EPITAXY

Edited by **Miao Zhong**

Epitaxy
http://dx.doi.org/10.5772/65173
Edited by Miao Zhong

Contributors

Ikai Lo, Chen-Chi Yang, Yu-Chi Hsu, Hong-Yi Yang, Celso Israel Fornari, Paulo H. De O. Rappl, Eduardo Abramof, Jerônimo Travelho, Gabriel Fornari, Hitoshi Habuka, Andra-Georgia Boni, Cristina Florentina Chirila, Luminita Hrib, Mihaela Botea, Raluca Negrea, Corneliu Ghica, Lucian Trupina, Iuliana Pasuk, Ioana Pintilie, Lucian Pintilie, Miguel Ciria, David Coffey, José I. Arnaudas, David Serrate, Yoon Hee Jeong, Abhijit Biswas, S.B. Krupanidhi, Basanta Roul, Shruti Mukundan, Greeshma Chandan, Xing-You Chen, Yi Gu, Yong-Gang Zhang, Ziheng Liu, Xiaojing Hao, Martin Green, Anita Ho-Baillie

Published by InTech

Janeza Trdine 9, 51000 Rijeka, Croatia

Publishing Process Manager Edi Lipovic
Technical Editor SPi Global
Cover InTech Design team

First published March, 2018
Printed in Croatia
Legal deposit, Croatia: National and University Library in Zagreb

Additional hard copies can be obtained from orders@intechopen.com

Epitaxy, Edited by Miao Zhong
p. cm.
Print ISBN 978-953-51-3889-1
Online ISBN 978-953-51-3890-7

For EU product safety concerns:
IN TECH d.o.o., Prolaz Marije Krucifikse Kozulić 3, 51000 Rijeka, Croatia,
info@intechopen.com or visit our website at intechopen.com.

Meet the editor

Dr. Miao Zhong is a Project Researcher at the School of Engineering of the University of Tokyo, where he received his PhD degree in 2012. His research interests include solid/liquid interface, chemistry and surface science, solid solution in the form of thin films and nanowire arrays, and epitaxial growth of high-quality films and nanostructures using (MO) CVD and MBE. His wide portfolio includes chapter publications, journal publications, conferences, and proceedings.

Contents

Preface

This edited volume is a collection of reviewed and relevant research chapters with a focus on the epitaxy field of study. This book includes scholarly contributions by accomplished authors and is edited by various experts in the field of materials science. Each contribution is available as a stand-alone chapter but is also directly related to the book's topics and objectives.

This book consists of nine chapters.

The target audience comprises scholars and specialists in the field.

InTechOpen

Section 1

High-Quality Epitaxy of Functional Heterostructures with Strain Engineering

Chapter 1

Strain Effect in Epitaxial Oxide Heterostructures

Abhijit Biswas and Yoon Hee Jeong

Additional information is available at the end of the chapter

http://dx.doi.org/10.5772/intechopen.70125

Abstract

In recent decades, extensive studies have been conducted on controlling and engineering novel functionalities in transition metal oxide (TMO) heterostructures by epitaxial strain. In this chapter, we discuss popular transition metal oxide thin films in the context of various research fields that are extensively studied in condensed matter physics. These materials include $La_{1.85}Sr_{0.15}CuO_4$ (a high temperature superconductor), $SrRuO_3$ (a highly conductive ferromagnetic metal), $La_{0.67}Sr_{0.33}MnO_3$ (a colossal magnetoresistive ferromagnetic metal), $BiFeO_3$ (a multiferroic oxide), $LaAlO_3$-$SrTiO_3$ (a conductive oxide interface), and $LaNiO_3$ (a strongly correlated metal). We focus on the appearance of novel functional properties from imposing epitaxial strain (compressive or tensile strain caused by the use of various lattice-mismatched substrates) on these films that cannot be observed in their bulk form. Subsequently, the intrinsic mechanisms for these novel phenomena are discussed based on experimental observations and theoretical modelling. We conclude that by using epitaxial strain, not only can thin film functionalities be tuned but many novel correlated phenomena can also be created. We believe that our collective efforts on the strain engineering of various transition metal oxide thin films will provide an insightful description of this emerging subject from a fundamental physics and nanoscale device applications point of view.

Keywords: oxides, substrates, thin films, strain effect, transport properties

1. Introduction

Matter can primarily be classified into three states: (1) gaseous, (2) liquid, and (3) solid (or condensed). In gaseous phases, the interactions among particles (atoms, molecules and ions) are very weak, and therefore they move freely. In liquid phases, the interactions between particles are comparatively strong. In solids, particles are closely packed or condensed, making the interactions between them the strongest compared to the other two states of matter.

Condensed matters (solids) are mostly crystalline; i.e., they have a periodic arrangement of atoms, ions or molecules. Depending on the periodicity of the atoms, they form different crystal structures. In nature, there are seven types of lattice structures: cubic (e.g., $SrTiO_3$), triclinic (e.g., $FeSiO_3$), monoclinic (e.g., $BiMnO_3$), orthorhombic (e.g., $GdFeO_3$), tetragonal (e.g., $BaTiO_3$), rhombohedral (e.g., $BiFeO_3$), and hexagonal (e.g., $YMnO_3$). Moreover, completely disordered systems displaying non-periodicity of atoms are called non-crystalline amorphous solids (e.g., glass). In reality, both types of systems show complex physics and chemistry with an immense number of functionalities [1].

Irrespective of their structural symmetry, three types of crystalline solids can be found either naturally or artificially, i.e., made in the laboratory. These include polycrystals, single crystals, and thin films. Polycrystalline materials are composed of many crystallites of various sizes that are oriented randomly (**Figure 1a**). Due to the random orientation of crystallites, they have many crystallite grains (~1 μm in size) with grain boundaries, twin boundaries and high porosity. As a result, polycrystals are occasionally considered dirty materials, and they show unusual behaviors at low temperature due to disorder. Polycrystals can be nearly several centimeters in size. Typically, the "solid-state reaction" method is used to synthesize polycrystalline materials [2].

Single crystals, in contrast, contain uniform orientations of their crystal lattices up to their edges, even at the macroscopic level; hence, there are no grain boundaries (**Figure 1b**). Therefore, it becomes easy to determine the various directions of a crystal and measure its properties along a particular direction. As a result, single crystals are regarded to be the cleanest and are very popular among the material science community as they reflect the exact properties of a material. Single crystals are nearly a few mm in size. For the most part, the floating-zone method, Czochralski method and Bridgman-Stockbarger method are used to grow high-quality single crystals [3]. A detailed analysis of these methods is beyond the scope of this chapter.

Thin films consist of very few layers of a solid material. They are typically deposited on structurally compatible metal oxide substrate surfaces (**Figure 1c**) by various thin-film deposition techniques [5]. Details about various thin-film deposition techniques are discussed later in

Figure 1. Schematics of a (a) polycrystal, (b) single crystal, and (c) thin film. Polycrystals have many grains, whereas the crystal orientation in single crystals is uniform. Moreover, thin films are grown on structurally compatible metal oxide substrates. Polycrystalline figure was taken from Ref. [4].

this chapter. Thin films are those with thicknesses ranging from a few angstroms to several nanometers (~4 Å to 1000 nm). Most thin films are made of oxides, particularly transition metal oxides (TMOs).

TMO thin films are one of the most investigated research topics in condensed matter physics as they show a variety of phenomena, e.g., metal insulator transitions (MIT), high-temperature superconductivity (HTSC), colossal magnetoresistance (CMR), and multiferrocity (coexistence of magnetism and ferroelectricity), as well as those exhibited by high-mobility two-dimensional electron gases (2DEGs), topological insulators (TI), and quantum spin-liquids (QSL) [6–20]. In TMOs, the *d*-orbital electrons of transition metal elements play a crucial role in determining the physical properties of a compound through the interplay between spin, lattice, charge, and orbital degrees of freedom (**Figure 2a**) [21–24]. Among the various types of TMOs, perovskites are a class of materials that shows almost all the properties mentioned above. They have been a deeply researched topic among physicists owing to their simple crystal structure (**Figure 2b**) [25, 26].

Perovskites, named after the Russian Mineralogist Count Lev Aleksevich von Perovski, have a general unit cell crystal structure of the ABO_3 type, where the A-site is an alkaline earth or rare earth metal and the B-site is a transition metal element (e.g., Fe, Co, Ni, Mn, Ti, Ru, and Ir). The structure can easily accommodate a wide range of valence states in both A- and B-sites (e.g. $A^{+1}B^{+5}O_3$, $A^{+2}B^{+4}O_3$, and $A^{+3}B^{+3}O_3$); so the variety of perovskite oxides is unlimited; for example, manganites, ruthenates, nickelates, titanates, and iridates [25]. $CaTiO_3$ was the first perovskite discovered by Gustav Rose in 1839 from samples found in the Ural Mountains in Russia. The B-site cation of ABO_3 perovskites is surrounded by six O anions, forming a corner-shared BO_6 octahedron (**Figure 2c**). This octahedral cage is the most important part of the crystal structure of these materials because the hopping of electrons from one *d*-orbital of the transition metal element to another *d*-orbital depends on the shape, size and position of this octahedron; thus, it affects the physics and chemistry of the material and the appearance of variety of phenomena [21–24].

(a) (b) (c)

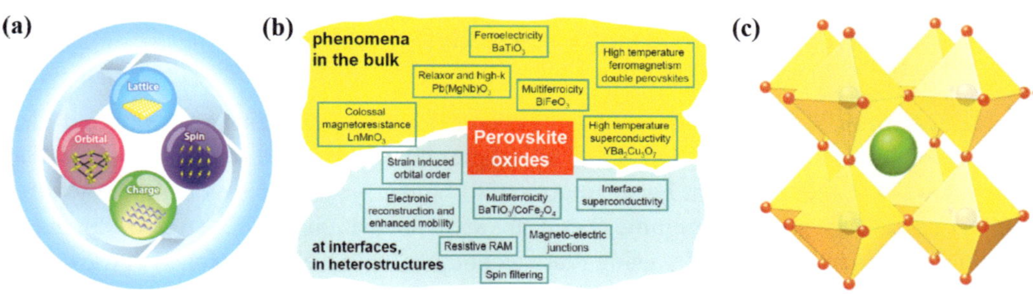

Figure 2. (a)-(b) Electronic and structural degrees of freedom in transition metal oxides and their interplay show a variety of correlated multifunctional phenomena in perovskite oxides, for both bulk and heterostructure thin films. (c) Schematic representation of a typical ABO_3 perovskite structure (e.g., $SrTiO_3$). Sr-cation is at the center; Ti-cation is surrounded by six O-anion forming TiO_6 octahedra cage. Reprinted with permission from [11, 20, 26]. Copyright 2012 Elsevier Ltd; Copyright 2014 Annual Reviews; Copyright 2008 IOP Publishing Group.

In bulk polycrystals and single crystals, the shape, size and position of the BO_6 octahedra can be manipulated externally by inducing chemical pressure (replacing A-site or B-site cations with other transition metal elements), or by partial oxygen pressure (changing the pressure from the atmospheric one) [6]. Ceramic materials fail structurally under modest strain (typically < 0.1% under strain) as these materials are brittle and thus will crack under this magnitude of strain, limiting the ability of routes involving chemical substitutions to control these materials. Cations with different sizes lead to the distortion of the crystal lattice, which is usually quantified as the Goldsmith tolerance factor [27] t_f, given by

$$t_f = \frac{r_A + r_O}{\sqrt{2}(r_B + r_O)} \tag{1}$$

where r_A, r_B, and r_O represent the ionic radii of ions A, B and O, respectively. The stability and distortion of a crystal structure is indicated by the value of t_f. For a perfect cubic structure, t_f is 1. Structure still remains cubic for $0.89 \leq t_f \leq 1.0$ [26]. For more lower value of t_f it forms other types of crystal structures, resulting in structural transitions to orthorhombic, or rhombohedral states that have lower symmetry than the cubic state.

However, as a result of chemical substitutions, disorder is introduced into the materials, which in most cases suppresses and even destroys the properties of a material. These difficulties can be overcome in a unique way in thin films by a disorder-free clean route approach. This can be achieved by growing thin films on substrates that are structurally compatible but have different cubic (pseudo-cubic) lattice constants. This is termed "strain engineering" in epitaxial thin films [28]. Once a strain effect is induced in a film, due to the change of energy scales of various degrees of freedoms (lattice, charge, spin, and orbital), it shows novel properties that cannot be found in parent bulk compound. This means that novel quantum-correlated phenomena can be obtained by the strain engineering of oxide heterostructures, which broadens the field and our understanding of condensed matter physics. In the next section, we will discuss how to grow such atomically controlled high-quality thin films and induce the strain effect in TMO heterostructure.

2. Thin film growth methods and substrates

In recent decades, significant advances have been made in synthesizing epitaxial thin films in the laboratory using various deposition techniques [5]. These methods are: (1) pulsed laser deposition (PLD), (2) molecular beam epitaxy (MBE), (3) off-axis radio frequency magnetron sputtering (RFMS), (4) metal-organic chemical vapor deposition (MOCVD), and (5) chemical solution deposition (CSD). By using these techniques, atomically controlled epitaxial thin films (*epi* means "above" and *taxy* means "in an ordered manner" in Greek), heterostructures and artificial superlattices can be grown. Among these techniques, PLD and MBE are the most popular ones adopted by the thin film community.

2.1. Pulsed laser deposition (PLD)

In 1986, the successful growth of HTSC $YBa_2Cu_3O_{7-\delta}$ (YBCO, T_{SC} ~90 K) thin films by the PLD technique by Dijkkamp et al., generated great interest among the material science community, as it provided an alternative method for making thin-film materials in the laboratory [29]. The PLD technique is probably the most commonly used method for growing oxide thin films [5, 30–32]. Films are grown inside a high-vacuum chamber. A homemade or commercially available polycrystalline target is ablated by an energy source (typically a KrF laser with a wavelength of 248 nm or a frequency-doubled Nd:YAG laser with a wavelength of 532 nm). When the target is ablated, it produces a highly energetic plasma plume from the target. This highly energetic plume contains ions and molecules that are then deposited onto the substrate surface, which is attached on a substrate holder and placed opposite the target along the same out-of-plane axis. The substrate temperature, which is controlled by a heater, is determined from outside the chamber using a pyrometer. The target-to-substrate distance is kept at ~40–50 mm as the dynamics and kinetics of the plume species are limited to a maximum critical distance from the target because of collisions. A schematic diagram of a PLD chamber is shown (**Figure 3a**) [33]. Gaseous atoms condense on a template created by the substrate to form a single crystal. During this process, one needs to fulfill the growing conditions, e.g., optimize the base pressure, gas (O_2, O_3, or Ar) pressure, substrate temperature, laser density, spot size, and substrate surface flatness. Since the whole process is a thermally non-equilibrium one, by tuning all these parameters and ultimately optimizing them, one can grow the highest quality atomically controlled epitaxial oxide thin films (**Figure 3b**) [15]. The *in-situ* growth process can be monitored in real time by using the reflection high-energy electron diffraction (RHEED) method.

The advantages of the PLD technique are: (1) *in-situ* stoichiometric transfer of composition from target to substrate; (2) compatible materials can be grown under oxygen pressures ranging from ultra-high vacuum (UHV) to atmospheric pressure; (3) materials ranging from ultra-thin homoepitaxial thin films to artificial superlattices can be grown with nanometer precision; (4) depending on the availability of the target material, a wide variety of films can be grown; and (5) materials are grown in a compact and inexpensive chamber. Furthermore, the disadvantages are: (1) sub-optimized growth conditions can lead to the non-stoichiometric films; and (2) due to the highly energetic plumes, macroparticles called "droplets" can be deposited on a substrate surfaces within a micrometer range [30]. Therefore, to grow the highest quality epitaxial thin films, one should be aware of these facts. There are many groups around the world who have been pioneers in growing artificial epitaxial single-crystalline thin films of the highest quality.

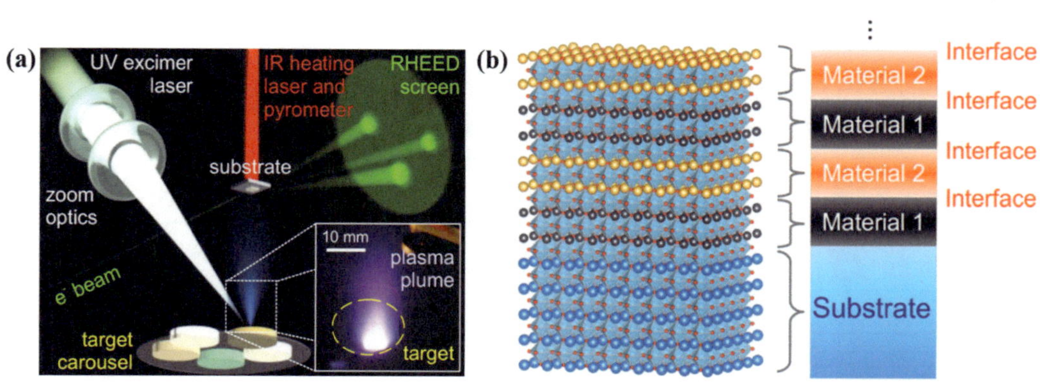

Figure 3. (a) Schematic diagram of a PLD chamber used for growing epitaxial oxide thin films. (b) Schematic of a layer-by-layer view of two different materials grown on a substrate. Reprinted with permission from Refs. [15, 33]. Copyright 2013 Materials Research Society; Copyright 2012 IOP Publishing Ltd.

2.2. Molecular beam epitaxy (MBE)

Molecular beam epitaxy (MBE) is also a method used to grow high-quality epitaxial thin films [5, 34]. It was invented in 1960s at Bell Labs by Arthur and Alfred Y. Cho [35]. The overall schematic of MBE is very similar to that of PLD thin film deposition. The only difference is the target material. Instead of a ceramic target, one uses "guns" called effusion cells (**Figure 4a**) [28, 36]. At the same time, one generates molecules from each cell using a highly intense laser beam (termed "atomic spray painting" by D. G. Schlom, a famous MBE thin film scientist) (**Figure 4b**) [28]. The spray duration is individually controlled for each beam by shutters. Once all the deposition conditions are satisfied, the ejected molecules travel to the substrate surface, condenses and form a single-crystalline thin film compatible with the substrate crystal structure. One of the main advantages of MBE thin film growth is its extreme cleanliness; i.e., no dirt particles (highly energetic species) or unwanted gas molecules can interfere with or contaminate the single-crystal thin film growth.

2.3. Substrate selection for epitaxial thin films

Substrates seem to be the basis of all thin film growth. Choosing a suitable metal oxide substrate is an important factor for growing high-quality epitaxial thin films as the structure and properties of a thin film depends on the underlying substrate and the interfacial interaction

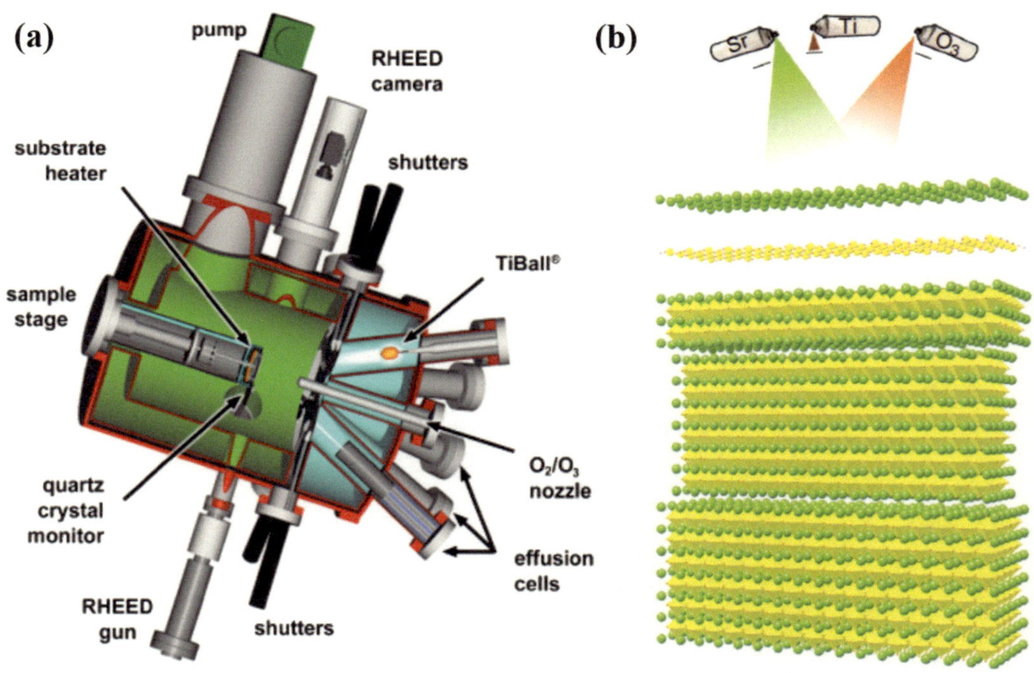

Figure 4. (a) Schematic diagram of a laser-MBE chamber for growing epitaxial thin films. (b) Schematic illustration of layer-by-layer MBE thin film growth, i.e., "atomic spray painting." Reprinted with permission from [28, 37]. Copyright 2008 The American Ceramic Society; Copyright 2014 Macmillan Publishers Limited.

between the substrate and film [38]. When choosing a metal oxide substrate for growing epitaxial films, one should consider the following factors:

1. Lattice matching between the substrate and film, which is important for the growth of most natural state films (structural compatibility).
2. No chemical reaction between the elements of the substrate and film (chemically compatibility).
3. Thermal-expansion matching between the substrate and film, as films are generally grown at high temperatures (good thermal-expansion match).
4. Surface quality of the substrate (e.g., free of cracks, unwanted particles, defects, and impure phases).

In most cases, the lattice constant and structure of a thin film should be compatible with those of the substrate to grow epitaxial films in their most natural state (**Figure 5a**). For most ABO_3 perovskites, their lattice constants range from 3.80 to 4.00 Å [39]. Fortunately, there are many perovskite single-crystal metal oxide substrates available commercially with lattice constants

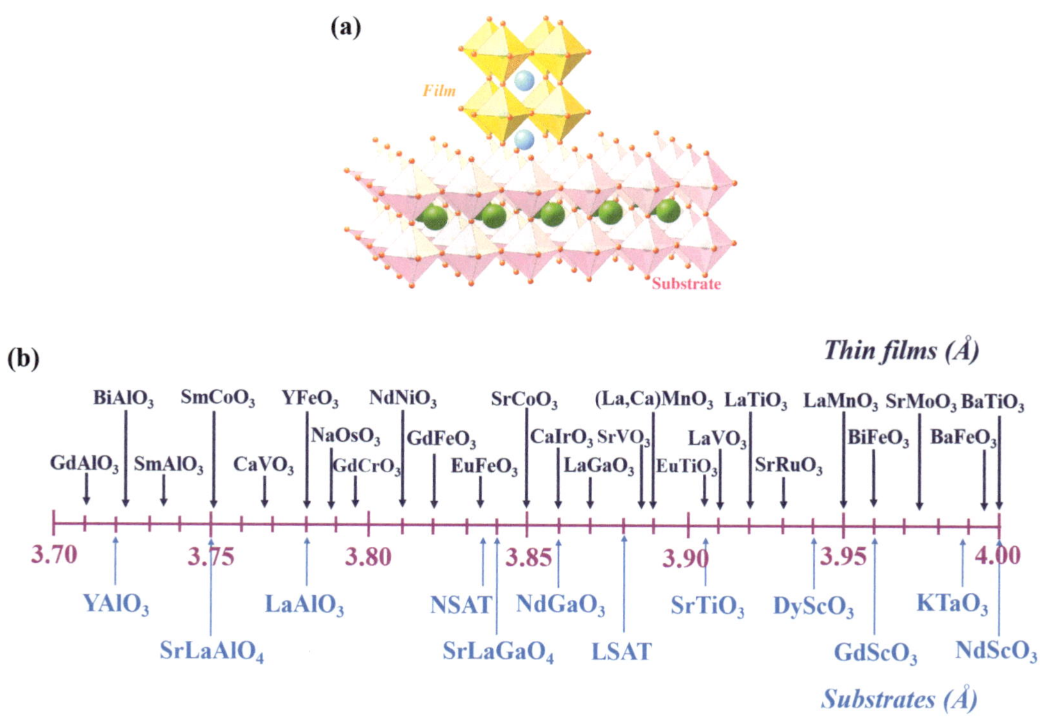

Figure 5. (a) Structural relationship between the substrate and film. For the most natural growth state of a film, a film's lattice constant (a_f) should be similar to the substrate's lattice constant (a_s), and the two should have structural compatibility. (b) List of cubic (pseudo-cubic) substrates and thin films within the lattice constant range from 3.70 to 4.00 Å. With a judicial choice of substrate, various atomically controlled high-quality thin films can be grown. Reprinted and adapted with permission from Ref. [41]. Copyright 2014 Materials Research Society.

ranging from 3.70 to 4.20 Å [28, 33, 40–42]. Among various available perovskite substrates, insulating $SrTiO_3$ is the most popular one. It has a cubic structure with a lattice constant of ~3.905 Å. There are also a broad range of substrates available with similar structures to that of $SrTiO_3$ while possessing different lattice constants and crystal orientations. These commercially available substrates include REScO$_3$ (RE = rare earths), $La_{0.18}Sr_{0.82}Al_{0.59}Ta_{0.41}O_3$ (LSAT), $NdGaO_3$, $SrLaAlO_4$, $LaAlO_3$, $SrLaAlO_4$, and $YAlO_3$ (**Figure 5b**). Thus, after careful consideration of all the important factors listed above and a film's lattice constant, it is easy to choose a substrate suitable for the epitaxial growth of thin films.

3. Strain in perovskite thin films

3.1. Strain engineering of perovskite thin films

Strain engineering is a unique way to create the novel functionalities in epitaxial oxide thin films [40–43]. From substrate-thin film relation point of view, when the lattice constant of a film (a_f) is dissimilar to the lattice constant a substrate (a_s), compression or elongation occurs within the film's crystal structure and thus elastic strain is induced in the film [41]. For cases of dissimilar lattice constants ($a_f \neq a_s$), the structure of the thin film tries to take the structure of the substrate, causing structural changes (mainly the change in BO_6 octahedron rotation, tilting, and distortion, and/or the change in B—O bond length) to occur from the original atomic position. This is defined as the typical strain effect in thin films. Quantitatively, the amount of strain (ε) induced in a film is defined as

$$\epsilon = \left(\frac{a_s - a_f}{a_f}\right) \quad (2)$$

where a_s is the substrate lattice constant and a_f is the film lattice constant.

Generally, compressive strain (**Figure 6a**) is induced in a film when $a_f > a_s$, whereas tensile strain (**Figure 6b**) is induced in a film when $a_f < a_s$. Under these epitaxial strain scenarios, the properties of functional oxide thin films can be drastically altered. Currently, the strain (ε) of ~2–3% is quite common in epitaxial oxide thin films, with highest strain to date of ~6.5% being imposed on multiferroic $BiFeO_3$ films grown on a highly lattice-mismatched (1 1 0) $YAlO_3$ substrate [43]. Thus, elastic strain is a viable route to observe materials with exceptional properties that cannot be observed in their bulk form by any other means [44, 45]. Although it looks simple, the intrinsic mechanism of the appearance of novel functionalities induced by the strain effect is quite complex to understand. In the next section, we briefly discuss about the intrinsic mechanism of the strain effect in perovskite thin films.

3.2. Mechanism of the strain effect in perovskite thin films

In post-Moore era, electronic devices with multifunctionality may offer a new alternative to replace the current silicon-based technology because the additional value the devices would

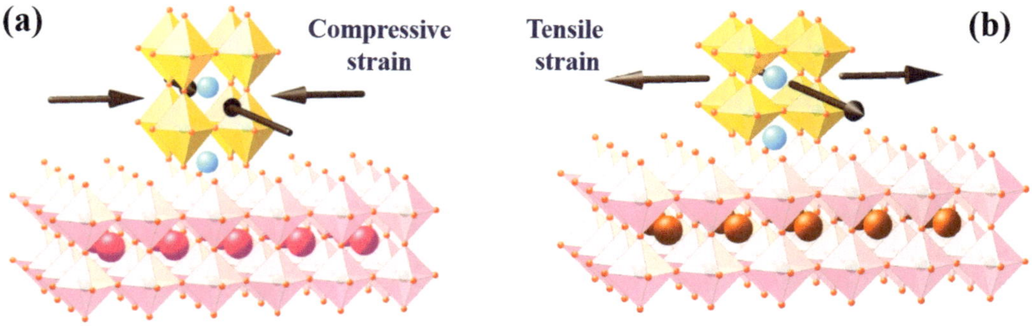

Figure 6. "*Strain engineering*" of perovskite heterostructures. (a) Compressive and (b) tensile strain is induced in a film through the use of various lattice-mismatched substrates. For compressive strain, $a_f > a_s$, whereas for tensile strain, $a_f < a_s$. In principle, for compressive strain, a thin film's lattice is compressed along the in-plane direction, and expanded along the out-of-plane direction. On the other hand, for tensile strained films, lattices expand along the in-plane direction and shrink along the out-of-pane direction. Reprinted with permission from Ref. [41]. Copyright 2014 Materials Research Society.

generate from multifunctionality may create an economically viable path superseding the miniaturization limit of silicon electronic devices. In this perspective, oxide electronics based on multifunctional properties of transition metal oxides looks promising [46, 47]. Even more exciting is the fact that advanced thin film growth techniques with atomic controllability provide further opportunities to design and synthesize artificial complex transition metal oxide heterostructures and superlattices to bring forth emergent physical properties, normally not seen in bulk states. However, despite the rapid progress and tremendous success in obtaining novel functionalities by the strain engineering of epitaxial oxide heterostructures, there is no general rule or theory available till date for predicting a material's electronic, magnetic, or other functional properties. This is perhaps due to a lack of knowledge about fully resolved atomic structures, especially the position of non-trivial oxygen atoms, as no experimental tool has yet been developed for the direct observation of oxygen atoms. In view of the lack of an experimental tool of this kind for transition metal oxides, electronic structure calculations could play a role instead. J. M. Rondinelli and N. A. Spaldin's recent article is particularly insightful in this regard, and we follow them briefly in this section [48].

The detailed structural distortion obtained by the movement of oxygen atoms due to strain is highly significant and it has a strong influence on the electronic properties of TMOs. In fact, just a small modification in an atomic structure would change the relevant energy scales (lattice, charge, spin, and orbital) and it is hard to predict functionalities as material properties are strongly dependent on the competition between these energy scales. To illustrate the effects of structural distortions in thin films, let us start with the ideal ABO_3 perovskite structure which is a simple cubic one with space group $Pm\bar{3}m$. The most important part of this structure is its corner-shared BO_6 (B = transition metal) octahedron, which results in an O—B—O—B chain with a 180° B—O—B bond angle (**Figure 7a**). There are very few systems that adopt this cubic structure (e.g., $SrTiO_3$); however, in practice, most perovskites show structural distortions that lower their symmetry from that of a highly symmetric cubic structure.

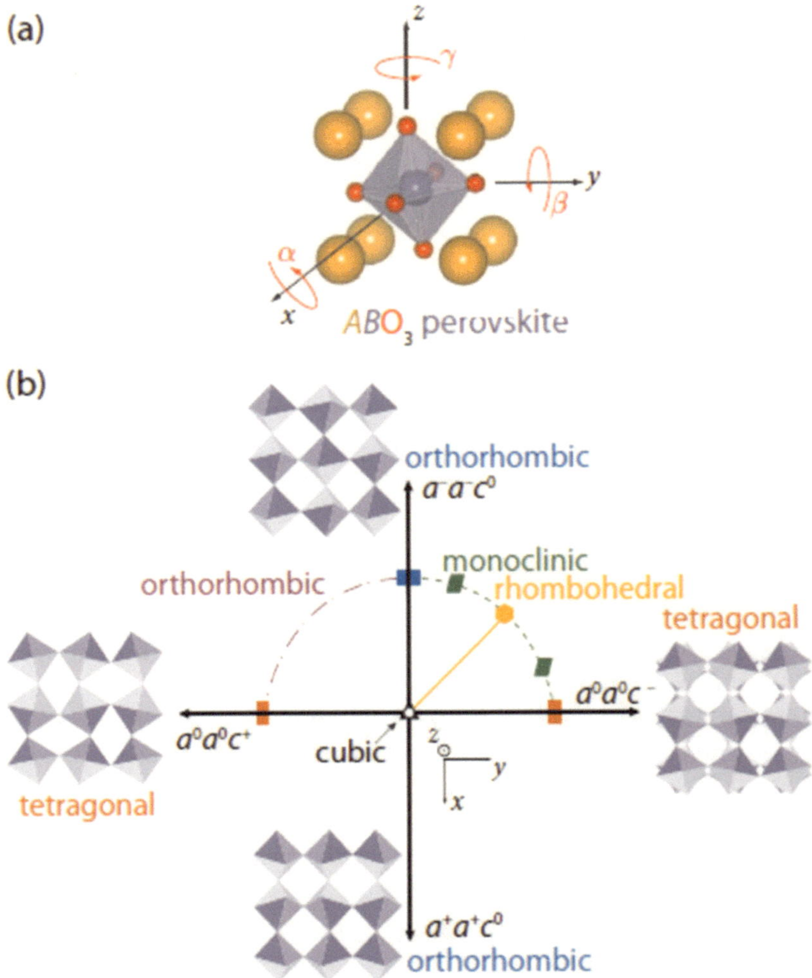

Figure 7. (a) The ideal ABO_3 perovskite crystal structure showing tilt in all three directions. (b) Distortion of BO_6 octahedra along various directions, lowering the symmetry of the cubic structure and forming other crystal structure. The +ve sign indicates in-phase rotation (c^+), and the −ve sign indicates out-of-phase rotation (c^-). Reprinted with permission from Ref. [48]. Copyright 2011 WILEY-VCH Verlag GmbH & Co. KGaA, Weinheim.

Rotations or tilts of BO_6 octahedra around the high-symmetry axes are the most common distortions in perovskite structures [49–51]. These are conveniently described by the Glazer notation [52, 53] and written as $a^{\#}b^{\#}c^{\#}$, where a, b, and c are the axes around which rotation occurs and the superscripts indicate whether the octahedral rotation is in phase (+) or out of phase (−). Thus, a, b, and c are not identical with the lattice constants; instead, they indicate that the nearest neighbor transition metal distances along that direction are equivalent. As perovskites are three-dimensional systems, a rotation or tilt in one direction restricts the rotation or tilt in other directions. Depending on the rotation or tilt of the BO_6 octahedra, the cubic structure can deform, leading to the symmetry lowering of other crystal structures (**Figure 7b**) [48–51].

This structural distortion is imposed on a thin film by the appropriate choice of lattice-mismatched substrates. It is widely believed that the strain imposed by film-substrate lattice mismatch generally changes the in-plane lattice parameter, but exactly what occurs still remains unclear and moreover is difficult to determine experimentally. Two possibilities remain: (1) changes in the in-plane lattice parameter are offset by changes in the in-plane metal-oxygen B—O bond lengths (**Figure 8a** and **b**), or (2) while keeping the B—O distance fixed, the lattice mismatch is offset by a change in magnitude of the tilt patterns through the rigid rotation of the BO_6 octahedron (**Figure 8c** and **d**). This is highly significant as, for example, the magnitude and symmetry of a crystal field are affected by changes in the B—O bond length, whereas the strength and sign of a superexchange interaction are affected by changes in the B—O—B bond angle [48].

Quantitatively, changes in the B—O bond length and B—O—B bond angles of octahedra affect the bandwidth (W) of ABO_3 perovskites as follows:

$$W \propto \frac{\cos\psi}{d^{3.5}} \tag{3}$$

where $\psi = (\pi - \phi)/2$ is the buckling deviation of the B—O—B bond angle ϕ from π and d is the B—O bond length (**Figure 8e**) [54]. Due to rigidity, it is hard to change the B—O bond-length. Thus, as a result of imposed strain, octahedral rotation and tilt angle changes, the electron hopping changes within the d-orbitals and thus changes a material's functionalities. Changes in the bandwidth also affect the effective correlation as in general changes in these energy scales cause the appearance of novel functionalities in oxide heterostructures under strain.

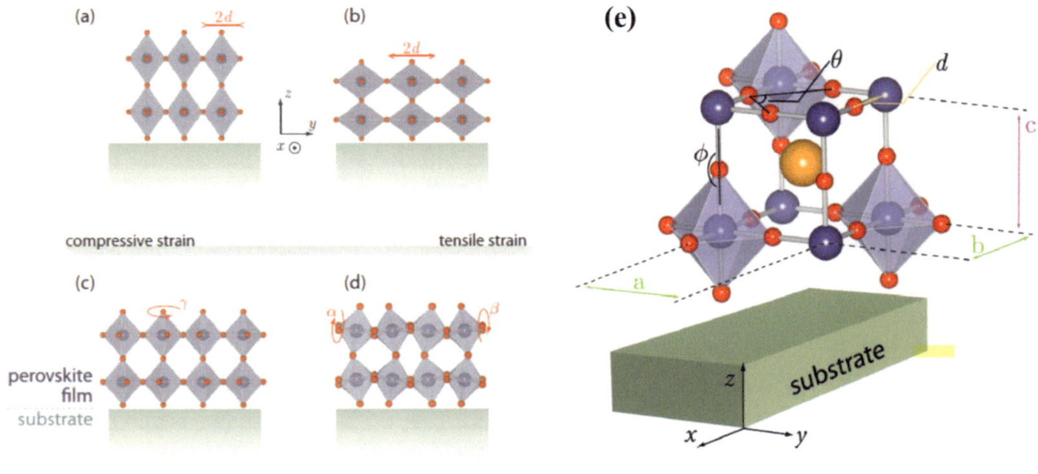

Figure 8. (a) Contraction and (b) elongation of B—O bond lengths, d, in a coherently strained perovskite film. Contraction is due to compressive strain, whereas elongation is due to tensile strain. Alternatively, change in the in-plane lattice parameters are due to the rotation of octahedra (c) perpendicular to the plane of the substrate or (d) about an axis parallel to the plane of the substrate. (e) Rotation (θ) and tilt (ϕ) angle used to describe the substrate-induced changes of octahedra. Reprinted with permission from Ref. [48]. Copyright 2011 WILEY-VCH Verlag GmbH & Co. KGaA, Weinheim.

Along with these changes in bond length, bond-angle and crystal symmetry that determine changes in the in-plane lattice parameters, another possibility for substrate-induced changes in lattice parameter remains: the defect stoichiometry or defect concentration of the material. Defect concentration, especially the oxygen concentration, is particularly important as films are generally grown under the high oxygen pressure or high vacuum. To accommodate the strain energy, it is easy to form oxygen vacancies. It is known that the higher the concentration of oxygen vacancies, the larger the lattice constants [55]. Since strain is induced by various lattice-mismatched substrates, it is difficult to establish whether changes in defect concentration are an intrinsic thermodynamic response due to strain or if they arise due to an extrinsic effect during the growth process.

4. Material properties tuned by epitaxial strain

4.1. A high temperature superconductor: $La_{1.85}Sr_{0.15}CuO_4$

In 1986, the discovery of HTSC in the cuprate oxide family by Bednorz and Müller generated considerable interest within the material science community, both in fundamental and applied research, due to the possibility obtaining oxides that are room-temperature superconductors [56]. $La_{1.85}Sr_{0.15}CuO_4$ (LSCO) is one example of these oxides. La_2CuO_4 is an antiferromagnetic insulator, but upon doping divalent Sr^{2+} ions in the trivalent La^{3+} site, magnetism is suppressed, and the compound makes a transition from an insulating to a superconducting state [57]. LSCO has a K_2NiF_4-type tetragonal structure with bulk lattice constants of $a = 3.777$ Å and $c = 13.226$ Å and space group *P42/ncm* (**Figure 9a**) [58]. It has quasi-two-dimensional copper-oxygen (Cu—O) planes, and superconductivity occurs within these planes [59]. In the bulk, around the optimal doping region, its maximum superconducting transition temperature is T_C ~25 K (**Figure 9b** and **c**) [60]. Its carrier doping remains within the Cu—O planes and the formation of electron pairs due to coupling between electrons and phonons seem to play a major role in achieving superconductivity [61]. Its critical temperature is controlled by either the density of electron pairs or the strength of electron pairing interactions [62].

In principle, external perturbation is applied to a material to enhance its functional properties, suggesting that substrate-induced strain might be a way to enhance the superconducting T_C of cuprates (**Figure 10a**) [63–70]. Indeed, Sato et al., and Locquet et al., grew $La_{2-x}Sr_xCuO_4$ (LSCO; $x = 0.1, 0.15$) thin films on two different substrates, such as (0 0 1) $SrTiO_3$ and (0 0 1) $SrLaAlO_4$ [63, 64]. Surprisingly, Locquet et al., observed that when films are grown on highly strained (0 0 1) $SrLaAlO_4$ substrates ($a_s = 3.75$ Å), which produces ~0.5% compressive strain onto these films, this amount of strain is enough to modify the superconducting T_C, making T_C almost double to its value found in the bulk, i.e., T_C ~49.1 K (**Figure 10b**) [64]. Independently, Sato, also reported the same study (**Figure 10c**) [69]. This is thought to be associated with strain-induced lattice deformation, which modifies the energy scales, leading to the formation and condensation of superconducting pairs. It was also observed that the residual resistivity value (ρ (0 K)) decreases as T_C increases [69]. More specifically, increasing T_C has a clear correlation with low residual resistivity. As stated by Sato, an increase in Cu—O bond length enhances the electrostatic potential at the Cu site relative to that at the oxygen site in the Cu—O plane.

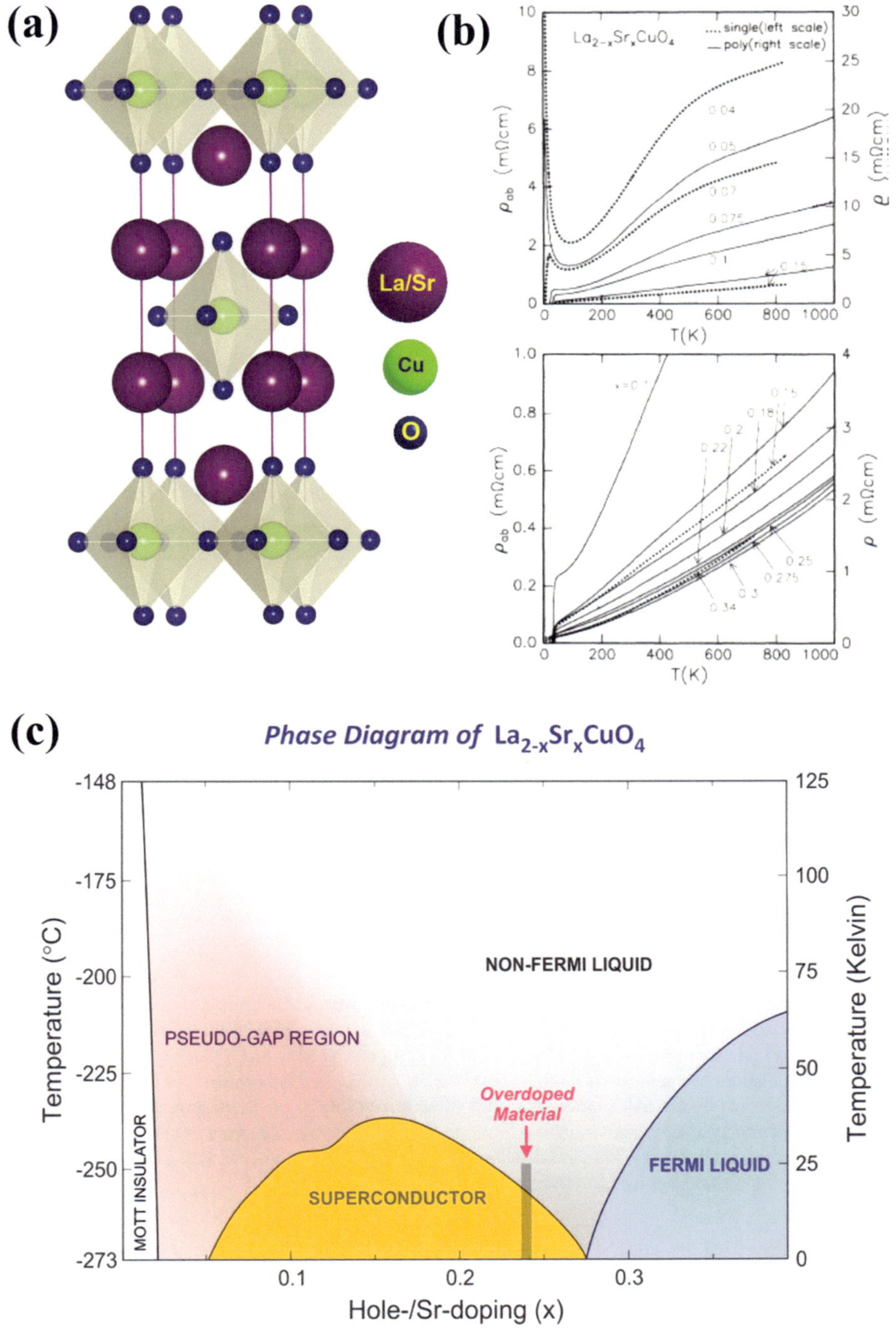

Figure 9. (a) Schematic representation of the K_2NiF_4-type tetragonal crystal structure of $La_{2-x}Sr_xCuO_4$ (LSCO) with lattice constants of $a = 3.777$ Å and $c = 13.226$ Å. (b) Resistivity of bulk single LSCO, showing the appearance of superconducting T_{SC} upon divalent Sr hole doping. (c) Sr hole doping dependence temperature vs. the material properties phase diagram of LSCO, showing that the system makes a transition from an antiferromagnetic insulator to a superconducting phase with the increase in Sr doping. Crystal structure was drawn using VESTA software. Reprinted with permission [57, 60]. Copyright 1992 American Physical Society; Copyright 2013 Macmillan Publishers Limited.

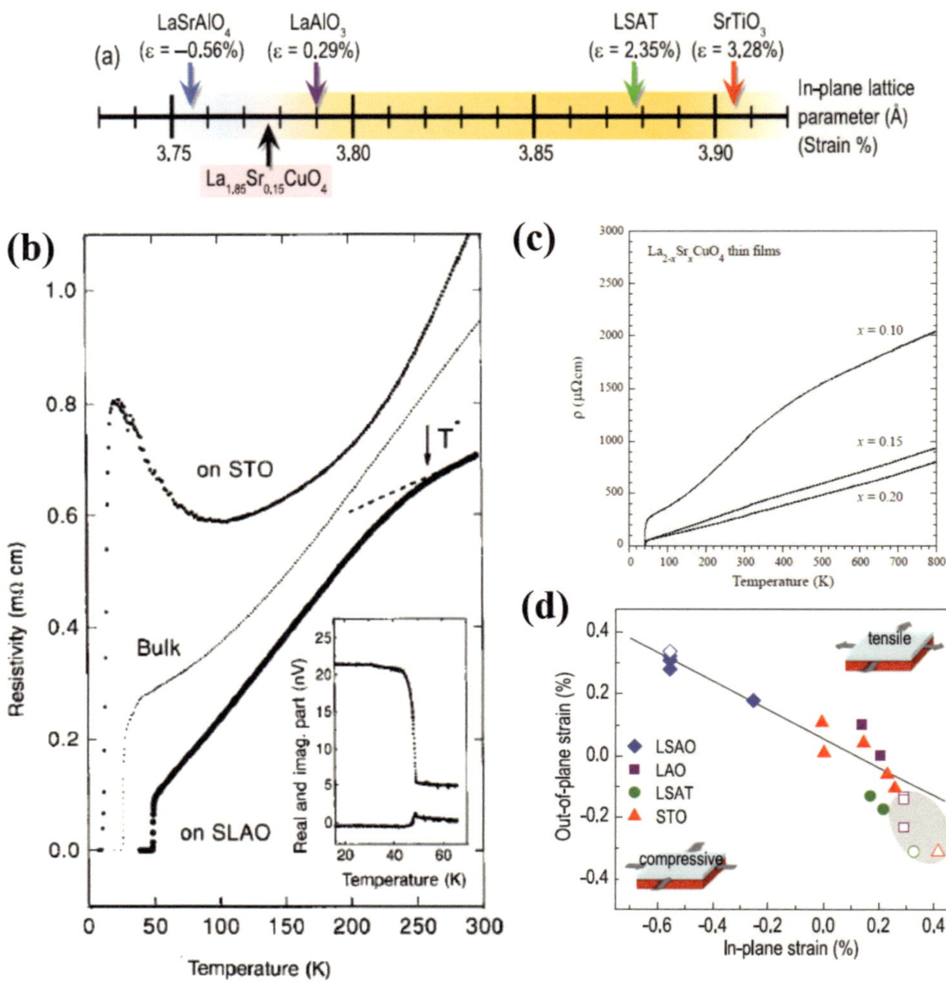

Figure 10. (a) In-plane strain and lattice parameters of the LSCO thin film and other perovskite substrates studied. (b) Strain-dependent resistivity (ρ) and superconductivity of LSCO films, showing that for tensile-strained films (on $SrTiO_3$ substrates), superconductivity occurs at T_C ~10 K, whereas for compressive-strained films (on $SrLaAlO_4$ substrates), the superconductivity transition temperature is doubled from the bulk value; i.e., T_C becomes ~49.1 K. (c) Superconductivity at T_C ~50 K for films grown on $LaSrAlO_4$ substrates with various amount of hole doping scenarios. (d) Changes in the *c*-axis lattice parameter and out-of-plane strain as a function of in-plane strain. The gray oval shape is the region where films do not show superconductivity. Reprinted with permission from Refs. [64, 69, 70]. Copyright 1998 Macmillan Publisher Ltd.; Copyright 2008 Elsevier Ltd.; Copyright 2015, AIP Publishing LLC.

Therefore, hole carriers are distributed more preferentially in itinerant states originating from the O 2p orbitals. As a result, antiferromagnetic spin fluctuation in the Cu—O plane is suppressed due to the reduction of the superexchange interaction between two adjacent Cu spins. Reduced spin fluctuation is the possible origin for reduction in ρ (0 K) and increase in T_C [69].

For compressively strained films grown on (0 0 1) $SrTiO_3$ substrate, which induced ~3% tensile strain on films, the T_C was found to be ~10 K [64]. Later, Božović et al., showed that with higher quality films (with much more oxygen intake as the films were annealed under an ozone atmosphere), T_C could reach up to ~40 K for tensile-strained films grown on (0 0 1)

$SrTiO_3$ substrates and ~51.5 K for compressive-strained films grown on (0 0 1) $LaSrAlO_4$ substrates [68]. Recently, Lee et al., also showed that oxygen vacancies and thickness-dependent strain relaxation indeed play a crucial role in increasing the superconducting T_C as lattice structures are highly sensitive to oxygen stoichiometry (**Figure 10d**) [70].

4.2. A highly conductive ferromagnetic metal: $SrRuO_3$

$SrRuO_3$ is one of the most promising oxide material among those used by thin-film researchers [71]. It is an itinerant ferromagnetic bad metal [72, 73]. Structurally, it has a $GdFeO_3$-type orthorhombic distorted perovskite structure at $T = 300$ K with lattice parameters of $a = 5.5670$ Å, $b = 5.5304$ Å, and $c = 7.8446$ Å and space group *Pbnm* (**Figure 11a**) [74]. This structure is converted into a pseudo-cubic (pc) lattice with a constant of $a_{pc} \sim 3.93$ Å $\left(a_c = b_c = \sqrt{a^2 + b^2/2} = d_{110}\right.$ and $\left.c_c = c/2 = d_{002}\right)$. Bulk $SrRuO_3$ shows structural phase transitions from orthorhombic to tetragonal at $T = 547$°C and then to cubic symmetry at $T = 677$°C, respectively [75]. This is associated with the occurrence of RuO_6 octahedral rotation, leading to a lowered structural symmetry from the ideal cubic perovskites.

In high-quality bulk single crystals, the ρ is ~200 μΩ cm with a mean free path of electrons of ~10 Å at $T = 300$ K (**Figure 11b**) [73]. $SrRuO_3$ shows a ferromagnetic transition at the Curie temperature T_C ~165 K (**Figure 11c**) with a magnetic moment of ~1.6 μ_B per Ru atom [76]. Although extensively studied, but the origin of its ferromagnetism, Stoner-type itinerant ferromagnetism vs. localized moment picture, is still under fierce debate because of contradicting experimental results and theoretical calculations [77, 78]. Due to its highly conductive nature and structural compatibility with other perovskite thin films, it is particularly interesting for its usefulness as a bottom electrode for $BiFeO_3$, $BaTiO_3$, or $Pb(Zr,Ti)O_3$ ferroelectric thin films used in electronic applications [79].

Eom et al. were the first to synthesize high quality metallic epitaxial $SrRuO_3$ thin films on various substrates [80]. The ferromagnetic transition temperature for these thin films were found be lower than the bulk value, T_C ~150 K, which is probably caused by the dimensionality and

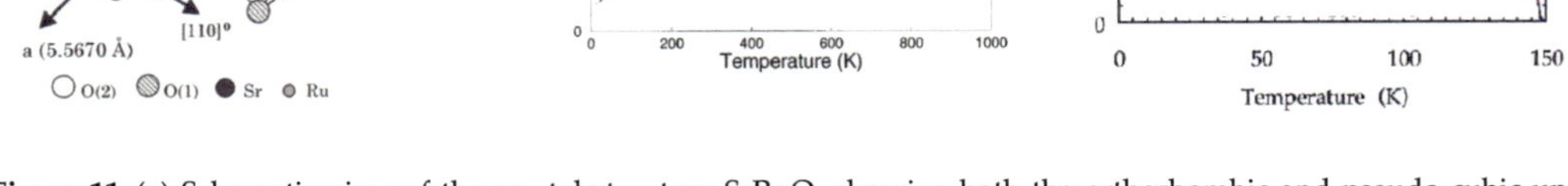

Figure 11. (a) Schematic view of the crystal structure $SrRuO_3$ showing both the orthorhombic and pseudo-cubic unit cell. Lattice parameters are $a = 5.5670$ Å, $b = 5.5304$ Å, and $c = 7.8446$ Å, with the pseudo-cubic (a_{pc}) one being ~3.93 Å. (b–c) Resistivity (ρ) and magnetization of $SrRuO_3$. The material was found be metal over the whole temperature range with ρ ~200 μΩ cm at $T = 300$ K. Magnetization measurement shows ferromagnetic ordering at $T_C = 165$ K. Reprinted with permission from Refs. [72–74]. Copyright 1996 IOP Publishing Ltd.; Copyright 1996, American Physical Society; Copyright 1999, American Institute of Physics.

strain effects [81]. Along with its electrode applications, due to its high metallicity upon ferromagnetic ordering, strain engineering of $SrRuO_3$ has become a popular research topic among thin-film scientists. Later, it has been found that the structural, metallic, and magnetic properties of $SrRuO_3$ thin films are highly sensitive to the substrate-induced strain (**Figure 12a**) [81–92].

The structural phase transition temperature (T_S) of $SrRuO_3$ is strongly affected by epitaxial strain. After imposing tensile strain on $SrRuO_3$ by growing films on (1 1 0) $DyScO_3$ substrates, an orthorhombic (stable in bulk phase) to tetragonal phase (stable at high temperature) transition could be observed at $T = 300$ K [84]. Strain imposed by substrates, induces additional rotation of RuO_6 octahedra, thus reducing the T_S [84, 86].

Transport wise, ρ at $T = 300$ K increases with the induced tensile strain (films grown on (1 1 0) $GdScO_3$ and (1 1 0) $DyScO_3$ substrates). The Ru—O—Ru bond angle is reduced when exerting tensile strain, which increases the effective correlation and thus reduces the bandwidth (W) (**Figure 12b**) [84]. Consequently, in the case of tensile-strained films, ρ has a higher value at $T = 300$ K compared to that of the most natural state of a film. On the other hand, it has also been found that for compressive-strained films (films grown on (0 0 1) $SrTiO_3$, (0 0 1) LSAT, and (0 0 1) $LaAlO_3$ substrates), ρ decreases, which is consistent with the increase in the Ru—O—Ru bond angle, decrease in effective correlation, and increase in bandwidth (W) (**Figure 12c**) [85]. The slight increase in the ρ of films grown on (0 0 1) $LaAlO_3$ substrates is associated with either the rough surface quality of the films caused by the twin structure of (0 0 1) $LaAlO_3$ or the films being fully relaxed on this substrate (**Figure 12d**) [85].

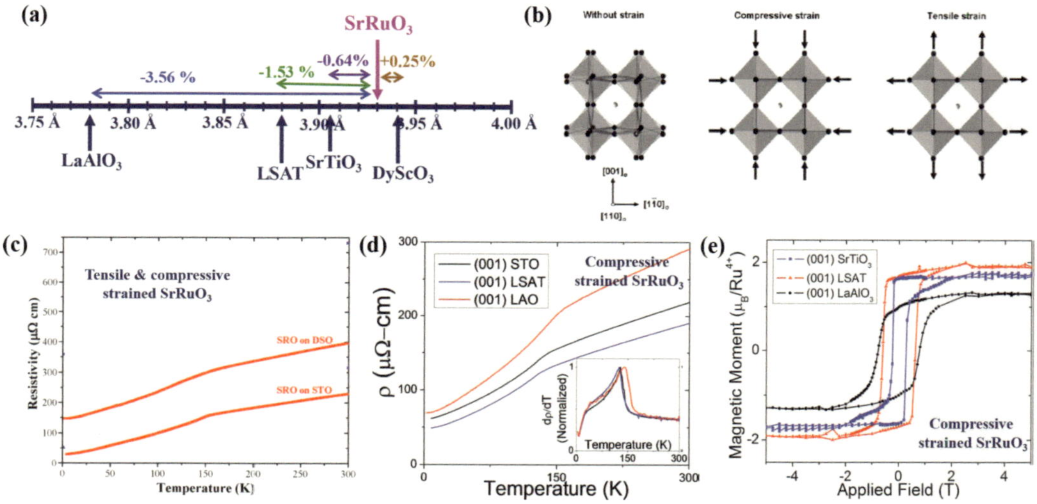

Figure 12. (a) Schematic illustration of the strain effect in epitaxial orthorhombic $SrRuO_3$ thin films. Films on $DyScO_3$ are subject to tensile strain, whereas films on $SrTiO_3$, LSAT, and $LaAlO_3$ undergo compressive strain. (b) Schematic representation of RuO_6 octahedral rotation from the bulk, due to both compressive and tensile strain. (c–d) Effect of tensile strain and compressive strain on the resistivity of $SrRuO_3$. For the tensile strain case, overall resistivity increases, while for the compressive-strained case, resistivity decreases from its bulk value. (e) Increase in saturation magnetization following the induction of compressive strain. The magnetic moment becomes ~2 μ_B per Ru atom, which is close to the theoretical value. Reprinted with permission from Refs. [84–86]. Copyright 2008 AIP Publishing LLC; Copyright 2010 AIP Publishing LLC; Copyright 2010 Wiley-VCH Gmbh & Co. KGaA.

Strain has definite effect on the magnetic properties of $SrRuO_3$. Inducing compressive strain on $SrRuO_3$ by using various lattice-mismatched substrates, e.g., (0 0 1) $SrTiO_3$, (0 0 1) LSAT, and (0 0 1) $LaAlO_3$, and its effect on magnetic properties has been investigated. The ferromagnetic T_C has a strong substrate dependence as it was found be T_C ~124 K for films grown on LSAT and T_C ~128 K for films grown on (0 0 1) $SrTiO_3$ [85]. It has been found that the saturated magnetic moment (M_S) increases from its bulk value and maximum of ~2 μ_B per Ru atom for films grown on (0 0 1) LSAT substrates (**Figure 12e**). For more compressive-strained films (those on (0 0 1) $LaAlO_3$), the M_S increases, possibly due to the deterioration in film quality caused by the twin structure of (0 0 1) $LaAlO_3$ substrates. The increase in M_S in compressive strained case is associated with the better alignment of moments in a low Ru^{4+} spin state. This higher value of magnetic moment was also expected based on the theoretical calculations [77].

Compressive and tensile strain-dependent physical properties have also been examined for the tetragonal phase of $SrRuO_3$. (1 1 0) $NaGaO_3$ substrates were used for imposing compressive strain, whereas (1 1 0) $GdScO_3$ substrates were used to induce tensile strain [90]. It has been found that tensile-strained films show low ferromagnetic ordering at T_C ~100 K, whereas compressively strained films show an almost bulk-like ferromagnetic transition at T_C ~155 K (**Figure 13**). Similar to the orthorhombic phase, compressive strain causes a lower residual resistivity ratio. The observations described above are associated with the deformation rather than the tilting of RuO_6 octahedra and thus the change in effective correlation [90–93].

4.3. A colossal magnetoresistive ferromagnetic metal: $La_{0.67}Sr_{0.33}MnO_3$

$La_{0.67}Sr_{0.33}MnO_3$ (LSMO) is an extremely important class of material in condensed matter physics [7, 94]. At T = 300 K, LSMO forms a rhombohedral crystal structure (a = 3.869 Å) with space group $R\bar{3}c$ (**Figure 14a**) [94]. Undoped $LaMnO_3$ is an antiferromagnetic insulator below T_N = 139.5 K [95]. Upon hole doping with divalent Sr^{2+} in place of La^{3+}, it becomes a ferromagnetic metal (**Figure 14b**) [96]. Doped LSMO is a mixed valence compound with Mn^{3+} ($3d^4$) and Mn^{4+} ($3d^3$). The change in its magnetism was well explained by the double-exchange hopping mechanism [96]. It is a highly conductive oxide that is useful as a bottom electrode for thin film device applications. It shows ferromagnetic ordering above room temperature with T_C ~360 K (**Figure 14c**) [97], having a magnetic moment of ~3.6 μ_B per Mn ion [98]. Its ρ changes greatly with the application of a magnetic field and shows colossal magnetoresistance (CMR; $\Delta\rho/\rho > 10^6\%$) (**Figure 14d**) [97], which is important in commercial applications, including magnetic field sensors, "read" heads of magnetic hard-disk drives and non-volatile magnetic random access memory (MRAM). Its half metallic behavior—i.e., spins are fully polarized within one band structure whereas others are empty—is highly important for spintronic applications [98]. It also shows compositional- and temperature-dependent MITs (**Figure 14e**) [99].

Due to its rich electronic and magnetic phase diagram (**Figure 14e**) [99], it is highly desirable to investigate the change in functionalities or appearance of novel states in LSMO by inducing the strain effect. Several groups have reported on the effect of strain on the electronic and magnetic properties of LSMO [100–111]. To observe the strain-dependent magnetic phase diagram, Tsui et al., grew epitaxial LSMO thin films on various substrates, such as (0 0 1) $LaAlO_3$, (1 1 0) $NdGaO_3$, (0 0 1) LSAT, and (0 0 1) $SrTiO_3$ [103]. By using four different substrates, the

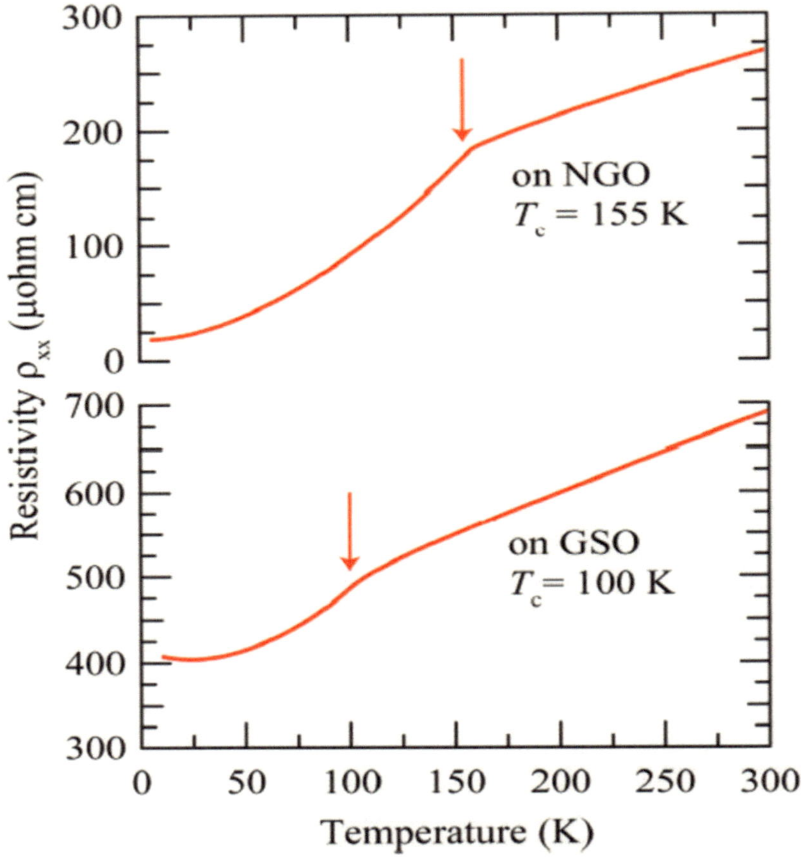

Figure 13. Compressive (on $NdGaO_3$) and tensile (on $GdScO_3$) strain-dependent resistivity of tetragonal phase $SrRuO_3$ thin films. A striking feature was obtained in magnetic ordering. For tensile-strained films (on $GdScO_3$), resistivity is higher than the bulk value and the magnetic ordering temperature is reduced with ferromagnetic T_C ~124 K. For compressive-strained film (on $NdGaO_3$), resistivity and magnetic ordering are very close to the bulk values with ferromagnetic T_C ~155 K. Reprinted with permission from Ref. [90]. Copyright 2013 AIP Publishing LLC.

strain ranged from compressive to tensile strain as follow; −2.0% compressive strain for films grown on (0 0 1) $LaAlO_3$, −0.25% compressive strain for films grown on (1 1 0) $NdGaO_3$, +0.25% tensile strain for films grown on (0 0 1) LSAT, and +0.85% tensile strain for films grown on (0 0 1) $SrTiO_3$. Compressively strained films on (0 0 1) $LaAlO_3$ substrates show in-plane compression and out-of-plane expansion in their lattice parameters. In contrast, tensile-strained films on (0 0 1) $SrTiO_3$ show the opposite effect. For films grown on the other two substrates with very low strain, they show very weak out-of-plane expansion in their lattice parameters. For compressively strained films on (0 0 1) $LaAlO_3$ substrates, there is a strong suppression of T_C from its bulk value; i.e., T_C is reduced from 360 to 300 K. There is an increase in in-plane magnetization compared to out-of-plane magnetization, which indicates the presence of easy-plane anisotropy. Films grown on (0 0 1) $SrTiO_3$ substrates show magnetic ordering below T_C ~320 K. On the other hand, films grown on the almost lattice-matched substrates, i.e., films

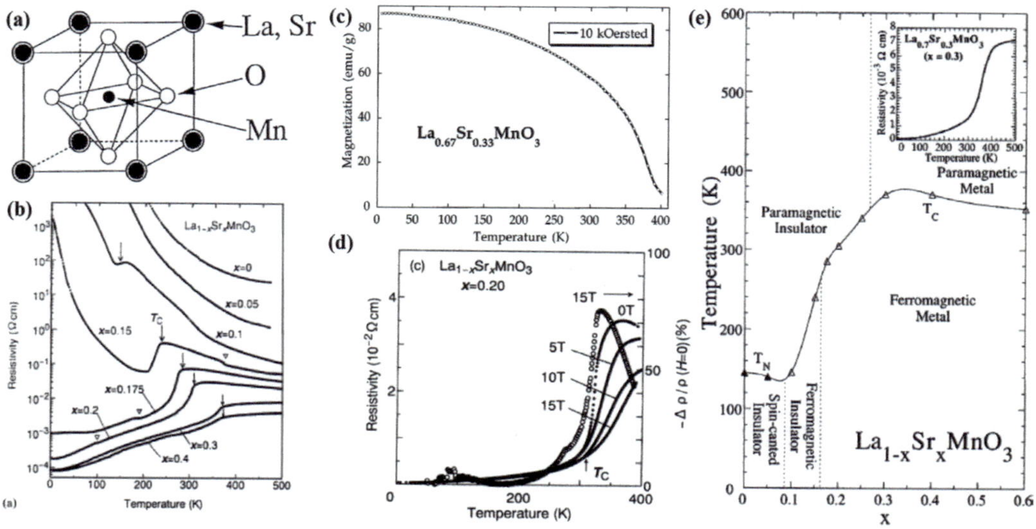

Figure 14. (a) Rhombohedral crystal structure of $La_{0.67}Sr_{0.33}MnO_3$ (LSMO) showing MnO_6 octahedra. (b) Sr hole doping-dependent resistivity of bulk LSMO single crystals showing insulator-to-metal transitions with the kink in resistivity; appearance of magnetic ordering. (c) Magnetization measurements show ferromagnetic ordering above room temperature at T_C ~360 K. (d) Magnetoresistance measurements, i.e., magnetic field-dependent resistivity, show a marked change in resistivity upon the application of a magnetic field, which is defined as the appearance of "colossal magnetoresistance" in LSMO. (e) Temperature- and compositional-dependent phase diagram of LSMO, showing that as Sr hole doping increases, various novel phases can appear in LSMO, making it a rich material that shows complex physics. Adapted and reprinted with permission from Refs. [94, 96, 97, 99]. Copyright 2001 Elsevier Ltd.; Copyright 1995 American Physical Society; Copyright 1996 American Physical Society; Copyright 1996 American Physical Society.

grown on (0 0 1) LSAT and (1 1 0) $NdGaO_3$ substrates, show magnetic ordering below T_C ~340 K. Field-dependent magnetization measurements confirm the presence of perpendicular magnetic anisotropy for films grown on (0 0 1) $LaAlO_3$ substrates, whereas films grown on (0 0 1) $SrTiO_3$ exhibit easy-plane magnetic anisotropy at low T. For the films grown on (0 0 1) LSAT and (1 1 0) $NdGaO_3$ substrates, the presence of distorted easy-pane anisotropy was confirmed by magnetization measurements [103].

Recently, Adamo et al., performed a comprehensive study on the strain effect on the electronic and magnetic properties of LSMO with biaxial strain ranging from −2.3 to +3.2% by using the following substrates: $(1\ 0\ 0)_{pc}$ $LaAlO_3$, (0 0 1) $LaSrGaO_4$, (1 1 0) $NdGaO_3$, (0 0 1) LSAT, (0 0 1) $SrTiO_3$, (1 1 0) $DyScO_3$, (1 1 0) $GdScO_3$, (1 1 0) $SmScO_3$ and (1 1 0) $NdScO_3$ (**Figure 15a–c**) [108]. ρ show a strong strain dependency on the MIT temperature. For low strain values, low temperature ρ is closer to the single-crystal value. The MIT value for films grown on (1 1 0) $NdGaO_3$ is $T_{MIT} > 390$ K, whereas the MIT transition value for films grown on (0 0 1) $SrTiO_3$ is around $T_{MIT} = 370 \pm 10$ K. Films with high compressive strain show fully insulating behavior over the entire T range (**Figure 15d**). For films under higher tensile strain, at $T = 300$ K, ρ is nearly 1 Ω cm. The magnetization behavior of these films are quite similar to that observed by Tsui et al., The observed magnetic behavior and change in T_C exhibit a strong strain dependence (**Figure 15e**), which is in good agreement with the theoretical predictions of Millis et al. [112].

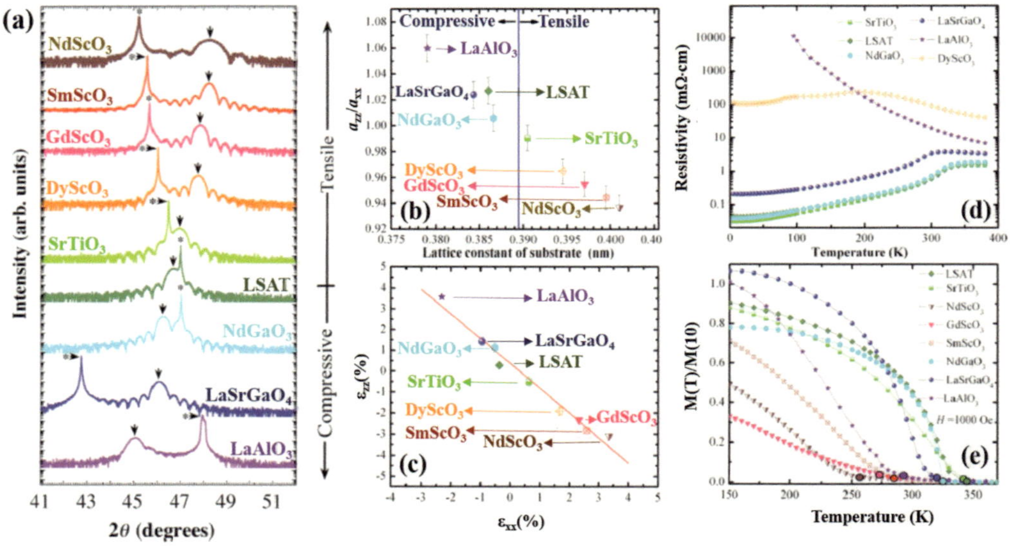

Figure 15. (a) Epitaxial growth of compressive- and tensile-strained $La_{0.67}Sr_{0.33}MnO_3$ (LSMO) films. Films grown on $LaAlO_3$, $LaSrGaO_4$, $NdGaO_3$, and LSAT are subject to compressive strain, whereas films grown on $SrTiO_3$, $DyScO_3$, $GdScO_3$, $SmScO_3$, and $NdScO_3$ are subject to tensile strain. The appearance of thickness fringes in the X-ray diffraction patterns shows the high crystallinity of each film. (b) Change in lattice parameter and (c) evolution of biaxial strain with the judicial choice of substrates for LSMO thin films. (d and e) Resistivity and magnetization of compressive- and tensile-strained films. Highly tensile- or compressive-strained films show insulating behavior (due to the strain and intrinsic atomic disorder effect), whereas metallic behavior was obtained for films that were exposed to moderate tensile or compressive strain. Magnetization measurements show that the ferromagnetic ordering temperature decreases as the tensile strain increases. Reprinted with permission from Ref. [108]. Copyright 2009 AIP Publishing LLC.

4.4. A multiferroic oxide: $BiFeO_3$

Bismuth ferrite, $BiFeO_3$ (BFO), is probably the most promising compound in condensed matter physics [9]. It possess a rhombohedral distorted perovskite structure ($a = b = c = 5.63$ Å, $\alpha = \beta = \gamma = 59.4°$) with space group $R\bar{3}c$ at $T = 300$ K (**Figure 16a**) [113–115]. There is a coexistence of its magnetism and ferroelectricity, that's why $BiFeO_3$ is called a multiferroic (**Figure 16b** and **c**) [115]. In principle, the coexistence of ferroelectricity and magnetism is a very rare phenomenon as ferroelectricity requires B-site ions with d^0 electronic configurations, whereas magnetism requires B-site ions with d^n ($n > 0$) electronic configurations [116]. Therefore, multiferrocity is a very unique phenomenon in condensed matter physics. Bulk single-crystal $BiFeO_3$ shows G-type antiferromagnetic ordering below the Néel temperature at $T_N = 643$ K [117]. In $BiFeO_3$, Fe moments are coupled ferromagnetically with the pseudo-cubic $\{1\ 1\ 1\}_C$ planes, whereas they are antiferromagnetically coupled between neighboring planes. In the bulk, an additional long-range cycloidal magnetic modulation is superimposed on the antiferromagnetic ordering, which results in a rotation of the spin axis through the crystal [43]. It also exhibits ferroelectricity below $T_C \sim 1103$ K with a polarization (P) value of ~3.5 μC/cm² along $(001)_C$ and ~6.1 μC/cm² along $[1\ 1\ 1]_C$ [118]. Recently, the polarization value of highly pure single-crystal $BiFeO_3$ was found to be ~100 μC/cm² along the $[1\ 1\ 1]_C$ [119]. Due to its

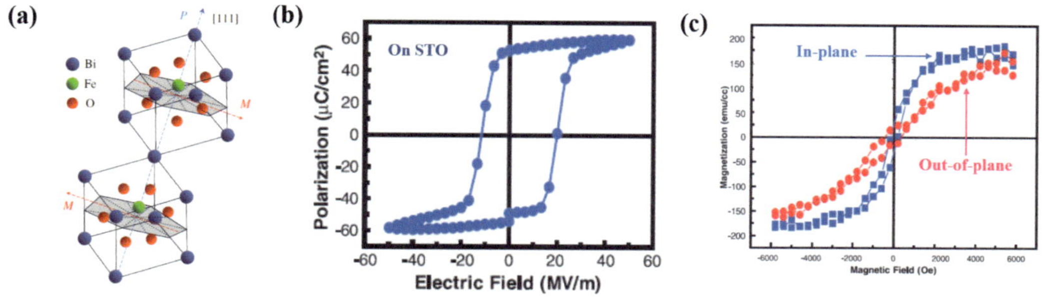

Figure 16. (a) Atomic and magnetic structure of rhombohedral $BiFeO_3$ showing rhombohedral distortion with easy-axis polarization along the $[1\ 1\ 1]_C$ and magnetization plane is perpendicular to the polarization direction. (b) Polarization-electric field hysteresis loop showing that $BiFeO_3$ is ferroelectric with a polarization value of ~60 μC/cm² along the $[1\ 1\ 1]_C$. (c) Magnetization vs. magnetic field hysteresis loop of $BiFeO_3$, showing that the system is antiferromagnetic with a saturation magnetization of ~150 emu/cm³. Blue: in-plane, and red: out-of-plane magnetization-magnetic field loop. Reprinted and adapted with permission [114, 115]. Copyright 2011 The Royal Society; Copyright 2003 American Association for the Advancement of Science.

multiferroic properties above $T = 300$ K, it is a very promising candidate for room-temperature magnetoelectric device applications [120].

Wang et al., were the first to synthesize high quality epitaxial $BiFeO_3$ heterostructures on (0 0 1) $SrTiO_3$ substrates by using the PLD method and observed multiferroic properties [115]. They observed that these films display a spontaneous P ~ 60 μC/cm² at $T = 300$ K, an order of magnitude higher than the bulk value. After this finding, it has been shown that the structural, polar, and magnetic behavior of $BiFeO_3$ thin films are highly sensitive to substrate-induced strain, i.e., the strain promoted by using different lattice-mismatched substrates [121–137].

As said, at $T = 300$ K, $BiFeO_3$ adopts rhombohedral symmetry (R-phase). With the availability of various substrates, compressive (−) or tensile (+) strain can be induced on the film, which changes its structural symmetry as octahedral tilt is highly sensitive to strain. Commercially available perovskite substrates such as (1 1 0) $YAlO_3$, (0 0 1) $LaAlO_3$, (0 0 1) LAST, (1 1 0) $DyScO_3$, (1 1 0) $GdScO_3$, (1 1 0) $SmScO_3$, (1 1 0) $NdScO_3$ and (1 1 0) $PrScO_3$ impose a strain on $BiFeO_3$ of −6.5%, −4.5%, −2.6%, −0.5%, −0.1%, +0.2%, +0.9% and +1.2%, respectively, -ve sign being the compressive strain and +ve sign corresponds to tensile strain. For cases of low compressive or tensile strain, $BiFeO_3$ shows monoclinic (M) phases (**Figure 17a–f**) [120]. In general, for R-phase $BiFeO_3$, the ratio of lattice parameters c/a is ~1. When grown on high lattice-mismatched (0 0 1) $LaAlO_3$ substrate, the ratio of lattice parameters c/a shows a large increase to a value of ~1.23 [124, 128]. This high c/a value can be regarded as being similar to a tetragonal (T_G)-phase or super T_G-phase of $BiFeO_3$. Upon imposing even higher compressive strain by using (1 1 0) $YAlO_3$ substrates, $BiFeO_3$ makes a transition to a fully T_G-phase (**Figure 17g**) [124, 134]. The transition from R-phase to T_G-phase was thought to be an isosymmetric monoclinic symmetry phase transition [128]. However, it was actually shown that octahedral tilt disappears, and the sudden jump in the c/a ratio can be attributed to structural relaxation through an out-of-plane shift. In contrast, imposing moderate tensile strain by using high lattice-mismatched (1 1 0) $NdScO_3$ substrates results in $BiFeO_3$ in a novel orthorhombic (O) phase [132].

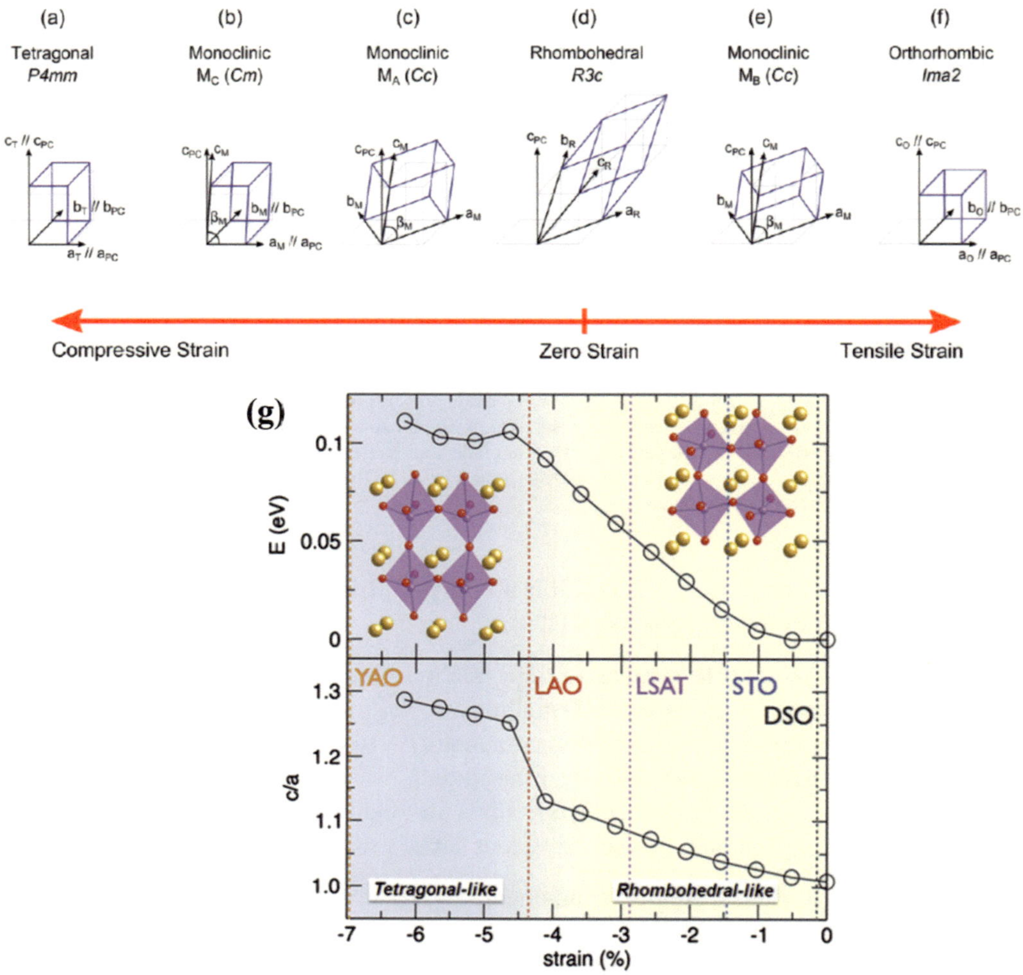

Figure 17. (a–f) Summary of the various crystal structures of $BiFeO_3$ thin films under epitaxial strain, i.e., both compressive and tensile strain. Under different amounts of strain, bulk rhombohedral (*R*) phase to monoclinic (*M*), tetragonal (T_G), and orthorhombic (*O*) phase transitions can be observed in $BiFeO_3$. (g) Calculated overall energy of the system and *c*/*a* ratio for strained $BiFeO_3$. Reprinted with permission from Refs. [120, 124]. Copyright 2014 IOP Publishing Ltd; Copyright 2009 American Association for the Advancement of Science.

First-principle calculations also suggested that with a tensile strain of 2% or more, the orthorhombic phase in $BiFeO_3$ can be stabilized. This is associated with the oxygen octahedral tilt exhibiting short atomic bonds and zig-zag cation displacement patterns. Consequently, the strain effect in $BiFeO_3$ induces very high structural flexibility, which changes the structure from *R*-phase to *M*-, T_G- and *O*-phases or even mixed (*R* + *O*) ones [43, 128, 131].

One of the key aspects of the strain-induced scenario is to increase the out-of-plane *P* in strained $BiFeO_3$ thin films. One of the goals is to induce strain and raise the *P* above 100 μC/cm^2 at T = 300 K. However, in the rhombohedral phase, whatever the growth direction is, the projected *P* along the $[1\ 1\ 1]_C$ increases by only up to ~20% compared to its initial value (**Figure 18a**) [133]. First-principle calculations suggested that suppressing octahedral

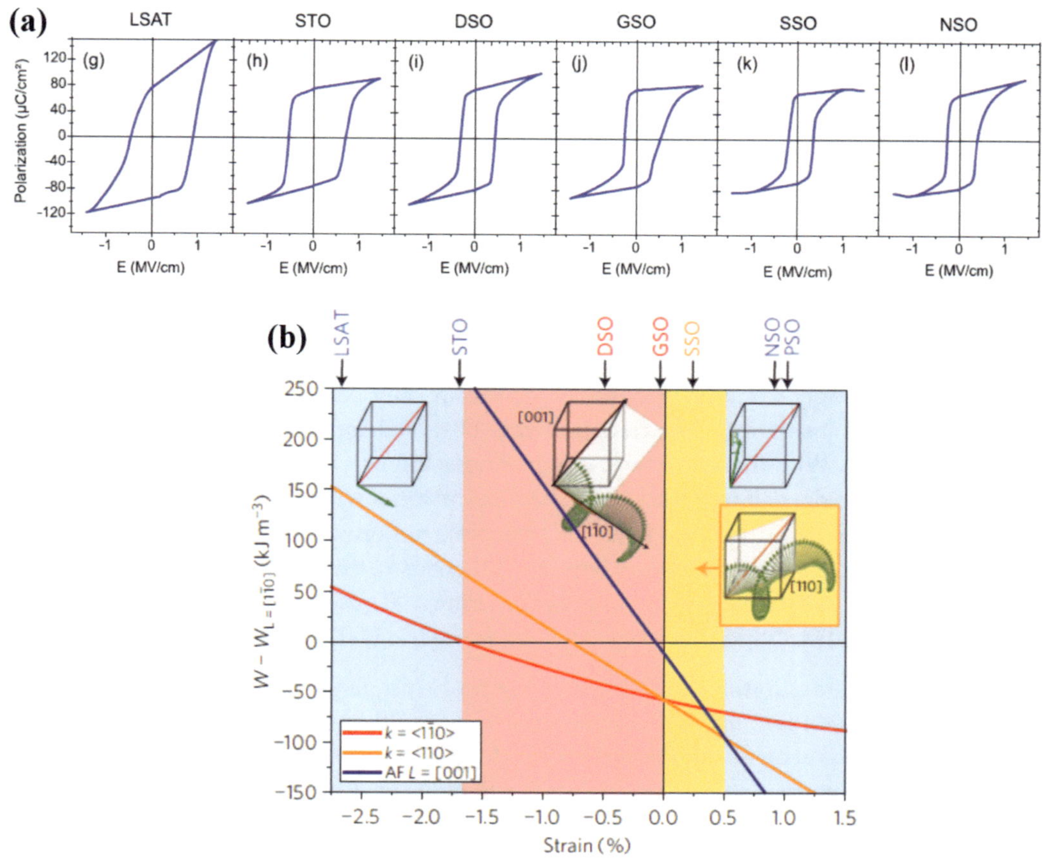

Figure 18. (a) Polarization values of strained $BiFeO_3$ thin films. The large value of polarization for compressively strained films grown on LSAT substrates was attributed to a leakage problem. (b) Magnetic phase diagram shows that with the evolution of strain, various novel magnetic phases can appear in $BiFeO_3$ thin films. The colors correspond to different stable magnetic states (blue: antiferromagnetic; red: type-1 cycloid; orange: type-2 cycloid). The different substrates used for the study are shown above with different colors corresponding to the different magnetic structures of $BiFeO_3$. The spins are denoted by green arrows. Reprinted with permission from Refs. [133, 135]. Copyright 2012 IOP Publishing Ltd; Copyright 2013 Macmillan Publisher Limited.

tilt favors an increase in *P*, whereas the presence of tilts instead favors a change in the direction without changing the *P* [130]. Recent theoretical calculations suggest that if highly strained $BiFeO_3$ is grown with *P4mm* structure, then it would show a high *P* of ~150 μC/cm² along the $[0\,0\,1]_C$ and ~100 μC/cm² along the $[1\,1\,1]_C$, which has been found experimentally in the super tetragonal-phase [137, 138].

To address how G-type antiferromagnetism is affected by strain effects, Sando et al., studied the strain effect within the range from −2.6% (compressive strain) to +1.0 (tensile strain) for the *R*-phase of $BiFeO_3$ (**Figure 18b**) [135]. By using Mössbauer and Raman spectroscopies combined with Landau–Ginzburg theory and effective Hamiltonian calculations, they observed different magnetic structures for different amounts of strain. For low compressive strain, there exists a bulk-like cycloidal spin modulation with non-collinear order, whereas for moderate tensile strain, they observed a new cycloidal phase with a propagation wave vector along [110]. For the high compressive- or tensile-strained case, the magnetic state was found to be a pseudo-collinear antiferromagnetic one.

4.5. A conductive oxide interface: $LaAlO_3$-$SrTiO_3$

$SrTiO_3$ and $LaAlO_3$ are both band insulators with band gaps of ~3.25 eV and ~5.6 eV, respectively [139, 140]. Along [0 0 1], $SrTiO_3$ unit cells consist of charge-neutral layers of SrO^0 and TiO_2^0, whereas $LaAlO_3$ consists of polar layers of LaO^{1+} and AlO_2^{1-} (**Figure 19a** and **b**) [14]. In 2004, Ohtomo and Hwang discovered that if band insulator $LaAlO_3$ (LAO) is grown on top of another band insulator, $SrTiO_3$ (STO), with atomic precision, the interface of LAO-STO can be highly conducting, which results in a two-dimensional electron gas (2DEG) at the interface [14]. Surprisingly, conductivity was observed for only one type of interface: LaO-TiO_2 (n-type), whereas insulating characteristics were observed for AlO_2-SrO (p-type) interfaces (**Figure 19c**) [141]. With this discovery, a tremendous interest has emerged among the thin-film community to deposit each and every thin film material with atomically controlled interfaces. Later, it was found that the LAO-STO interface not only is highly conductive but can also show coexistence of superconductivity and magnetism, the quantum Hall effect, and the Rashba effect [18, 142–145]. Based on various experimental observations, the origin of 2DEG, interfacial charge distribution vs. oxygen vacancy scenario is still under debate [146, 147].

Since the atomic configuration of a substrate's topmost layer plays a key role in forming 2DEGs at interfaces between two non-conducting oxides, it would be highly desirable to study the formation of 2DEGs at oxide interfaces with the strain effects [148–155]. Bark et al., grew LAO-STO interfaces on various substrates, such as (1 1 0) $NdGaO_3$, (0 0 1) LSAT, (1 1 0) $DyScO_3$, and (1 1 0) $GdScO_3$ (**Figure 20a**) [148]. By using four different substrates, the strain ranged from compressive to tensile strain as follow; −1.21% compressive strain for films grown on (1 1 0) $NdGaO_3$, −0.96% compressive strain for films grown on (0 0 1) LSAT, +1.11% tensile strain for films grown on (1 1 0) $DyScO_3$, and +1.59% tensile strain for films grown on (1 1 0) $GdScO_3$. They have shown that imposing tensile strain on $SrTiO_3$ destroys the 2DEG, whereas exerting compressive strain leads to the 2DEG nature being retained but with a reduction of carrier concentration compared to that of unstrained LAO-STO interfaces (**Figure 20b**). Using theoretical calculations, they suggested that this behavior is associated with the built-in polarization in $SrTiO_3$, as it was observed that with induced strain, polarization can have built up in pure $SrTiO_3$ as well [44]. This polarization is directed away from the interface and creates a negative polarization opposing that of the polar $LaAlO_3$ layer. It has also been calculated that the distortion of interfacial Ti—O octahedra enhances with increases in in-plane compressive strain, which also modulates the carrier concentration [152]. Applied in-plane compressive strain also reduces the carrier concentration. On the other hand, under tensile strain, the interfacial charge carrier density increases, which is consistent with the theoretical calculations [155]. It has been found that when moving from the compressive-strained to the tensile-strained scenario at the LAO-STO interface, the Ti—O bond length becomes elongated, which confines the Ti d_{xy} orbital electrons at the interface, thus increasing the sheet carrier concentration (n_s) at the interface. For the compressively strained scenario, the Ti—O bond length decreases and the Ti d_{xy} orbital cannot hold all the electrons at the interface; hence, the remaining electrons are transferred to a deeper layer, reducing the carrier concentration at the interface.

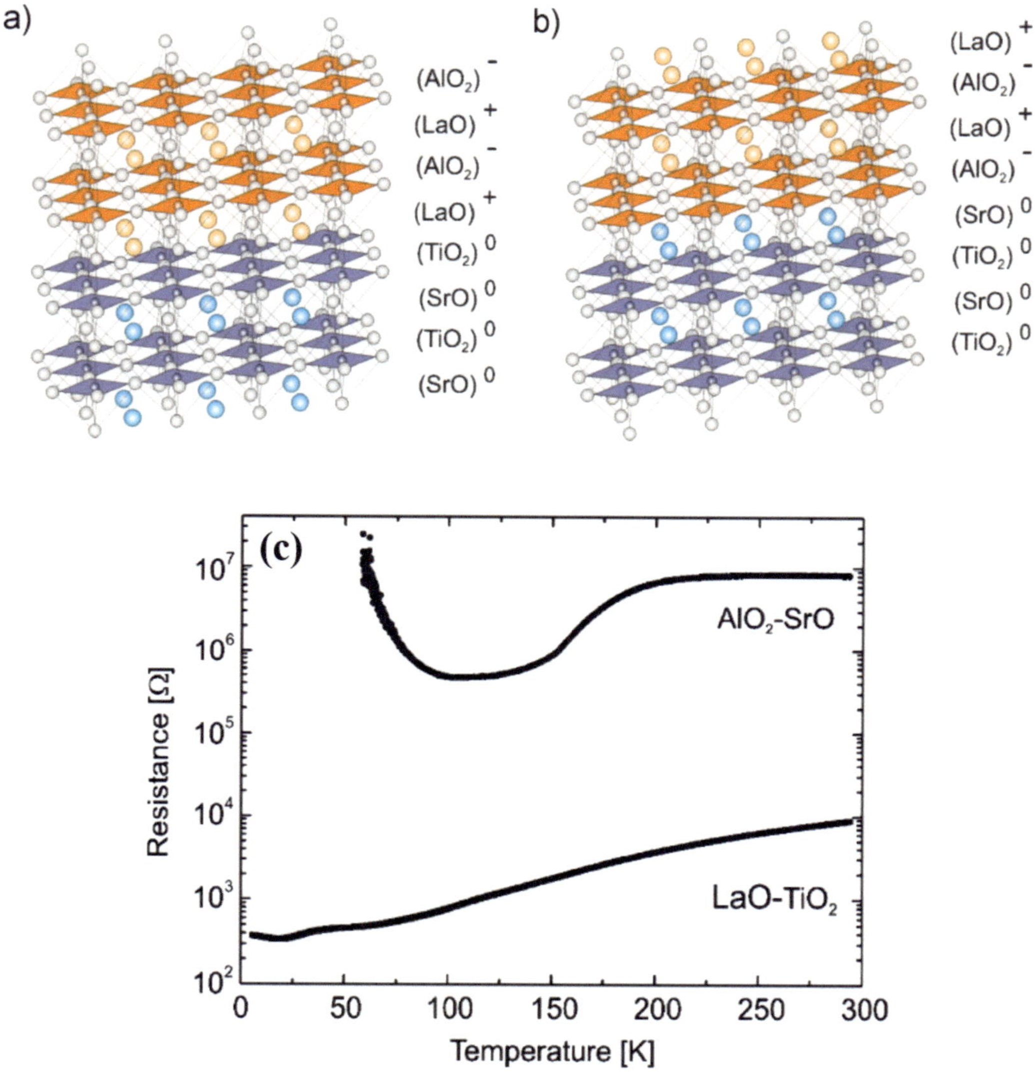

Figure 19. (a) and (b) Schematic representation of two possible interfaces between change neutral $SrTiO_3$ (band insulator) and polar $LaAlO_3$ (band insulator) materials showing the composition and ionic distribution. Depending on the topmost layer termination of the (0 0 1) $SrTiO_3$ substrate, two possible interfaces can form: (1) LaO^{1+}-TiO_2^{0}, and (2) AlO_2^{1-}-SrO^{0}. (c) Temperature-dependent resistivity of these two types of interfaces. LaO^{1+}-TiO_2^{0} interfaces show metallicity and the formation of 2-dimesnioanl electron gas (2DEG) with n-type charge carriers at the interface, whereas AlO_2^{1-}-SrO^{0} interfaces show insulating characteristics at the interface. Reprinted with permission from Refs. [14, 141]. Copyright 2004 Nature Publishing Group; Copyright 2009 WILEY-VCH Verlag GmbH & Co. KGaA, Weinheim.

To observe the correlation and strain effect induced by the topmost polar layer, Ariando et al., grew various combination of polar/non-polar ABO_3 perovskites, e.g., $NaAlO_3/SrTiO_3$, $PrAlO_3/SrTiO_3$, and $NdGaO_3/SrTiO_3$ (**Figure 20c**) [150]. They found that these interfaces are also conducting and form 2DEGs. They also claimed that the interfacial strain and electron correlation caused by the polar layers seem to control the carrier density and mobility at the interface (**Figure 20d**). The presence of large octahedral distortion due to strain also plays an important role in observing

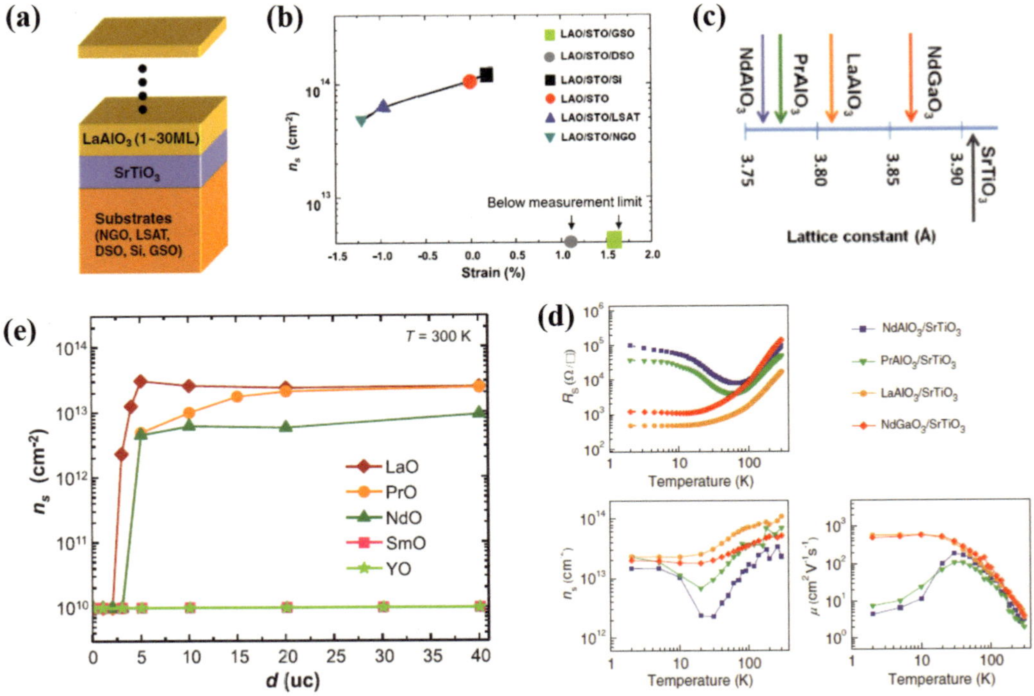

Figure 20. (a) Schematic representation of a $LaAlO_3$-$SrTiO_3$ interface grown on various substrates (i.e., on $GdScO_3$, $DyScO_3$, LSAT and $NdGaO_3$). (b) Carrier concentration at room temperature at the LAO-STO interface under various epitaxial strains. For tensile-strained films, the carrier concentration is below the measurement limit. (c) The strain effect at the interface and lattice parameters of various perovskites. (d) Sheet resistance (R_S), carrier concentration (n_s), and mobility (μ) of various polar/non-polar interfaces. (e) Carrier concentration at the interface of LAO-STO with monolayer growth of rare earth oxides. For some rare earth cases, the interfaces are conductive, whereas insulating interfaces are formed in some cases involving rare-earth oxides, showing the effect of correlation on the formation of 2DEGs at interfaces. Reprinted with the permission from Refs. [148–150]. Copyright 2011 National Academy of Sciences; Copyright 2011 American Association for the Advancement of Science; Copyright 2012 American Physical Society.

these novel phases. Instead of growing various rare earth-based polar ABO_3 perovskites, Jang et al., directly grew monolayers of rare-earth oxides (*R*O, R = La, Pr, Nd, Sm, and Y) to observe the correlation effect at interfaces (**Figure 20e**) [149]. Surprisingly, they found that some oxides (La, Pr, and Nd) forms 2DEGs at the interface, whereas an insulating interface is formed in the case of SmO and YO. Based on the observations and theoretical calculations, they claimed that this is due to the correlation effect at the interface. Independent theoretical calculation also suggest that in-plane strain can induce metal-insulator transitions at oxide interfaces [155].

4.6. A strongly correlated metal: $LaNiO_3$

The physics of strongly correlated materials, i.e., those having strong electronic correlations, is remarkably rich and complex and cannot be understood within the framework of conventional theories of metals and insulators [21, 22]. For example, in strongly correlated nickelate materials, spin, lattice charge, and orbital degrees of freedom result in competing interactions. Due to this, these materials show exotic phases [156–158]. Among strongly correlated materials, nickelates, more specifically $LaNiO_3$ is a strongly correlated Mott metal [159, 160]. $LaNiO_3$ has a highly distorted rhombohedral structure with lattice parameters of $a = b = 5.4573$ Å, $c = 13.1462$ Å, and $\alpha = \beta = \gamma = 60.49°$ and with space group $R\bar{3}c$ (**Figure 21a**) [161]. Its pseudo-cubic lattice constant is $a_{pc} = 3.84$ Å. It is highly metallic over the entire temperature range (**Figure 21b**) [162]. It is the only member in the perovskite nickelate family ($RENiO_3$, RE = rare earths) which shows no long-range magnetic ordering [156–158].

The functionalities of strongly correlated rare-earth nickelates are highly sensitive to external perturbation, e.g., chemical pressure and atmospheric pressure [156–158]. Thus, it would be very interesting to investigate the ability of strain-based perturbation (i.e., by using various lattice-mismatched substrates) to obtain novel functionalities in $LaNiO_3$. Several groups have reported the strain dependent transport properties in epitaxial $LaNiO_3$ thin films [163–172]. Moon et al., investigated in detail the transport properties of tensile- and compressive-strained $LaNiO_3$ systems. They grew 10-unit cell ultra-thin $LaNiO_3$ films on (1 1 0) $YAlO_3$, (0 0 1) $SrLaAlO_4$, (0 0 1) $LaAlO_3$, (0 0 1) $SrLaGaO_4$, (0 0 1) LSAT, (0 0 1) $SrTiO_3$, and (1 1 0) $GdScO_3$ substrates (**Figure 22a**) [167–169]. They investigated ρ of tensile-strained films and observed that when the tensile strain increases, $LaNiO_3$ gradually evolves from the Mott to the Mott-Anderson regime; i.e., correlation and disorder play a crucial role at low temperature (**Figure 22b**) [163, 167].

The authors have also investigated the Hall effect ($1/R_H \propto T$; R_H = Hall coefficient) in both the compressive- and tensile-strained cases (**Figure 22c**) [168]. They claimed that the evolution of the linear T dependent transport coefficient is quite similar to that of hole-doped cuprate superconductors. By using density functional theoretical (DFT) calculations, they claimed that strain-induced changes in transport properties arise from changes in the low-energy electronic band structure that induces self-doping, i.e., a transfer of charge between the O p and Ni d states. Using detailed quantitative structural analysis and theoretical calculations May et al., found that strain systematically modifies both the Ni–O–Ni bond angles and Ni–O lengths in this functional perovskite oxide (**Figure 22d**) [164, 172], which has strong effect on its strain-dependent transport properties.

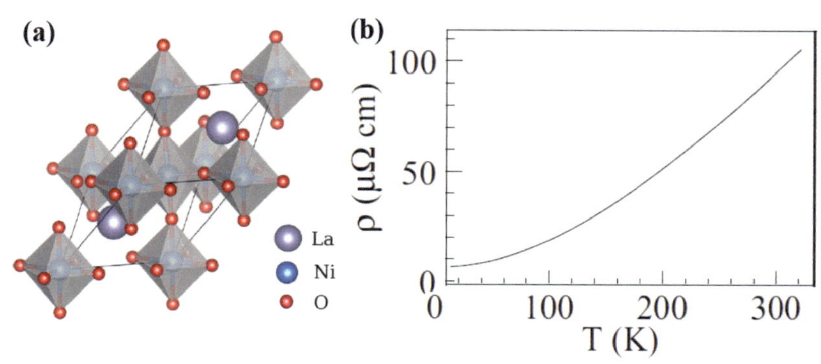

Figure 21. (a) Schematic representation of the distorted rhombohedral crystal structure of $LaNiO_3$ showing highly distorted NiO_6 octahedra with lattice parameters of $a = b = 5.4573$ Å and $c = 13.1462$ Å. The pseudo-cubic unit cell is ~3.84 Å. (b) Resistivity of bulk polycrystalline $LaNiO_3$ showing it is a paramagnetic metal without any indication of magnetic ordering over the entire temperature range. Reprinted with permission from Refs. [161, 162]. Copyright 2015 Elsevier B. V. Ltd.; Copyright 2014 American Physical Society.

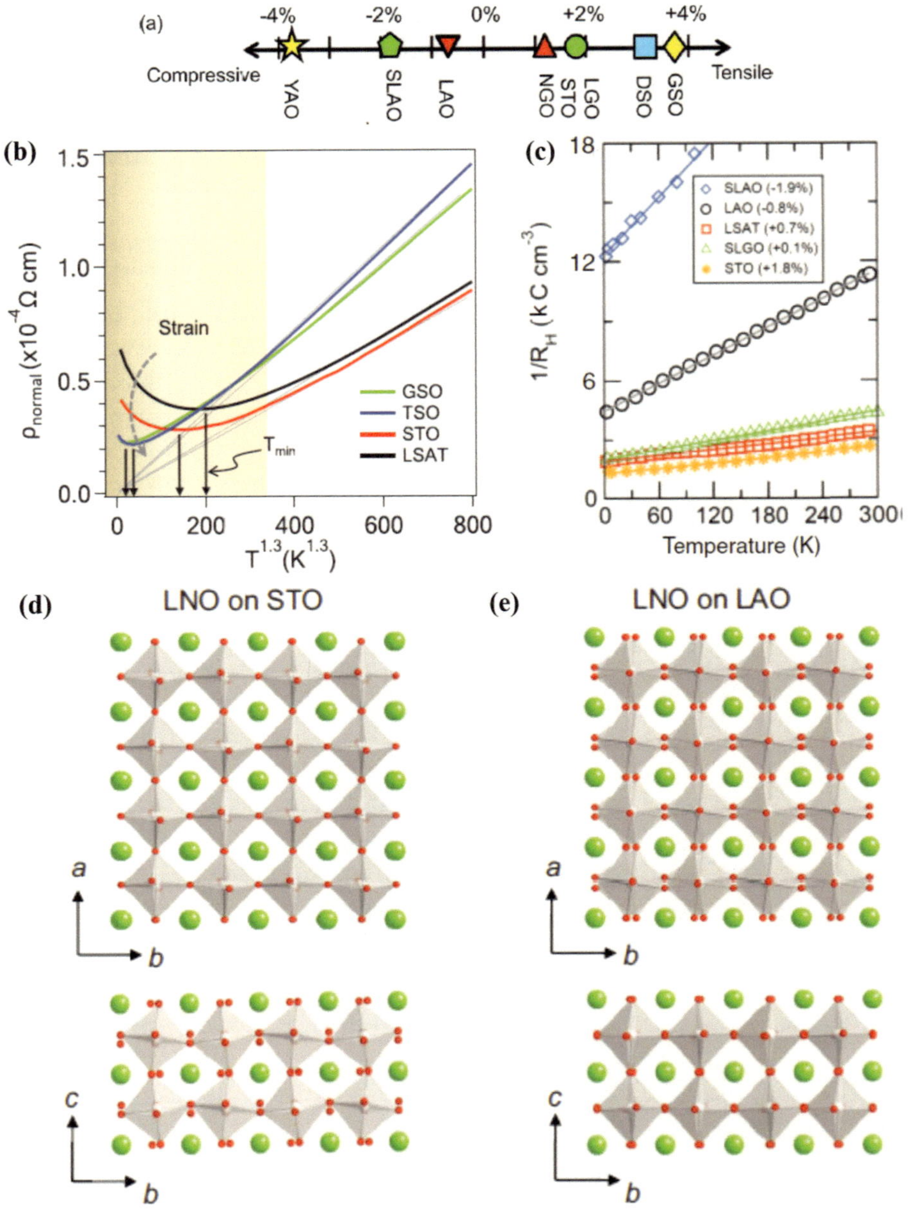

Figure 22. (a) Schematic representation of compressive and tensile strain imposed on $LaNiO_3$ by various metal oxide substrates. (b) Low-temperature resistivity of strained $LaNiO_3$ films. With an increase in tensile strain, resistivity at room temperature increases, although the resistivity minima also decrease. (c) The temperature-dependent Hall coefficient ($1/R_H$) for strained $LaNiO_3$ films grown on various substrates. With an increase in compressive strain, $1/R_H$ becomes almost linearly dependent on temperature ($1/R_H \propto T$), thus bearing a striking resemblance to the behavior of cuprate superconductors. (d) Tensile strain increases rotation of octahedra along the [1 0 0] and [0 1 0], and decreases it along the [0 0 1]. (e) Compressive strain reduces the rotation along the [1 0 0] and [0 1 0], and increases it along the [0 0 1]. Reprinted with permission from Refs [164, 167, 168]. Copyright 2010 American Physical Society; Copyright 2013 American Chemical Society; Copyright 2011 IOP Publishing Ltd and Deutsche Physikalische Gesellschaft.

5. Summary

In principle, material functionalities arise from the coupling between spin, lattice, charge, and orbital degrees of freedom. Lattice strain is thus found to be a unique way to engineering the functionalities of many TMOs, which modifies above energy scales. Here, we presented the effect of strain-dependent functionalities of various TMO-based thin films: (1) a high temperature superconductor, $La_{1.85}Sr_{0.15}CuO_4$, (2) a highly conductive oxide, $SrRuO_3$, (3) a colossal magnetoresistive metal, $La_{0.67}Sr_{0.33}MnO_3$, (4) a multiferroic oxide, $BiFeO_3$, (5) a conducting oxide interface, $LaAlO_3$-$SrTiO_3$, and (6) a strongly correlated metal, $LaNiO_3$. The aforementioned materials all show rich and complex structural, electronic, magnetic and polar phase diagrams that are dependent on epitaxial strain, which is mainly caused by modifications of their crystal structures and the effects of these modification on the coupling of their various degrees of freedoms (relevant energy scales). More specifically, tailoring the shape, size, and position of BO_6 octahedra by strain give rise to new functionalities in ABO_3 perovskite oxides. Strain-dependent MITs, increase in magnetic transition temperature and ferroelectric polarization can be observed in these materials. Obtaining novel properties by designing artificial oxide heterostructures and the subsequent new physics resulting from the strain effect have always been an interesting topic of research among the thin-film community as cheap and environment friendly oxide thin film-based electronic devices are highly in demand in industry. Although there has been vast progress in the strain-dependent tuning of material properties, there is still long way to go to fully understand the intrinsic mechanisms and theoretical developments behind these strain-dependent phenomena. Nevertheless, strain has provided an avenue to explore materials with novel functionalities. We believe that our experimental investigations along with insightful explanations will provide readers with an easier way to understand the strain effect in epitaxial oxide heterostructures and utilize it to explain the fundamental physics and to commercialize oxide-based electronic devices.

Acknowledgements

YHJ was supported by the National Research Foundation (NRF) of Korea (SRC-2011-0030786 and 2015R1D1A1A02062239).

Author details

Abhijit Biswas and Yoon Hee Jeong*

*Address all correspondence to: yhj@postech.ac.kr

Department of Physics, Pohang University of Science Technology, Pohang, Republic of Korea

References

[1] Ashcroft NW, Mermin DN. Solid State Physics. 1st ed. New York: Holt, Rinehart and Winston; 1976. 826 p. DOI: 10.1002/piuz.19780090109

[2] Zachariev ZT, editor. Polycyrstalline Materials—Theoretical and Practical Aspects. 1st ed. Croatia: InTech; 2012. DOI: 10.5772/1391

[3] Rudolph P, editor. Handbook of Crystal Growth: Bulk Crystal Growth. 2nd ed. North-Holland: Elsevier; 2015. 1418 p. ISBN: 978-0-444-63303-3

[4] http://wolfweb.unr.edu/homepage/yjiang/jixi_zhang.html

[5] Kuech T, editor. Handbook of Crystal Growth: Thin Films and Epitaxy. 2nd ed. North-Holland: Elsevier; 2015. ISBN: 978-0-444-63304-0

[6] Imada M, Fujimori A, Tokura Y. Metal-insulator transitions. Reviews of Modern Physics. 1998;**70**:1039-1263. DOI: 10.1103/RevModPhys.70.1039

[7] Prellier W, Lecoeur P, Mercey B. Colossal-magnetoresistive manganite thin films. Journal of Physics: Condensed Matter. 2001;**13**:R915. DOI: 10.1088/0953-8984/13/48/201

[8] Norton DP. Synthesis and properties of epitaxial electronic oxide thin-film materials. Materials Science and Engineering Reports. 2004;**43**:139-247. DOI: 10.1016/j.mser.2003.12.002

[9] Ramesh R, Spaldin NA. Multiferroics: Progress and prospects in thin films. Nature Materials. 2007;**6**:21-29. DOI: 10.1038/nmat1805

[10] Blamire G, MacManus-Driscoll JL, Mathur ND, Barber ZH. The materials science of functional oxide thin films. Advanced Materials. 2009;**21**:3827-3840. DOI: 10.1002/adma.200900947

[11] Martin LW, Schlom DG. Advanced synthesis techniques and routes to new single-phase multiferroics, Current Opinion in Solid State & Materials Science. 2012;**16**:199-215. DOI: 10.1016/j.cossms.2012.03.001

[12] Monkman EJ, Adamo C, Mundy JA, Shai DE, Harter JW, Shen D, Burganov B, Muller DA, Schlom DG, Shen KM. Quantum many-body interaction in digital oxide superlattices. Nature Materials. 2012;**11**:855-859. DOI: 10.1038/nmat3405

[13] Yu P, Chu YH, Ramesh R. Oxide interfaces: Pathway to novel phenomena. Materials Today. 2012;**15**:320-327. DOI: 10.1016/S1369-7021(12)70137-2

[14] Ohtomo A, Hwang HY. A high-mobility electron gas at the $LaAlO_3/SrTiO_3$ heterointerface. Nature (London). 2004;**427**:423-426. DOI: 10.1038/nature02308

[15] Dawber M, Bousquet E. New developments in artificially layered ferroelectric oxide superlattices. MRS Bulletin. 2013:**38**:1048-1055. DOI: 10.1557/mrs.2013.263

[16] Nagaosa N, Sinova J, Onoda A, MacDonald AH, Ong NP. Anomalous Hall effect. Reviews of Modern Physics. 2010;**82**:1539-1592. DOI: 10.1103/RevModPhys.82.1539

[17] Zubko P, Gariglio S, Gabay M, Ghosez P, Triscone JM. Interface physics in complex oxide heterostructures. Annual Review of Condensed Matter Physics. 2011;**2**:141-165. DOI: 10.1146/annurev-conmatphys-062910-140445

[18] Hwang HY, Iwasa Y, Kawasaki M, Keimer B, Nagaosa N, Tokura Y. Emergent phenomena at oxide interfaces. Nature Materials. 2012;**11**:103-113. DOI: 10.1038/nmat3223

[19] Krempa WW, Chen G, Kim YB, Balents L. Correlated quantum phenomena in the strong spin-orbit regime. Condensed Matter Physics. 2014;**5**:57-82. DOI: 10.1146/annurev-conmatphys-020911-125138

[20] Zhang J, Averitt RD. Dynamics and control in complex transition metal oxides. Annual Review of Materials Research. 2014;**44**:19-43. DOI: 10.1146/annurev-matsci-070813-113258

[21] Tokura Y. Correlated-electron physics in transition-metal oxides. Physics Today. 2003;**56**:50-55. DOI: 10.1063/1.1603080

[22] Dagotto E. Complexity in strongly correlated electronic systems. Science. 2005;**309**:257-262. DOI: 10.1126/science.1107559

[23] Tokura Y, Nagaosa N. Orbital physics in transition-metal oxides. Science. 2000;**288**:462-468. DOI: 10.1126/science.288.5465.462

[24] Cheong SW. Transition metal oxides: The exciting world of orbitals. Nature Materials. 2007;**6**:927-928. DOI: 10.1038/nmat2069

[25] Pan L, Zhu G, editors. Perovskite Materials - Synthesis, Characterisation, Properties, and Applications, 1st ed. Croatia: InTech; 2016. DOI: 10.5772/60649. ISBN: 978-953-51-2245-6

[26] Johnsson M, Lemmens P. Perovskites and thin films-crystallography and chemistry. Journal of Physics: Condensed Matter. 2008;**20**:264001. DOI: 10.1088/0953-8984/20/26/264001

[27] Goldschmidt VM. Die Gesetze der Krystallochemie. Naturwissenschaften. 1926;**14**:477-485. DOI: 10.1007/BF01507527

[28] Schlom DG, Chen LQ, Pan X, Schmehl A, Zurbuchen MA. A thin film approach to engineering functionality into oxides. Journal of the American Ceramic Society. 2008;**91**:2429-2454. DOI: 10.1111/j.1551-2916.2008.02556.x

[29] Dijkkamp D, Venkatesan T, Wu XD, Shaheen SA, Jisrawi N, Min-Lee YH, Mclean WL, Croft M. Preparation of Y-Ba-Cu oxide superconductor thin films using pulsed laser evaporation from high T_c bulk material. Applied Physics Letters. 1987;**51**:619-621. DOI: 10.1063/1.98366

[30] Chrisey DB, Hubler GK. Pulsed Laser Deposition of Thin Films. 1st ed. John Wiley & Sons, New York; 1994. 648 p. ISBN: 0-471-59218-8

[31] Christen HM, Eres G. Recent advances in pulsed-laser deposition of complex oxides. Journal of Physics: Condensed Matter. 2008;**20**:264005-264020. DOI: 10.1088/0953-8984/20/26/264005

[32] Koster G, Huijben M, Rijnders G, editors. Epitaxial Growth of Complex Metal Oxides. 1st ed. Elsevier Science & Technology, United Kingdom; 2015. 504 p. ISBN: 978-1-78242-245-7

[33] Opel M. Spintronic oxide grown by laser-MBE. Journal of Physics D: Applied Physics. 2012;**45**:033001-033031. DOI: 10.1088/0022-3727/45/3/033001

[34] McCray WP. MBE deserves a place in the history books. Nature Nanotechnology. 2007;**2**:259-261. DOI: 10.1038/nnano.2007.121

[35] Cho AY, Arthur JR Jr. Molecular beam epitaxy. Progress in Solid State Chemistry. 1975;**10**:157-192. DOI: 10.1016/0079-6786(75)90005-9

[36] Moustakas TD. Molecular beam epitaxy: Thin film growth and surface studies. MRS Bulletin. 1988;**13**:29-36. DOI: 10.1557/S0883769400063892

[37] Nie YF, Zhu Y, Lee CH, Kourkoutis LF, Mundy JA, Junquera J, Ghosez P, Baek DJ, Sung S, Xi XX, Shen KM, Muller DA, Schlom DG. Atomically precise interfaces from non-stoichiometric deposition. Nature Communications. 2014;**5**:4530-4537. DOI: 10.1038/ncomms5530

[38] Phillips JM. Substrate selection for high-temperature superconducting thin films. Journal of Applied Physics. 1996;**79**:1829-1848. DOI: 10.1063/1.362675

[39] Jiang LQ, Guo JK, Liu HB, Zhu M, Zhou X, Wu P, Li CH. Prediction of lattice constant in cubic perovskites. Journal of Physics and Chemistry of Solids. 2006;**67**:1531-1536. DOI: 10.1016/j.jpcs.2006.02.004

[40] Schlom DG, Chen LQ, Eom CB, Rabe KM, Streiffer SK, Triscone JM. Strain tuning of ferroelectric thin films. Annual Review of Materials Research. 2007;**37**:589-626. DOI: 10.1146/annurev.matsci.37.061206.113016

[41] Schlom DG, Chen LQ, Fennie CJ, Gopalan V, Muller DA, Pan X, Ramesh R, Uecker R. Elastic strain engineering of ferroic oxides. MRS Bulletin. 2014;**39**:118-130. DOI: 10.1557/mrs.2014.1

[42] Damodaran AR, Agar JC, Pandya S, Chen Z, Dedon L, Xu R, Apgar B, Saremi S, Martin LW. New modalities of strain-control of ferroelectric thin films. Journal of Physics: Condensed Matter. 2016;**28**:263001-263036. DOI: 10.1088/0953-8984/28/26/263001

[43] Yang Y, Infante IC, Dkhil B, Bellaiche L. Strain effect on multiferroic $BiFeO_3$ films. Comptes Rendus Physique. 2015;**16**:193-203. DOI: 10.1016/j.crhy.2015.01.010

[44] Haeni JH, Irvin P, Chang W, Uecker R, Reiche P, Li YL, Choudhury S, Tian W, Hawley ME, Craigo B, Tagantsev AK, Pan XQ, Streiffer SK, Chen LQ, Kirchoefer SW, Levy J, Schlom DG. Room-temperature ferroelectricity in strained $SrTiO_3$. Nature (London). 2004;**430**:758-761. DOI: 10.1038/nature02773

[45] Lee JH, Fang L, Vlahos LFE, Ke X, Jung YW, Kourkoutis LF, Kim JW, Ryan PJ, Heeg T, Roeckerath M, Goian Bernhagen M, Uecker R, Hammel PC, Rabe KM, Kamba S, Schubert J, Freeland JW, Muller DA, Fennie CJ, Schiffer P, Gopalan V, Halperin EJ, Schlom DG. A strong ferroelectric ferromagnet created by means of spin–lattice coupling. Nature (London). 2010;**466**:954-958. DOI: 10.1038/nature09331

[46] Schlom DG, Mannhart J. Oxide electronics: Interface takes charge over Si. Nature Materials. 2011;**10**:168-169. DOI: 10.1038/nmat2965

[47] Lorentz M, Ramachandra Rao MS, Venkatesan T, Fortunato E, Barquinha P, Branquinho R, Salgueiro D, Martins R, Carlos E, Liu A, Shan FK, Grundmann M, Boschker H, Mukherjee J, Priyadarshini M, DasGupta N, Rogers DJ, Teherani FH, Sandana EV, Bove P, Rietwyk Zaban A, Veziridis A, Weidenkaff A, Muralidhar M, Murakami M, Abel S, Fompeyrine J, Zuniga-Perez J, Ramesh R, Spaldin NA, Ostanin S, Borisov V, Mertig I, Lazenka V, Srinivasan G, Prellier W, Uchida M, Kawasaki M, Pentcheva R, Getenwart P, Miletto Granozio FM, Fontcuberta J, Pryds N. The 2016 oxide electronic materials and oxide interfaces roadmap. Journal of Physics D: Applied Physics. 2016;**49**:433001. DOI: 10.1088/0022-3727/49/43/433001

[48] Rondinelli JM, Spaldin NA. Structure and properties of functional oxide thin films: Insights from electronic-structure calculations. Advanced Materials. 2011;**23**:3363-3381. DOI: 10.1002/adma.201101152

[49] Rondinelli JM, Coh S. Large issosymmetric reorientation of oxygen octahedra rotation axes in epitaxially strained perovskites. Physical Review Letters. 2011;1**06**:235502-235505. DOI: 10.1103/PhysRevLett.106.235502

[50] Rondinelli JM, May SJ, Freeland JW. Control of octahedral connectivity in perovskite oxide heterostructures: An emerging route to multifunctional materials discovery. MRS Bulletin. 2012;**37**:261-270. DOI: 10.1557/mrs.2012.49

[51] Rondinelli JM, Fennie CJ. Octahedral rotation-induced ferroelectricity in cation ordered perovskites. Advanced Materials. 2012;**24**:1961-1968. DOI: 10.1002/adma.201104674

[52] Glazer AM. The classification of tilted octahedra in perovskites. Acta Crystallographica Section A 1972;**B28**:3384-3392. DOI: 10.1107/S0567740872007976

[53] Glazer AM. Simple ways of determining perovskite structures. Acta Crystallographica Section A 1975;**A31**:756-762. DOI: 10.1107/S0567739475001635

[54] Harrison WA. Electronic Structure and the Physical Properties of Solids: The Physics of the Chemical Bonds. 1st ed. New York: Dover; 1989. 608 p. ISBN: 0486660214

[55] Brooks CM, Kourkoutis LF, Heeg T, Schubert J, Muller DA, Schlom DG. Growth of homoepitaxial $SrTiO_3$ thin films by molecular-beam epitaxy. Applied Physics Letters. 2009;**94**:162905-162907. DOI: 10.1063/1.3117365

[56] Bednorz JG, Müller KAZ. Possible high T_c superconductivity in the Ba–La–Cu–O system. Physik B–Condensed Matter. 1986;**64**:189-193. DOI:10.1007/BF01303701

[57] Takagi H, Batlogg B, Kao HL, Kwo J, Cava RJ, Krajewski JJ, Pech WF Jr. Systematic evolution of temperature-dependent resistivity in $La_{2-x}Sr_xCuO_4$. Physical Review Letters. 1992;**69**:2975-2978. DOI: 10.1103/PhysRevLett.69.2975

[58] Cava RJ, Santoro A, Johnson DW, Rhodes WW. Crystal structure of the high-temperature superconductor $La_{0.85}Sr_{0.15}CuO_4$ above and below Tc. Physical Review B. 1987;**35**:6716. DOI: 10.1103/PhysRevB.35.6716

[59] Emery VJ. Theory of high-T_C superconductivity in oxides. Physical Review Letters. 1987;**58**:2794-2797. DOI: 10.1103/PhysRevLett.58.2794

[60] Chang J, Mansson M, Pailhès S, Claesson T, Lipscombe OI, Hayden SM, Patthey L, Tjernberg O, Mesot J. Anisotropc breakdown of Fermi liquid quasiparticle excitations in overdped $La_{2-x}Sr_xCuO_4$. Nature Communications. 2013;**4**:2559-2563. DOI: 10.1038/ncomms3559

[61] Dagotto E. Correlated electrons in high-temperature superconductors. Reviews of Modern Physics. 1994;**66**:763-840. DOI: 10.1103/RevModPhys.66.763

[62] Božović I, He X, Wu J, Bollinger AT. Dependence of the critical temperature in overdoped copper oxides on superfluid density. Nature. 2016;**536**:309-313. DOI: 10.1038/nature19061

[63] Sato H, Naito M. Increase in the superconducting transition temperature by anisotropic strain effect in (0 0 1) $La_{1.85}Sr_{0.15}CuO_4$ thin films on $LaSrAlO_4$ substrates. Physica C. 1997;**274**:221-226. DOI: 10.1016/S0921-4534(96)00675-2

[64] Locquet JP, Perret J, Fompeyrine J, Mächler E, Seo JW, Van Tendeloo G. Doubling the critical temperature of $La_{1.9}Sr_{0.1}CuO_4$ using epitaxial strain. Nature. 1998;**394**:453-456. DOI: 10.1038/28810

[65] Si W, Li HC, Xi XX. Strain and oxygenation effects on superconductivity of $La_{1.85}Sr_{0.15}CuO_4$ thin films. Applied Physics Letters. 1999;**74**:2839-2841. DOI: 10.1063/1.124031

[66] Sato H, Tsukada A, Naito M, Matsuda A. $La_{2-x}Sr_xCuO_y$ epitaxial thin films (x = 0 to 2): Structure, strain, and superconductivity. Physical Review B. 2000;**61**:12447-12456. DOI: 10.1103/PhysRevB.61.12447

[67] Sato H, Tsukada A, Naito M, Matsuda A. La-214 thin films under epitaxial strain. Physica C. 2000;**341-248**;1767-1770. DOI: 10.1016/j.physc.2004.03.218

[68] Božović I, Lonvenov G, Belca I, Narimbetov B, Sveklo I. Epitaxial strain and superconductivity in $La_{2-x}Sr_xCuO_4$ thin films. Physical Review Letters. 2002;**89**:107001. DOI: 10.1103/PhysRevLett.89.107001

[69] Sato H. Enhanced superconductivity by reducing magnetism in strained $La_{2-x}Sr_xCuO_4$ films. Physica C. 2008;**468**:2366-2368. DOI: 10.1016/j.physc.2008.09.001

[70] Meyers TL, Jiang L, Park S, Egami T, Lee HN. Strain-relaxation and critical thickness of epitaxial $La_{1.85}Sr_{0.15}CuO_4$ films. APL Materials. 2015;**3**:126102. DOI: 10.1063/1.4937170

[71] Koster G, Klein L, Siemons W, Rijnders G, Steven Dodge J, Eom CB, Blank DHA, Beasley MR. Structure, physical properties, and applications of $SrRuO_3$ thin films. Reviews of Modern Physics. 2012;**84**:253-298. DOI: 10.1103/RevModPhys.84.253

[72] Klein L, Dodge JS, Ahn CH, Reiner JW, Geballe TH, Beasley MR, Kapitulnik A. Transport and magnetization in the badly metallic itinerant ferromagnet $SrRuO_3$. Journal of Physics: Condensed Matter. 1996;**8**:10111-10126. DOI: 10.1088/0953-8984/8/48/026

[73] Allen PB, Berger H, Chauvet O, Forro L, Jarlborg T, Junod A, Revaz B, Santi G. Transport properties, thermodynamics properties, and electronic structure of $SrRuO_3$. Physical Review B. 1996;**53**:4393-4398. DOI: 10.1103/PhysRevB.53.4393

[74] Gan Q, Rao RA, Eom CB, Wu L, Tsui F. Lattice distortion and uniaxial magnetic anisotropy in single domain epitaxial (1 1 0) films of $SrRuO_3$. Journal of Applied Physics. 1999;**85**:5297-5299 DOI: 10.1063/1.369859

[75] Kennedy BJ, Hunter BA. High-temperature phases of $SrRuO_3$. Physical Review B. 1998;**58**:653-658. DOI: 10.1103/PhysRevB.58.653

[76] Cao G, McCall S, Shepard M, Crow JE, Guertin RP. Thermal, magnetic, and transport properties of single-crystal $Sr_{1-x}Ca_xRuO_3$ ($0 \le x \le 1.0$). Physical Review B. 1997;**56**:321-329. DOI: 10.1103/PhysRevB.56.321

[77] Mazin II, Singh DJ. Electronic structure and magnetism in Ru-based perovskites. Physical Review B. 1997;**56**:2556. DOI: 10.1103/PhysRevB.56.2556

[78] Kim M, Min BI. Nature of itinerant ferromagnetism in $SrRuO_3$: A DFT + DMFT study. Physical Review B. 2015;**91**:205116. DOI: 10.1103/PhysRevB.91.205116

[79] Funakubo H, Oikawa T, Higashi N, Saito K. Metal organic chemical vapour deposition growth of epitaxial $SrRuO_3$ and $CaRuO_3$ thin films with different orientations as the bottom electrode for epitaxial ferroelectric thin films. Journal of Crystal Growth. 2002;**235**:401-406. DOI: 10.1016/S0022-0248(01)01921-2

[80] Eom CB, Cava RJ, Fleming RM, Phillips JM, van Dover RB, Marshall JH, Hsu JWP, Krajewski JJ, Pecj WF Jr. Single-crystal epitaxial thin films of the isotropic metallic oxides $Sr_{1-x}Ca_xRuO_3$ ($0 \le x \le 1$). Science. 1992;**258**:1766-1769. DOI: 10.1126/science.258.5089.1766

[81] Gan Q, Rao RA, Eom CB, Garrett JL, Lee M. Direct measurement of strain effects on magnetic and electrical properties of epitaxial $SrRuO_3$ films. Applied Physics Letters. 1998;**72**:978-980. DOI: 10.1063/1.120603

[82] Chodekar RV, Takamura Y, Suzuki Y. Disorder-induced carrier localization in ultrathin strained $SrRuO_3$ epitaxial films. Journal of Applied Physics. 2006;**99**:08F503. DOI: 10.1063/1.2168437

[83] Zayak AT, Huang X, Neaton JB, Rabe KM. Manipulating magnetic properties of $SrRuO_3$ and $CaRuO_3$ with epitaxial and uniaxial strains. Physical Review B. 2008;**77**:214410. DOI: 10.1103/PhysRevB.77.214410

[84] Vailionis A, Siemons W, Koster G. Room temperature epitaxial stabilization of a tetragonal phase in $ARuO_3$ (A = Ca and Sr) thin films. Applied Physics Letters. 2008;**93**:051909. DOI: 10.1063/1.2967878

[85] Grutter A, Wong F, Arenholz E, Liberati M, Vailinois A, Suzuki Y. Enhanced magnetism in epitaxial $SrRuO_3$ thin films via substrate-induced strain. Journal of Applied Physics. 2010;**107**:09E138. DOI: 10.1063/1.3360345

[86] Choi KJ, Baek SH, Jang HW, Belenky LJ, Lyubchenko M, Eom CB. Phase-transition temperature of strained single-crystal $SrRuO_3$ thin films. Advanced Materials. 2010;**22**:759-762. DOI: 10.1002/adma.200902355

[87] Dirsyte R, Schwarzkopf J, Schmidbauer M, Wagner G, Irmsccher K, Anooz SB, Fornari R. Impact of epitaxial strain on the ferromagnetic transition temperature of $SrRuO_3$ thin films. Thin Solid Films. 2011;**519**:6264-6268. DOI: 10.1016/j.tsf.2011.03.132

[88] Vailionis A, Boschker H, Siemons W, Houwman EP, Blank DHA, Rijnders G, Koster G. Misfit strain accommodation in epitaxial ABO_3 perovskites: Lattice rotations and lattice modulations. Physical Review B. 2011;**83**:064101. DOI: 10.1103/PhysRevB.83.064101

[89] Kan D, Shimakawa Y. Strain effects on structural transition in $SrRuO_3$ epitaxial thin films. Crystal Growth & Design. 2011;**11**:5483-5487. DOI: 10.1021/cg201070n

[90] Kan D, Aso R, Kurata H, Shimakawa Y. Epitaxial strain effect in tetragonal $SrRuO_3$ thin films. Journal of Applied Physics. 2013;**113**:173912. DOI: 10.1063/1.4803869

[91] Herklotz A, Kataja M, Nenkov K, Biegalski MD, Christen HM, Deneke C, Schultz L, Dorr K. Magnetism of the tensile-strain-induced tetragonal state of $SrRuO_3$ films. Physical Review B. 2013;**88**:144412. DOI: 10.1103/PhysRevB.88.144412

[92] Aso R, Kan D, Fujiyoshi Y, Shimakawa Y, Kurata H. Strong dependence of oxygen octahedral distortions in $SrRuO_3$ films on types of substrate-induced epitaxial strain. Crystal Growth & Design. 2014;**14**:6478-6485. DOI: 10.1021/cg501340e

[93] Kan D, Aso R, Kurata H, Shimakawa Y. Phase control of a perovskite transition-metal oxide through oxygen displacement at the heterointerface. Dalton Transactions. 2015;**44**:10594-10607. DOI: 10.1039/c4dt03749a

[94] Dagotto E, Hotta T, Moreo A. Colossal magnetoresistive materials: The key role of phase separation. Physics Reports. 2001;**344**:1-153. DOI: 10.1016/S0370-1573(00)00121-6

[95] Moussa F, Hennion M, Rodriguez-Carvajal J, Moudden H, Pinsard L, Revcolevschi A. Spin waves in the antiferromagnet perovskite $LaMnO_3$: A neutron-scattering study. Physical Review B. 1996;**54**:15149. DOI: 10.1103/PhysRevB.54.15149

[96] Urushibara A, Moritomo Y, Arima T, Asamitu A, Kido G, Tokura Y. Insulator-metal transition and giant magnetoresistance in $La_{1-x}Sr_xMnO_3$. Physical Review B. 1995;**51**:14103-14109. DOI: 10.1103/PhysRevB.51.14103

[97] Snyder GJ, Hiskes R, DiCarolis S, Beasley MR, Geballe TH. Intrinsic electrical and magnetic properties of $La_{0.67}Ca_{0.33}MnO_3$ and $La_{0.67}Sr_{0.33}MnO_3$ MOCVD thin films and bulk material. Physical Review B. 1996;**53**:14434-14444. DOI: 10.1103/PhysRevB.53.14434

[98] Haghiri-Gosnet AM, Renard JP. CMR manganites: Physics, thin films and devices. Journal of Physics D: Applied Physics. 2003;**36**:R127-R150. DOI: 10.1088/0022-3727/36/8/201

[99] Martin MC, Shirane G, Endoh Y, Hirota K, Moritomo Y, Tokura Y. Magnetism and structural distortion in the $La_{0.7}Sr_{0.3}MnO_3$ metallic ferromaganet. Physical Review B. 1996;**53**:14285-14290. DOI: 10.1103/PhysRevB.53.14285

[100] Suzuki Y, Hwang HY, Cheong SW, van Dover RB. The role of strain in magnetic anisotropy of manganite thin films. Applied Physics Letters. 1997;**71**:140-142. DOI: 10.1063/1.119454

[101] Wu Y, Suzuki Y, Rüdigger U, Yu J, Kent AD, Nath TK, Eom CB. Magnetotransport and magnetic domain structure in compressively strained colossal magnetoresistance films. Applied Physics Letters. 1999;**75**:2295-2297. DOI: 10.1063/1.124995

[102] Haghiri-Gosnet AM, Wolfman J, Mercey B, Simon C, Lecoeur P, Korzenski M, Hervieu M, Desfeux R, Baldinozzi G. Microstructure and magnetic properties of strained $La_{0.7}Sr_{0.3}MnO_3$ thin films. Journal of Applied Physics. 2000;**88**:4257-4264. DOI: 10.1063/1.1309040

[103] Tsui F, Somak MC, Nath TK, Eom CB. Strain-dependent magnetic phase diagram of epitaxial $La_{0.67}Sr_{0.33}MnO_3$ thin films. Applied Physics Letters. 2000;**76**:2421-2423. DOI: 10.1063/1.126363

[104] Dho J, Kim YN, Hwang YS, Kim JC, Hur NH. Strain-induced magnetic stripe domains in $La_{0.7}Sr_{0.3}MnO_3$ thin films. Applied Physics Letters. 2003;**82**:1434-1436. DOI: 10.1063/1.1556967

[105] Tebano A, Aruta C, Medaglia PG, Tozzi F, Balestrino G, Sidorenko AA, Allodi G, De Renzi R, Ghiringhelli G, Dallera C, Braicovich L, Brookes NB. Strain-induced phase separation in $La_{0.7}Sr_{0.3}MnO_3$ thin films. Physical Review B. 2006;**74**:245116. DOI: 10.1103/PhysRevB.74.245116

[106] Dey P, Nath TK, Taraphder A. Effects of substrate-induced strain on transport and magnetic properties of epitaxial $La_{0.66}Sr_{0.33}MnO_3$ thin films. Applied Physics Letters. 2007;**91**:012511. DOI: 10.1063/1.2750399

[107] Takamura Y, Chopdekar RV, Arenholz E, Suzuki Y. Control of the magnetic and magnetrotrasport properties of $La_{0.67}Sr_{0.33}MnO_3$ thin films. Applied Physics Letters. 2008;**92**:162504. DOI: 10.1063/1.2908051

[108] Adamo C, Ke X, Wang HQ, Xin HL, Heeg T, Hawley ME, Zander W, Schubert J, Schiffer P, Muller DA, Maritato L, Schlom DG. Effect of biaxial strain on the electrical and magnetic properties of (0 0 1) $La_{0.7}Sr_{0.3}MnO_3$ thin films. Applied Physics Letters. 2009;**95**:112504. DOI: 10.1063/1.3213346

[109] Ovsyannikov GA, Petrzhik AM, Borisenko IV, Klimov AA, Ignatov YA, Demodov VV, Nikitov SA. Magnetotransport characteristics of strained $La_{0.7}Sr_{0.3}MnO_3$ epitaxial manganite films. Journal of Experimental and Theoretical Physics. 2009;**108**:48-55. DOI: 10.1134/S1063776109010075

[110] Španková M, Štrbík V, Dobročka E, Chromik Š, Sojková M, Zheng DN, Li J. Characterization of epitaxial LSMO thin films with high Curie temperature prepared on different substrates. Vacuum. 2016;**126**:24-28. DOI: 10.1016/j.vacuum.2016.01.009

[111] Liu GZ, Yang YY, Qiu J, Chen XX, Jiang YC, Yao JL, Zhao M, Zhao R, Gao J. Substrate-related structural, electrical, magnetic and optical properties of $La_{0.7}Sr_{0.3}MnO_3$ films. Journal of Physics D: Applied Physics. 2016;**49**:075309. DOI: 10.1088/0022-3727/49/7/075304

[112] Millis AJ, Darling T, Migliori A. Quantifying strain dependence in "colossal" magnetoresistance manganites. Journal of Applied Physics. 1998;**83**:1588-1591. DOI: 10.1063/1.367310

[113] Kubel F, Schmid H. Structure of a ferroelectric and ferroelastic monodomain crsytal of the perovskite $BiFeO_3$. Acta Crystallographica Section B 1990;**B46**:698-702. DOI: 10.1107/S0108768190006887

[114] Velev JP, Jaswal SS, Tsymbal EY. Multi-ferroic and magnetoelectric materials and Interfaces. Philosophical Transactions of the Royal Society A. 2011;**369**:3069-3097. DOI: 10.1098/rsta.2010.0344

[115] Wang J, Neaton JB, Zheng H, Nagarajan V, Ogale SB, Liu B, Viehland D, Vaithyanathan V, Schlom DG, Waghmare UV, Spaldin NA, Rabe KM, Wuttig M, Ramesh R. Epitaxial $BiFeO_3$ multiferroic thin film heterostructure. Science. 2003;**299**:1719-1722. DOI: 10.1126/science.1080615

[116] Hill NA. Why are there so few magnetic ferroelectrics?. Journal of Physical Chemistry B. 2000;**104**:6694-6709. DOI: 10.1021/jp000114x

[117] Moreau JM, Michel C, Gerson R, James WJ. Ferroelectric $BiFeO_3$ X-ray and neutron diffraction study. Journal of Physics and Chemistry of Solids. 1971;**32**:1315-1320. DOI: 10.1016/S0022-3697(71)80189-0

[118] Teague JR, Gerson R, James WJ. Dielectric hysteresis in single crystal $BiFeO_3$. Solid State Communications. 1970;**8**:1073-1074. DOI: 10.1016/0038-1098(70)90262-0

[119] Lebeugle D, Colson D, Forget Viret M. Very large spontaneous electric polarization in $BiFeO_3$ single crystals at room temperature and its evolution under cycling fields. Applied Physics Letters. 2007;**91**:022907. DOI: 10.1063/1.2753390

[120] Sando D, Barthélémy A, Bibes M. $BiFeO_3$ epitaxial thin films and devices: Past, present and future. Journal of Physics: Condensed Matter. 2014;**26**:473201-473223. DOI: 10.1088/0953-8984/26/47/473201

[121] Zhang JX, Li YL, Wang Y, Liu ZK, Chen LQ, Chu YH, Zavaliche F, Ramesh R. Effect of substrate-induced strains on the spontaneous polarization of epitaxial $BiFeO_3$ thin films. Journal of Applied Physics. 2007;**101**:114105. DOI: 10.1063/1.2743733

[122] Kim DH, Lee HN, Biegalski MD, Christen HM. Effect of epitaxial strain on ferroelectric polarization in multiferroic $BiFeO_3$ films. Applied Physics Letters. 2008;**92**:012911. DOI: 10.1063/1.2830799

[123] Jang HW, Baek SH, Ortiz D, Folkman CM, Das RR, Chi YH, Shafer P, Zhnag JX, Choudhusry S, Vaithyanathan V, Chen YB, Felker DA, Biegalski MD, Rzchowski MS, Pan XQ, Schlom DG, Chen LQ, Ramesh R, Eom CB. Strain-induced polarization rotation in epitaxial (0 0 1) $BiFeO_3$ thin films. Physical Review Letters. 2008;**101**:107602. DOI: 10.1103/PhysRevLett.101.107602

[124] Zeches RJ, Rossell MD, Zhang JX, Hatt AJ, He Q, Yang CH, Kumar A, Wang CH, Melville A, Adamo A, Sheng G, Chu YH, Ihlefeld JF, Erni R, Ederer C, Gopalan V, Chen LQ, Schlom DG, Spaldn NA, Martim LW, Ramesh R. A strain-driven morphotropic phase boundary in $BiFeO_3$. Sicence. 2009;**326**:977-980. DOI: 10.1126/science.1177046

[125] Infante IC, Lisenkov S, Dupé B, Bibes M, Fusil S, Jacquet E, Geneste G, Petit S, Courtial A, Juraszek J, Bellaiche L, Barthélémy A, Dkhil B. Bridging multiferroic phase transitions by epitaxial strain in $BiFeO_3$. Physical Review Letters. 2010;**105**:057601. DOI: 10.1103/PhysRevLett.105.057601

[126] Dupe B, Infante IC, Geneste G, Janolin PE, Bibes M, Berthelemy A, Lisenkov S, Bellaiche L, Ravy S, Dkhil B. Competing phases in $BiFeO_3$ thin films under compressive epitaxial strain. Physical Review B. 2010;**81**:144128. DOI: 10.1103/PhysRevB.81.144128

[127] Huang CW, Chu YH, Chen ZH, Wang J, Sritharan T, He Q, Ramesh R, Chen L. Strain-driven phase transitions and associated dielectric/piezoelectric anomalies in $BiFeO_3$ thin films. Applied Physics Letters. 2010;**97**:152901. DOI: 10.1063/1.3499658

[128] Hatt AJ, Spaldin NA, Ederer C. Strain-induced isosymmetric phase transition in $BiFeO_3$. Physical Review B. 2010;**81**:054109. DOI: 10.1103/PhysRevB.81.054109

[129] Woo CS, Lee JH, Chu K, Jang BK, Kim YB, Koo TY, Yang P, Qi Y, Chen Z, Chen L, Choi HC, Shim JH, Yang CH. Suppression of mixed-phase areas in highly elongated $BiFeO_3$ thin films on $NdAlO_3$ substrates. Physical Review B. 2012;**86**:054417. DOI: 10.1103/PhysRevB.86.054417

[130] Dupé B, Prosandeev S, Geneste G, Dkhil B, Bellaiche L. $BiFeO_3$ films under tensile epitaxial strain from first principles. Physical Review Letters. 2011;**106**:237601. DOI: 10.1103/PhysRevLett.106.237601

[131] Christen HM, Nam JH, Kim HS, Hatt AJ, Spaldin NA. Stress-induced R-M_A-M_C-T symmetry changes in $BiFeO_3$ films. Physical Review B. 2011;**83**:144107. DOI: 10.1103/PhysRevB.83.144107

[132] Yang JC, He Q, Suresha SJ, Kuo CY, Peng CY, Haislmaier RC, Motyka MA, Sheng G, Adamo C, Lin HJ, Hu Z, Chnag L, Tjeng LH, Arenholz E, Podraza NJ, Bernhagen M, Uecker R, Schlom DG, Gopalan V, Chen LQ, Chen CT, Ramesh R, Chu YH. Orthorhombic $BiFeO_3$. Physical Review Letters. 2012;**109**:247606. DOI: 10.1103/PhysRevLett.109.247606

[133] Daumont C, Ren W, Infante IC, Lisenkov S, Allibe J, Carŕeťero C, Fusil S, Jacquet E, Bouvet T, Bouamrane F, Prosandeev S, Geneste G, Dkhil B, Bellaiche L, Barthélémy A, Bibes M. Strain dependence of polarization and piezoelectric response in epitaxial $BiFeO_3$ thin films. Journal of Physics: Condensed Matter. 2012;**24**:162202. DOI: 10.1088/0953-8984/24/16/162202

[134] Liu HJ, Chen HJ, Liang WI, Lee HY, Lin SJ, Chu YH, Structural study in highly compressed $BiFeO_3$ epitaxial thin films on $YAlO_3$. Journal of Applied Physics. 2012;**112**:052002. DOI: 10.1063/1.4746036

[135] Sando D, Agbelele A, Rahmedov D, Liu J, Rovillain P, Toulouse C, Infante IC, Pyatakov AP, Fusil S, Jacquet E, Carrétéro C, Deranlot C, Lisenkov S, Wang D, LeBreton JM, Cazayous M, Sacuto A, Juraszek J, Zvezdin AK, Bellaiche L, Dkhil B, Barthélémy A, Bibes M. Crafting the magnonic and spintronic response of $BiFeO_3$ films by epitaxial strain. Nature Materials. 2013;**12**:641-646. DOI: 10.1038/nmat3629

[136] Sando D, Xu B, Bellaiche L, Nagarajan V. A multiferroic on the brink: Uncovering the nuances of strain-induced transitions in $BiFeO_3$. Applied Physics Reviews. 2016;**3**:011106. DOI: 10.1063/1.4944558

[137] Ederer C, Spadin NA. Effect of epitaxial strain on the spontaneous polarization of thin film ferroelectrics. Physical Review Letters. 2005;**95**:257601. DOI: 10.1103/PhysRevLett.95.257601

[138] Zhen Fan Z, Juanxiu Xiao J, Huajun Liu H, Ping Yang P, Qingqing Ke Q, Wei Ji W, Kui Yao K, Khuong P, Ong KP, Kaiyang Zeng K, John Wang J. Stable ferroelectric perovskite structure with giant axial ratio and polarization in epitaxial $BiFe_{0.6}Ga_{0.4}O_3$ thin films.. ACS Applied Materials and Interfaces. 2015;**7**:2648-2653. DOI: 10.1021/am509016w

[139] van Benthem K, Elsässer C, French RH. Bulk electronic structure of $SrTiO_3$: Experiment and theory. Journal of Applied Physics. 2001;**90**:6156-6164. DOI: 10.1063/1.1415766

[140] Lim SG, Kriventsov S, Jackson TN, Haeni JH, Schlom DG, Balbashov AM, Uecker R, Riche P, Freeouf JL, Lucovsky G. Dielectric functions and optical bandgaps of high-K dielectrics for metal-oxide semiconductor field-effect transistors by far ultraviolet spectroscopic ellipsometry. Journal of Applied Physics. 2002;**91**:4500-4505. DOI: 10.1063/1.1456246

[141] Huijben M, Brinkman A, Koster G, Rijnders G, Hilgenkamp H, Blank DHA. Structure-property relation of $SrTiO_3$/$LaAlO_3$ interfaces. Advanced Materials. 2009;**21**:1665-1677. DOI: 10.1002/adma.200801448

[142] Li L, Richter C, Mannhart J, Ashoori RC. Coexistence of magnetic order and two-dimensional superconductivity at $LaAlO_3$/$SrTiO_3$ interfaces. Nature Physics. 2011;**7**:762-766. DOI: 10.1038/nphys2080

[143] Bert JA, Kalisky, Bell C, Kim M, Hikita Y, Hwang HY, Moler KA. Direct imaging of the coexistence of ferromagnetism and superconductivity at the $LaAlO_3$/$SrTiO_3$ interfaces. Nature Physics. 2011;**7**:767-771. DOI: 10.1038/nphys2079

[144] Trier F, Prawiroatmodjo GEDK, Zhong Z, Christensen DV, von Soosten M, Bhowmik A, Lastra JMG, Chen Y, Jespersen TS, Pryds N. Quantization of hall resistance at the metallic interface between an oxide insulator and $SrTiO_3$. Physical Review Letters. 2016;**117**:096804. DOI: 10.1103/PhysRevLett.117.096804

[145] Caviglia AD, Gabay M, Gariglio S, Reyren N, Cancellieri C, Triscone JM. Tunable rashba spin-orbit interaction at oxide interfaces. Physical Review Letters. 2012;**104**:126803. DOI: 10.1103/PhysRevLett.104.126803

[146] Nakagawa N, Hwang HY, Muller DA. Why some interfaces cannot be sharp. Nature Materials. 2006;**5**:204-209. DOI: 10.1038/nmat1569

[147] Liu ZQ, Li CJ, Lu WM, Huang XH, Huang Z, Zeng SW, Qiu XP, Huang LS, Annadi A, Chen JS, Coey JMD, Ariando VT. Origin of the two-dimensional electron gas at $LaAlO_3$/$SrTiO_3$ interfaces: The role of oxygen vacancies and electronic reconstruction. Physical Review X. 2013;**3**:021010. DOI: 10.1103/PhysRevX.3.021010

[148] Bark CW, Felker DA, Wnag Y, Zhnag Y, Jang HW, Folkman CM, Park JW, Baek SH, Zhou H, Fong DD, Pan XQ, Tsymbal EY, Rzchowski MS, Eom CB. Tailoring a two-dimensional electron gas at the $LaAlO_3/SrTiO_3$ (001) interfaces by epitaxial strain. PNAS. 2011;**108**:4720-4724. DOI: 10.1073/pnas.1014849108

[149] Jang HW, Felker DA, Bark CW, Wang Y, Niranjan MK, Nelson CT, Zhnag Y, Su D, Folkman CM, Baek SH, Lee S, Janicka K, Zhu Y, Pan XQ, Fong DD, Tsymnal EY, Rzchowski MS, Eom CB. Metallic and insulating oxide interfaces controlled by electronic correlations. Science. 2011;**331**:886-889. DOI: 10.1126/science.1198781

[150] Anandi A, Putra A, Liu ZQ, Wang X, Gopinadhn K, Huang Z, Dhar S, Ariando VT. Electronic correlation and strain effects at the interfaces between polar and nonpolar complex oxides. Physical Review B. 2012;**86**:085450. DOI: 10.1103/PhysRevB.86.085450

[151] Huang Z, Liu ZQ, Yang M, Zeng SW, Anandi A, Lu WM, Tan XL, Chn PF, Sun L, Wang XR, Zhao YL, Li CJ, Zhou J, Han K, Wu WB, Feng YP, Coey JMD. Biaxial strain-induced transport property changes in atomically tailored $SrTiO_3$-based systems. Physical Review B. 2014;**90**:125156. DOI: 10.1103/PhysRevB.90.125156

[152] Nazir S, Behtash M, Yang K. Enhancing interfacial conductivity and spatial charge confinement of $LaAlO_3/SrTiO_3$ heterostructures via strain engineering. Applied Physics Letters. 2014;**105**:1416020. DOI: 10.1063/1.4897626

[153] Fister TT, Zhou H, Luo Z, Seo SSA, Hruszkewycz SO, Proffit DL, Esatman JA, Fuoss PH, Baldo PM, Lee HN, Fong DD. Octahedral rotations in strained $LaAlO_3/SrTiO_3$ (001) heterostructures. APL Materials. 2014;**2**:021102. DOI: 10.1063/1.4865160

[154] Du Y, Wang C, Li J, Zhang X, Wang F, Zhu Y, Yin N, Mei L. The effect of in-plane strain on the electronic properties of $LaAlO_3/SrTiO_3$ interfaces. Computational Materials Science. 2015;**99**:57-61. DOI: 10.1016/j.commatsci.2014.11.039

[155] Nazir S, Behtash M, Yang K. The role of uniaxial strain in tailoring the interfacial properties of $LaAlO_3/SrTiO_3$ heterostructures. RSC Advances. 2015;**5**:15682-15689. DOI: 10.1039/c4ra15866k

[156] Medarde ML, Structural, magnetic and electronic properties of $RNiO_3$ perovskites (R = rare earth). Journal of Physics: Condensed Matter. 1997;**9**:1679-1707. DOI: 10.1088/0953-8984/9/8/003

[157] Catalan G. Progress in perovskite nickelate research. Phase Transitions: A Multinational Journal. 2008;**81**:729-749. DOI: 10.1080/01411590801992463

[158] Middey S, Chakhalian J, Mahadevan P, Freeland JW, Millis AJ, Sarma DD. Physics of ultrathin films and heterostructures of rare-earth nickelates. Annual Review of Materials Research. 2016;**46**:11.1-11.30. DOI: 10.1146/annurev-matsci-070115-032057

[159] Rajeev KP, Shivashankr GV, Raychaudhuri AK. Low-temperature electronic properties of a normal conducting perovskite oxide ($LaNiO_3$). Solid State Communications. 1991;**79**:591-595. DOI: 10.1016/0038-1098(91)90915-I

[160] Sreedhar K, Honig JM, Darwin M, McElfresh M, Shand PM, Xu J, Crooker BC, Spalek J. Electronic properties of the metallic perovskite $LaNiO_3$: Correlated behavior of 3*d* electrons. Physical Review B. 1992;**46**:6382-6386. DOI: 10.1103/PhysRevB.46.6382

[161] Masys Š, Jonauskas V. Elastic properties of rhombohedral, cubic, and monoclinic phases of $LaNiO_3$ by first principle calculations. Computational Materials Science. 2015;**108**:153-159. DOI: 10.1016/j.commatsci.2015.06.034

[162] Zhou JS, Marshall LG, Goodenough JB. Mass enhancement versus Stoner enhancement in strongly correlated metallic perovskites: $LaNiO_3$ and $LaCuO_3$. Physical Review B. 2014;**89**:245138. DOI: 10.1103/PhysRevB.89.245138

[163] Son J, Moetakef P, LeBeau JM, Ouellette D, Balents L, Allen SJ, Stemmer S. Low-dimensional Mott materials: Transport in ultrathin epitaxial $LaNiO_3$ films. Applied Physics Letters. 2010;**96**:062114. DOI: 10.1063/1.3309713

[164] May SJ, Kim JW, Rondinelli JM, Karapetrova E, Spaldin NA, Bhattacharya A, Ryan PJ. Quantifying octahedral rotations in strained perovskite oxide films. Physical Review B. 2010;**82**:014110. DOI: 10.1103/PhysRevB.82.014110

[165] Chakhalian J, Rondinelli JM, Liu J, Gray BA, Kareev M, Moon EJ, Prasai N, Cohn JL, Varela M, Tung IC, Bedzyk MJ, Altendorf SG, Strigari F, Dabrowski B, Tjeng LH, Ryan PJ, Freeland JW. Asymmetric orbital-lattice interactions in ultrathin correlated oxide films. Physical Review Letters. 2011;**107**:11680. DOI: 10.1103/PhysRevLett.107.116805

[166] Stewart MK, Yee CH, Liu J, Kareev M, Smith RK, Chapler BC, Varela M, Ryan PJ, Haule K, Chakhalian J, Basov DN. Optical study of strained ultrathin films of strongly correlated $LaNiO_3$. Physical Review B. 2011;**83**:075125. DOI: 10.1103/PhysRevB.83.075125

[167] Moon EJ, Gray BA, Kareev M, Liu J, Altendorf SG, Strigari F, Tjeng LH, Freeland JW, Chakhalain J. Strain-dependent transport properties of the ultra-thin correlated metal, $LaNiO_3$. New Journal of Physics. 2011;**13**:073037. DOI: 10.1088/1367-2630/13/7/073037

[168] Moon EJ. Rondinelli JM, Prasai N, Gray BA, Kareev M, Chakhalain J, Cohn JL. Strain-controlled band engineering and self-doping in ultrathin $LaNiO_3$ films. Physical Review B. 2012;**85**:121106(R). DOI: 10.1103/PhysRevB.85.121106

[169] Moon EJ, Gray BA, Pimpinelli A, Kareev M, Meyers D, Chakhalain J. Strain-controlled epitaxial stabilization in ultrathin $LaNiO_3$ films grown by pulsed laser deposition. Crystal Growth & Design. 2013;**13**:2256-2259. DOI: 10.1021/cg300958z

[170] Zhu M, Komissinsky P, Radetinac A, Wang Z, Alff L. Joint effect of composition and strain on the anomalous transport properties of $LaNiO_3$ films. Journal of Applied Physics. 2015;**117**:155306. DOI: 10.1063/1.4918661

[171] Weber MC, Gunnou, Dix N, Pesquera D, Sanchez F, Herranz G, Fontcuberta J, López-Conesa L, Estradé S, Peiró F, Iñiguez J, Kreisel J. Multiple strain-induced phase transitions in $LaNiO_3$ thin films. Physical Review B. 2016;**94**:014118. DOI: 10.1103/PhysRevB.94.014118

[172] Misra D, Kundu TK. Strain-controlled transport mechanism in strongly correlated $LaNiO_3$. Journal of Electronic Materials. 2017;**46**:150-157. DOI: 10.1007/s11664-016-4889-3

Chapter 2

Epitaxial Growth of Ge on Si by Magnetron Sputtering

Ziheng Liu, Xiaojing Hao, Anita Ho-Baillie and
Martin A. Green

Additional information is available at the end of the chapter

http://dx.doi.org/10.5772/intechopen.73554

Abstract

Epitaxial growth of Ge on Si has received considerable attention for its compatibility with Si process flow and the scarcity of Ge compared with Si. Applications that drive the efforts for integrating Ge with Si include high mobility channel in metal-oxide-semiconductor field-effect transistors, infrared photodetector in Si-based optical devices, and template for III-V growth to fabricate high-efficiency solar cells. Epitaxy Ge on Si can be used as a virtual Ge substrate for fabrication of III-V solar cells, which has advantages of superior mechanical properties and low cost over Ge wafers. This work investigates the epitaxial growth of Ge on Si using magnetron sputtering, which is an environment-friendly, inexpensive, high throughput, and simple deposition technique. The effects of substrate temperature on the properties of Ge are analyzed. A novel method to epitaxially grow Ge on Si by magnetron sputtering at low temperature is developed using one-step aluminum-assisted crystallization. By applying an *in-situ* low temperature (50–150°C) heat treatment in between Al and Ge sputter depositions, the epitaxial growth of Ge on Si is achieved. This method significantly lowers the required temperature for and therefore the cost of epitaxial growth of Ge on Si.

Keywords: germanium, epitaxy, silicon, magnetron sputtering, substrate temperature, one-step aluminum-assisted crystallization

1. Introduction

Epitaxial growth of Ge on Si has received considerable attention for its compatibility with Si process flow and the scarcity of Ge compared with Si. Applications driving the efforts for integrating Ge with Si include: high mobility channel in metal-oxide-semiconductor field-effect transistors [1], infrared photodetector in Si-based optical devices [2], and template for III-V growth to fabricate high efficiency solar cells [3].

Ge wafers are the commonly used substrates for the fabrication of high efficiency III-V tandem solar cells [4–6]. Though cheaper than III-V materials, Ge wafers are over 100 times more expensive than Si accounting for more than 50% of the cell cost [3]. Compared with Ge wafer, Si wafer is an alternative with low cost, superior mechanical properties, and higher band gap more desirable for the bottom cell in a double or triple stack [7]. However, the lattice constant of Si is too small to match that of the III-V materials as shown in **Figure 1**. The lattice mismatch can induce large densities of defects negating the advantages of Si substrate. Several approaches have been investigated to control the defect density in this mismatched heterostructure including the insertion of various III-V intermediate layers, strained layer super-lattices, and the use of thermal annealing [8]. The obtained material qualities through these methods are not high enough to yield high efficiency III-V cells. A promising alternative is growing a Ge buffer layer to engineer the lattice constant of substrate surface to match that of III-V materials. Ge epitaxial film on Si can be used as a "virtual Ge substrate" for III-V solar cells. The virtual Ge substrate has advantages of superior mechanical properties and low cost over Ge wafer.

This work investigates the epitaxial growth of Ge on Si by magnetron sputtering, which is an environment-friendly, economical, high throughput, and simple deposition technique. Molecular beam epitaxy (MBE) and chemical vapor deposition (CVD) are widely used for Ge epitaxial growth on Si [10–13]. The MBE and CVD systems require higher vacuum (5×10^{-11} and 1.5×10^{-9} mbar, respectively) than magnetron sputtering (5×10^{-7} mbar) used in this work [14]. While MBE is the most expensive of the three and toxic gases such as germane and silane are used in a CVD system, magnetron sputtering offers a lower cost and safer alternative in

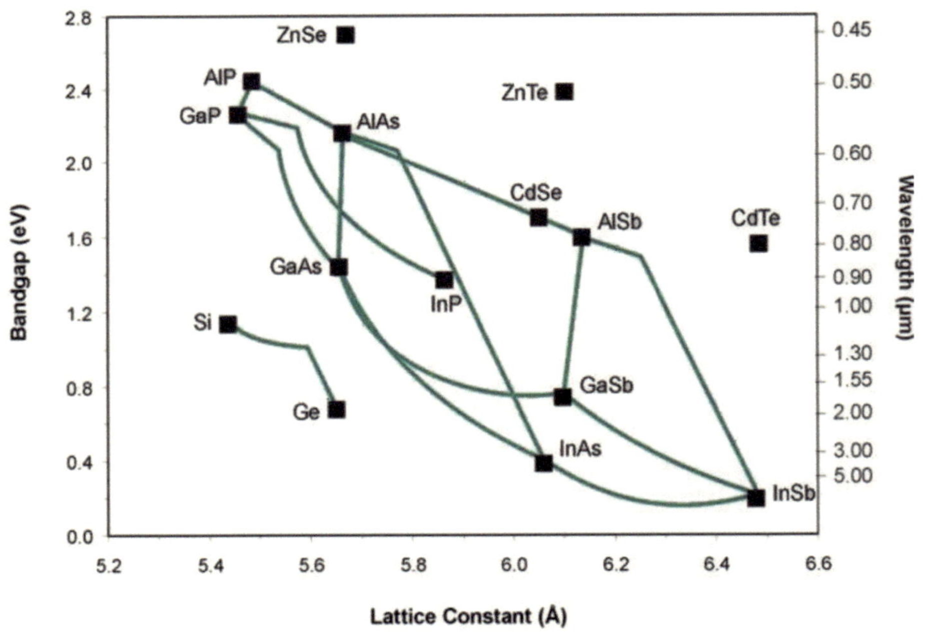

Figure 1. Lattice/band gap diagram for tetrahedrally coordinated semiconductors and their alloys [9].

supplying epitaxial Ge films on Si, which is potentially capable of large-scale production with good uniformity. This chapter presents the successful epitaxial growth of Ge on Si by magnetron sputtering, investigation on the effects of substrate temperature, and the development of a novel method to grow epitaxial Ge on Si by magnetron sputtering at low temperature through one-step aluminum-assisted crystallization.

2. Growth mechanism of Ge epitaxy on Si

2.1. Stranski-Krastanow growth

Three modes are possible in epitaxial growth: Frank-van der Merwe [15], Volmer-Weber [16], and Stranski-Krastanow [17], as shown in **Figure 2**. Frank-van der Merwe and Volmer-Weber modes are pure 2D layer-by-layer growth and 3D island growth, respectively. Stranski-Krastanow (SK mode) is a unique mode of 2D growth plus 3D island formation.

The interfacial free energy and the lattice mismatch determine which growth mode will be adopted in a given system [18]. In lattice-matched systems, the epitaxial film grows either in layer-by-layer mode or island growth mode depending on the interface energy and surface energy of the epitaxial film. In systems with large lattice mismatch, the growth mode may transit from 2D to island growth (SK mode) to relax strain in the epitaxial film. The early stage of the growth could be layer-by-layer due to the small interface energy. With epitaxial film growing thicker, strain energy is accumulated. The island formation is triggered to lower the total energy by introducing misfit dislocations.

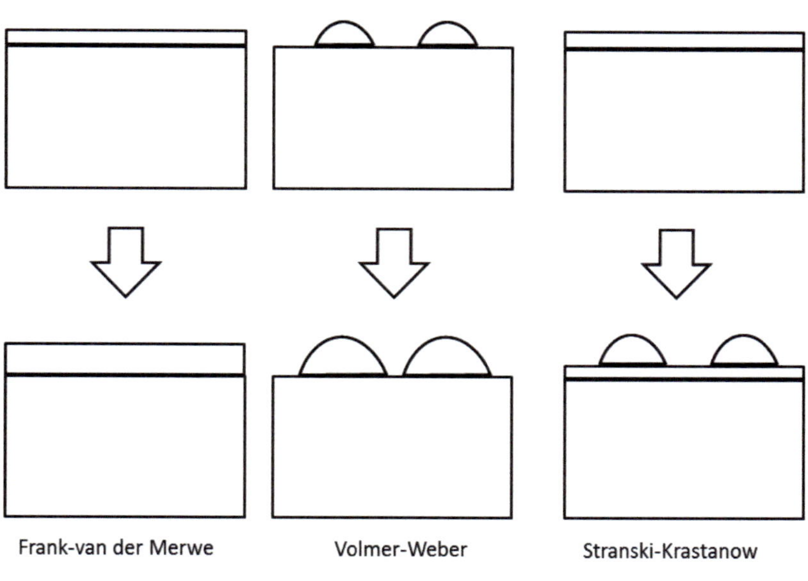

Figure 2. Illustrations of three possible epitaxial growth modes: Frank-van der Merwe, Volmer-Weber, and Stranski-Krastanow [18].

The Ge epitaxial growth on Si can be described as SK growth mode due to the 4.2% lattice mismatch between Si and Ge [19, 20]. In order to be used as a virtual substrate for the III-V deposition, smooth Ge surface is required [21, 22]. The island formation kinetics can be suppressed by shortened atomic surface migration length [23]. The reduced diffusion length can forbid the mass transport over a certain distance, which is required to form islands.

The diffusion length of Ge atoms could be reduced by using surfactant [24, 25] or low growth temperature [26, 27]. Sb has been used as a surfactant to suppress the Ge island formation. The energy barrier to diffuse is higher on the surfactant-covered surface than that of pure Si surface. In addition, the Ge atoms may be incorporated below the surfactant layer due to the site exchange process and therefore it is difficult for the Ge on top of the surfactant layer to diffuse as a relatively high diffusion barrier that has to be overcome. However, the use of surfactant also induces the incorporation of Sb leading to a n-type doping in the Ge film [28]. The effect of substrate temperature on the Ge surface roughness will be investigated in this work.

2.2. Lattice mismatch

Due to the 4.2% lattice mismatch between Si and Ge, the Ge epitaxial growth on Si is defect-free only below the critical thickness. The thin wetting layer is compressively strained in plane to adapt its lattice constant to that of the underlying Si substrate. In the meanwhile, a tensile strain is introduced in perpendicular inducing a tetragonal distortion to the Ge lattice. The biaxial strain compensates the lattice mismatch and therefore no defect is formed. The critical thickness for the defect-free growth of strained Ge on Si is in the range of several nanometers, which is also affected by the growth temperature [29].

When Ge growing above the critical thickness, misfit dislocations will nucleate at the interface and thread segments of dislocations run through the layer to the surface as threading dislocations. The misfit dislocations are incorporated to relax the strain arising from the lattice mismatch between Si and Ge by introducing extra half plane of atoms [30]. The misfit dislocations are energetically stable at the Si and Ge interface when Ge layer is above the critical thickness. As byproduct of misfit dislocations, threading dislocations thread either form a dislocation loop or terminate at the film surface. The threading dislocations are detrimental for the electrical devices because they lie cross the whole film reducing the carrier mobility, carrier lifetime, and device reliability [31]. In this work, the epitaxial Ge layer has to be above the critical thickness to achieve a fully relaxed Ge surface matching the lattice of overlying III-V materials. The effect of substrate temperature on the threading dislocation density (TDD) of Ge will be investigated.

3. Epitaxial growth of Ge on Si by magnetron sputtering

3.1. Magnetron sputtering

Sputtering is a physical vapor deposition method. The target and the substrate are put on the cathode and anode, respectively. An inert gas such as argon (Ar) is introduced to create gaseous plasma by applying a voltage between the cathode and anode. The produced ions (Ar^+)

are accelerated toward the source target to sputter neutral atoms of the target. The ejected neutral atoms will travel to the substrate in a straight line unless they have collision with particles such as Ar atoms. The sputtered atoms, which arrive at the substrate may implant, bounce, diffuse, or simply stick onto the substrate, depending on their kinetic energies. As a result, the substrate will be coated by a thin film composed of target materials.

In conventional RF sputtering, most electrons lose their energy in nonionizing collisions are collected by the anode. The efficiency of ionization from energetic collisions between the electrons and gas atoms is low. Magnets are used to increase the percentage of electrons that participate in the ionization process. Large magnets are formed behind the target by applying a magnetic field at right angles to the electric field. The electrons are trapped near the target surface and kept in spiral motion until they collide with gas atoms. The increased probability of ionization significantly improves the efficiency of target materials sputtering and therefore increases the deposition rate at the substrate. Moreover, this allows the use of lower gas pressure, which may improve the film quality.

Figure 3 shows a schematic diagram of the RF magnetron sputtering system employed in this work. Four-inch intrinsic Ge target was used for the depositions of Ge film on Si and 4-inch SiO_2 target was used for capping layer deposition. The RF powers were supplied to the Ge and SiO_2 targets by two independent RF generators. The RF reverse power was reduced to zero by tuning the variable capacitors in the impedance matching network. Each target had a shutter to isolate the substrate from the plasma. The tilt angle of the targets and the distance between the targets and the substrate could be adjusted to achieve good uniformity.

Gas inlet with mass flow meter was used to supply argon into the main chamber. The vacuum in the main chamber was established by a mechanical rotary pump and a turbo molecular pump. Moreover, a load lock chamber was employed to protect the vacuum condition in the main chamber. Quartz halogen lamps were used to heat the substrate. The deposition rate was controlled by varying the RF power applied on the target and measured by a crystal monitor. The substrate was rotated during deposition to improve uniformity of the films.

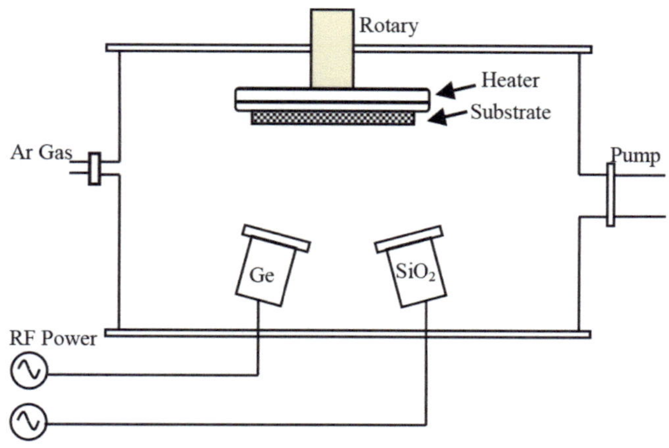

Figure 3. Schematic diagram of RF magnetron sputtering system used in this work.

3.2. Experimental details

In this work, the epitaxial growth of Ge on Si is demonstrated by sputtering Ge target using the AJA ATC-2200 magnetron sputtering system. The base pressure of the chamber was 5×10^{-7} mbar. N-type Si (100) wafers were used as the substrates. The Si substrates were cleaned using RCA solutions [32] followed by a HF dip. The Si substrate was immediately loaded into a load lock chamber after cleaning to minimize the oxidation of the Si surface.

The Ge films were sputter-deposited from a 4-inch intrinsic Ge target (99.999% purity) at a process pressure of 1.5×10^{-3} mbar. Rotation of 30 revolutions per minute was applied to the substrate during deposition to ensure the uniformity of the films. The Ar flow was kept at 15 sccm and the RF power applied to the Ge target was 150 W. The Ge deposition rate was 5 nm/min examined by a quartz crystal deposition rate monitor. 300 nm thick Ge films were sputter-deposited on Si at various substrate temperatures of 300, 400, and 500°C to investigate the effects of substrate temperature. The temperature calibration data was supplied and measured with a Si wafer by the manufacturer of the sputter system.

The surface morphology of Ge films was examined by atomic force microscopy (AFM) with Bruker Icon using the tapping mode. The scan area was 2 × 2 μm. The crystalline quality of the annealed Ge films was analyzed by high resolution X-ray diffraction (XRD), Raman spectroscopy, and transmission electron microscopy (TEM). The XRD measurements were performed with Bruker D8 at a voltage of 45 kV and a current of 100 mA, using Cu $K_{\alpha 1}$ radiation ($\lambda = 1.5406$ Å). The diffractometer was calibrated by making the Si (400) diffraction peak from the substrate maximized and at its theoretical position. Raman spectra of the Ge films were measured with Renishaw inVia Raman microscope using Ar^+ laser with wavelength of 514 nm as the excitation source. The beam power was limited to 6 mW to prevent the locally induced crystallization of Ge films during the measurement. Static mode with 20 times accumulation was employed to improve the signal to noise ratio. TEM measurements were conducted with Phillips CM200 microscope operating at 200 kV. The TEM samples were prepared by focused ion beam milling using Nova Nanolab 200.

3.3. Results and discussions

XRD 2θ-Ω scans were conducted on the Ge films deposited on Si at 300, 400, and 500°C in the 2θ range between 20 and 75° to examine the crystallinity of the Ge films. As shown in **Figure 4(a)**, apart from the strong Si (400) peak at 69.2° attributed to the substrate, only one peak at around 66° is observed which corresponds to Ge (400). The absence of any other Ge peaks indicates the Ge films might be single-crystalline Ge (100) which requires further examination by XRD Phi scans. **Figure 4(b)** shows Si (220) and Ge (220) Phi scan patterns collected from the sample deposited at 300°C by rotating the specimen with respect to the [110] axis. Only the four (220) reflections are observed in the Ge Phi scan pattern suggesting the film is with fourfold symmetry about an axis normal to the substrate [33]. In addition, the Ge (220) reflections align with the Si substrate (220) reflections indicating the Ge is single-crystalline epitaxy film.

The interface of the Ge film and Si substrate is investigated by high-resolution TEM to confirm the epitaxial growth of Ge on Si. As shown in the atomic-resolution image at the interface in

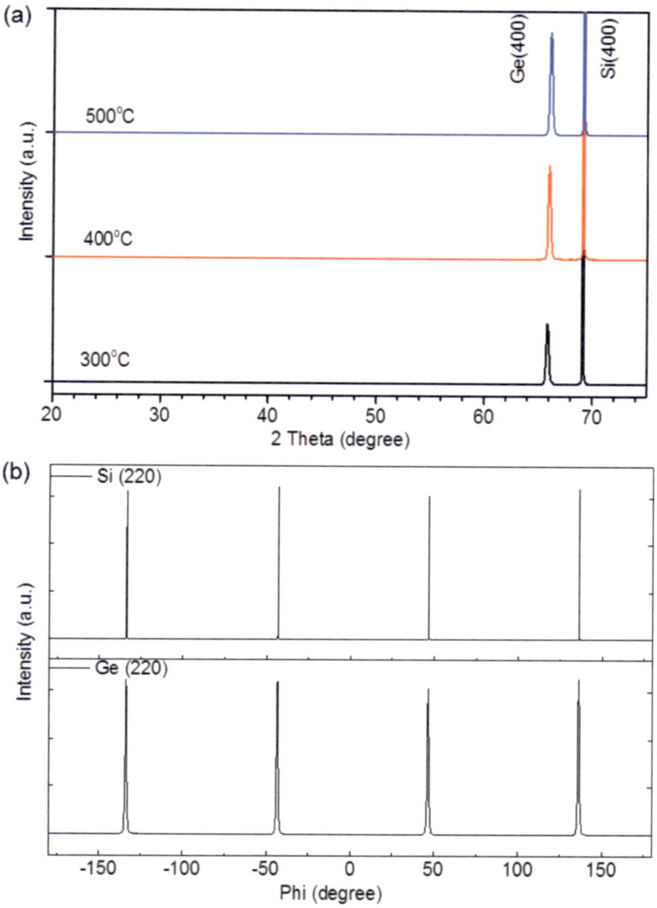

Figure 4. (a) XRD 2θ-Ω diffraction patterns of the Ge films deposited on Si at 300°C, 400°C, and 500°C, (b) Si (220) and Ge (220) phi scan patterns collected from the sample deposited at 300°C showing the epitaxial relationship between the Ge film and Si substrate.

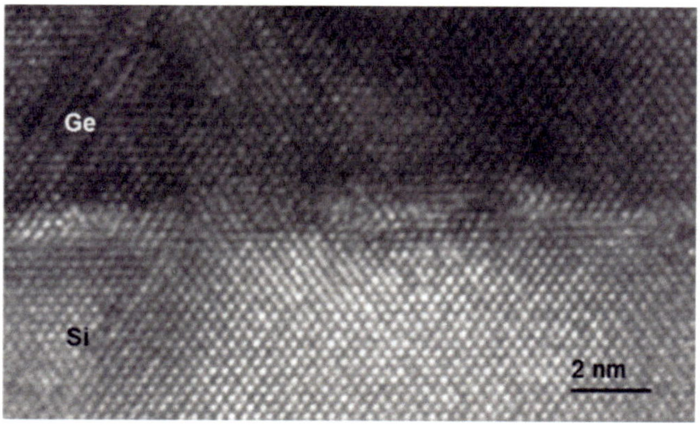

Figure 5. Atomic-resolution cross-sectional TEM image of Ge/Si interface on the sample deposited at 300°C.

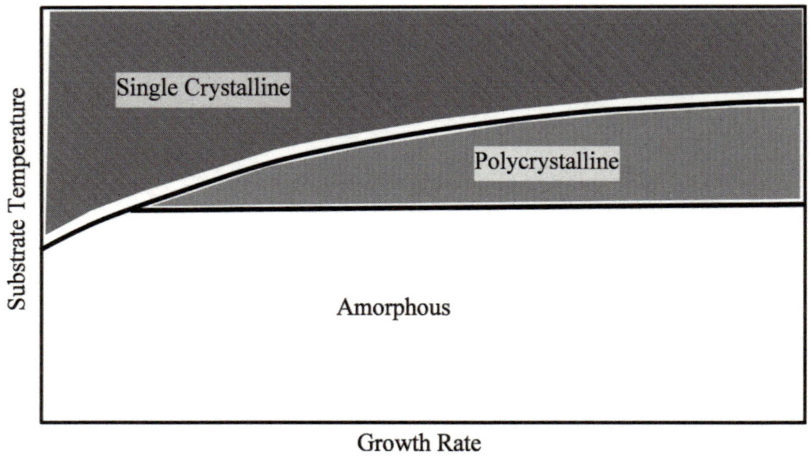

Figure 6. A schematic phase map of the crystallinity of as-deposited semiconductor films as function of growth rate and temperature [34].

Figure 5, the atoms are continuously aligned from the Si substrate to the grown Ge film suggesting successful epitaxy. This results is in good agreement with the XRD measurements.

The crystallinity of the as-deposited film depends on both the substrate temperature and growth rate as indicated in the schematic diagram shown in **Figure 6** [34]. The crystallinity can be improved by increasing the substrate temperature and reducing the growth rate. The XRD and TEM results suggest that substrate temperature of 300°C is enough to obtain single-crystalline Ge epitaxial growth on Si at the growth the rate of 5 nm/min. The effects of substrate temperature on the quality of Ge films are investigated in the following section.

4. Effects of substrate temperature

As reviewed in the previous section, the substrate temperature may play an important role in determining the growth mode. The effects of substrate temperature on the properties of sputter-deposited epitaxial Ge films are discussed in this section. 300 nm thick Ge films were sputter-deposited on Si at various substrate temperatures of 300, 400, and 500°C.

The effect of substrate temperature on surface morphology of the Ge films is investigated using tapping mode AFM. **Figure 7** shows the 2D and 3D AFM images of the Ge films deposited at (a) 300°C, (b) 400°C, and (c) 500°C. It can be seen from the 3D AFM images that the surface morphology varies significantly among the Ge films deposited at different temperatures. The root mean square (RMS) surface roughness of the Ge films increases from 0.49 to 6.87 nm with substrate temperature increasing from 300 to 500°C. The increase in surface roughness with increasing substrate temperature indicates the growth switching from layer-by-layer mode to islanding mode with increasing substrate temperature.

In general, the epitaxial growth of Ge on Si follows the Stranski-Krastanow mode due to the lattice mismatch [18]. The growth initially follows layer-by-layer mode and progresses into island

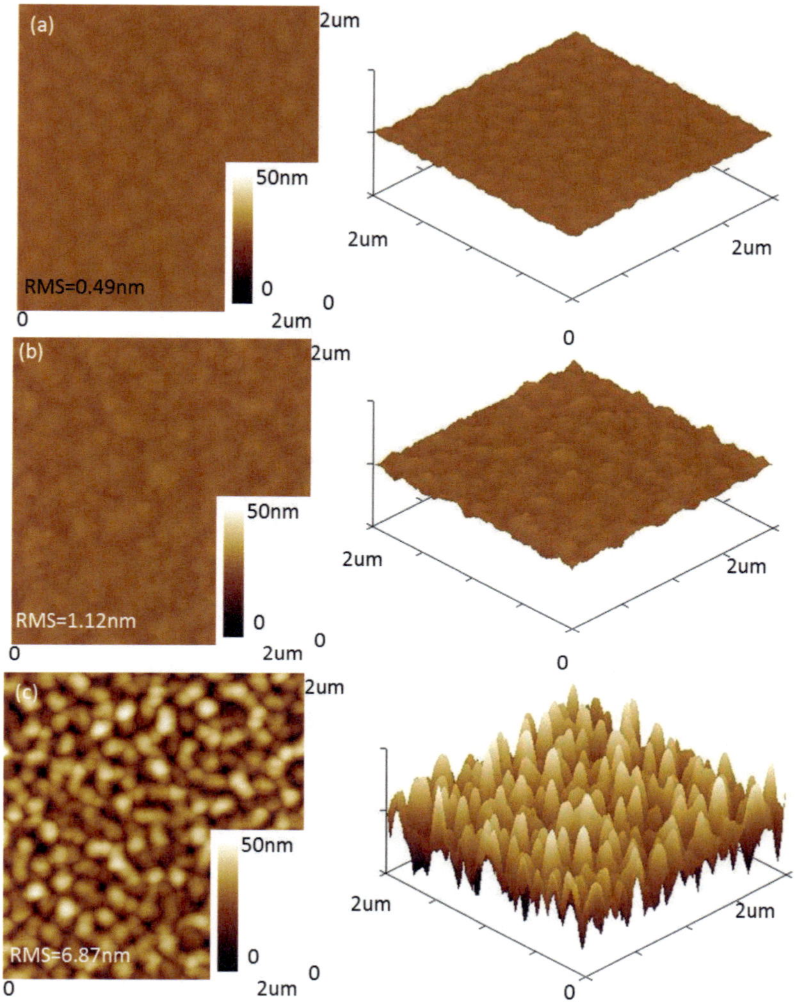

Figure 7. 2D and 3D AFM images showing the surface morphology of the Ge films deposited at (a) 300°C, (b) 400°C, and (c) 500°C.

mode when the layer becoming thicker. The thicker layer has large strain energy, which can be lowered by forming isolated thick islands. The island formation can be avoided by reducing the diffusion length of Ge. The reduced diffusion length hinders the mass transport of Ge over large distances which is necessary for the formation of islands [23]. Since the diffusion length of Ge decreases with reducing substrate temperature, the islanding is suppressed at low substrate temperature. As shown in **Figure 7**, layer-by-layer growth can be obtained at low temperature of 300°C to achieve smooth Ge surface, which is favored for the following III-V deposition. However, the low substrate temperature might induce the degradation of crystallinity simultaneously which will be investigated by the following XRD and TEM measurements.

The XRD reciprocal space mappings (RSM) were conducted to investigate the effect of substrate temperature on the crystallinity of the Ge films. **Figure 8** shows the (004) RSM of the Ge films deposited at (a) 300°C, (b) 400°C, and (c) 500°C. **Figure 8(a)** demonstrates that the Ge

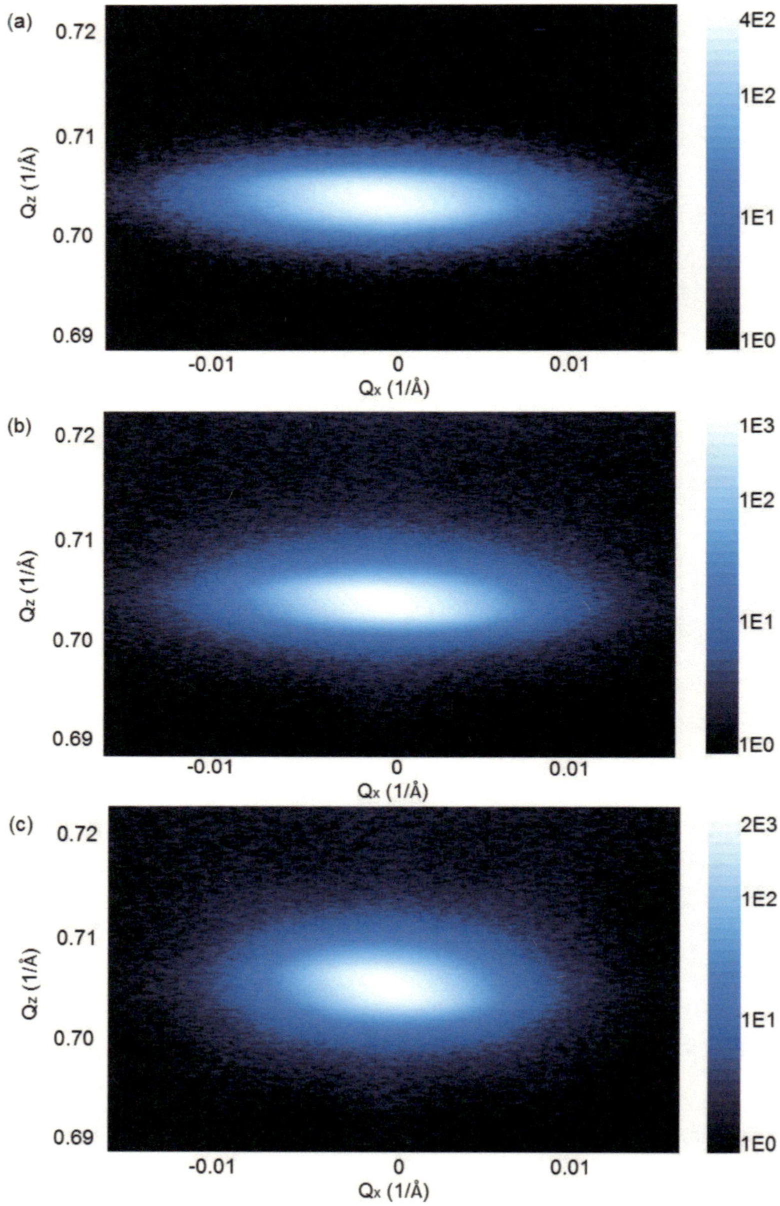

Figure 8. XRD (004) reciprocal space maps of the Ge films deposited at (a) 300°C, (b) 400°C, and (c) 500°C.

diffraction spots are elongated along the Q_x direction, which is due to the deteriorated crystal quality [35]. With the substrate temperature increasing from 300 to 500°C, the Ge peak exhibits steeper decay in the Q_x direction and the Ge peak position shows a slight upwards shift along the Q_z direction. These results indicate that the Ge film deposited at higher temperature has lower defect density and reduced compressive strain.

Micro-Raman spectra were used to investigate the structural property of the surface layer in the Ge samples deposited on Si at 300°C, 400°C, and 500°C. The penetration depth of the laser in the Ge layer was limited within the top 20 nanometers by using the wavelength of 514 nm excitation source [36]. As shown in **Figure 9**, the Ge films exhibit peaks centered around 300 cm^{-1} corresponding to the Ge-Ge optical vibration modes [37]. All the Ge films deposited at various temperatures exhibit peaks positioned at a higher wavenumber than the bulk unstrained Ge, suggesting compressive strains in the films. With increasing substrate temperatures, the peak positions of the Ge films shift to lower wavenumbers toward that of the bulk Ge indicating decreased compressive strain in the films [38], which is in agreement with the XRD results.

The reduction of compressive strain with increasing substrate temperature might be due to the difference in linear thermal expansion coefficients between Si and Ge. The thermal expansion coefficient of Ge is $\Delta a/a$ (Ge) = 5.8×10^{-6} ΔT (°C), which is larger than that of Si, $\Delta a/a$ (Si) = 2.6×10^{-6} ΔT (°C) [39]. The Ge films, which are nearly fully lattice-matched to the Si substrate at the growth temperature experience tensile strain when cooling to room temperature [40]. This is because the perpendicular lattice parameter of the Ge films shrinks more easily during cooling process than the in-plane lattice which is influenced by the underneath Si substrate with lower thermal expansion coefficient.

Figure 10 shows the cross-sectional TEM images of Ge samples deposited at 300°C in (a) bright and (b) dark field, deposited at 400°C in (c) bright and (d) dark field, and deposited at 500°C in (e) bright and (f) dark field. As shown in **Figure 10(a)** and **(b)**, the Ge film deposited at 300°C exhibits very high TDD which is estimated to be of the order of 10^{10} cm^{-2}. The high TDD might be owing to the reduced diffusion length of Ge at low temperature. With increasing substrate temperature, the TDD decreases and some planar defects are observed in the Ge film deposited at 500°C as shown in **Figure 10(c)–(f)**. The density of the planar defects is particularly high in the vicinity area of the Ge/Si interface and most of them are restricted to that region and do not extend to the film

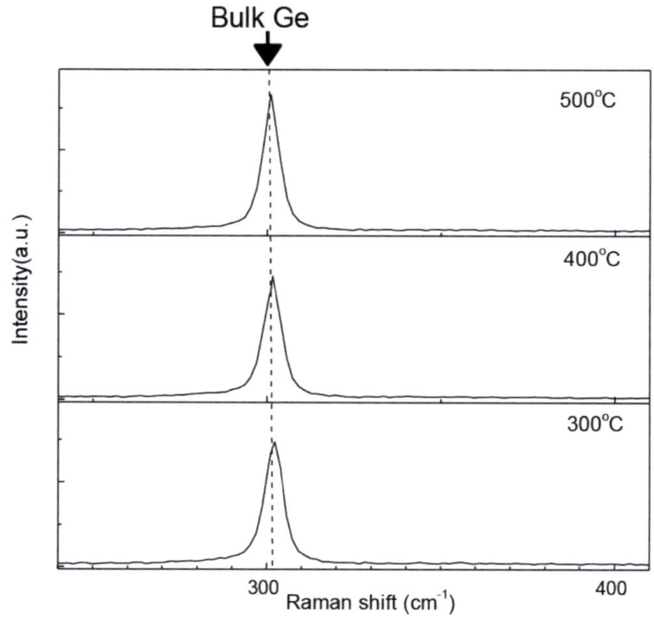

Figure 9. Raman spectra of the Ge films deposited on Si at 300, 400, and 500°C.

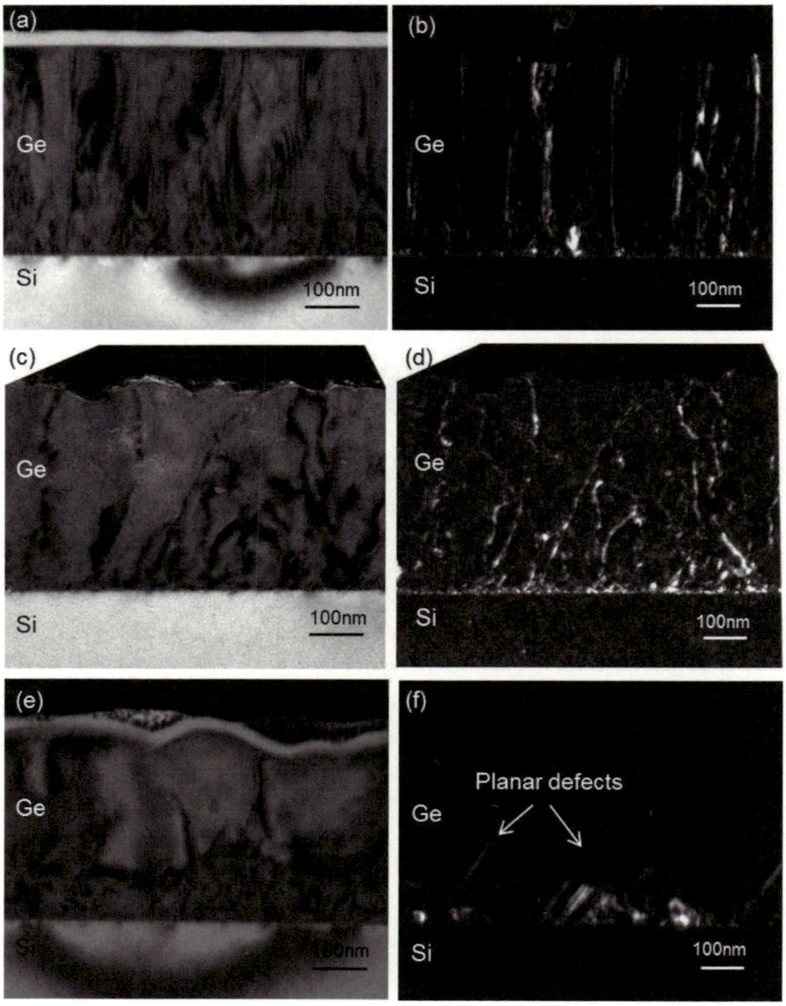

Figure 10. Cross-sectional TEM images of Ge samples deposited at 300°C in (a) bright and (b) dark field, deposited at 400°C in (c) bright and (d) dark field, and deposited at 500°C in (e) bright and (f) dark field.

surface which is consistent with previous report [41]. The TDD of the Ge film deposited at 500°C is in the order of 10^9 cm^{-2}, one magnitude order lower than that deposited at 300°C. The improved crystallinity with increasing substrate temperature agrees well with the XRD results.

5. Epitaxial growth of Ge on Si at low temperatures by one-step aluminum-assisted crystallization

The aluminum-induced crystallization (AIC) of Si, Ge and SiGe on foreign substrates has been extensively studied by several groups to obtain polycrystalline material at a low temperature [42–45]. The a-Ge/Al/c-Si structure has been investigated and epitaxial SiGe alloys were obtained [46]. The aforementioned conventional AIC includes two steps: (1) depositing a stacked Al and amorphous Ge layer on the Si substrate, (2) postdeposition annealing to induce the layer exchange process. The postdeposition annealing introduces the diffusion of Si into the Ge layer resulting in formation of Si_xGe_{1-x} alloy. In order to achieve epitaxial growth of pure Ge on Si through Al at low temperature, one-step aluminum-assisted crystallization is developed.

The novelty of one-step aluminum-assisted crystallization of Ge epitaxy on Si lies in the elimination of the postdeposition annealing step [47]. This process simply requires sequential depositions of Al and Ge films via magnetron sputtering in the same chamber without breaking the vacuum. By applying an *in-situ* low temperature (50–150°C) heat treatment in between Al and Ge sputter depositions, the epitaxial growth of Ge on Si is achieved. This low temperature process has a low thermal budget and can fabricate pure Ge layer compared with Si_xGe_{1-x} alloy as obtained in the conventional process. The effects of Al heating temperature on the properties of the epitaxial Ge films are investigated and the mechanism of epitaxial growth of Ge on Si by one-step aluminum-assisted crystallization is discussed based on observations on samples with various Ge deposition times.

The Al films were sputter-deposited onto Si substrates at room temperature using a 2 inch intrinsic Al target (99.999% purity) at a deposition rate of 3 nm/min. The samples then underwent an in-situ heat treatment for 10 minutes prior to the Ge deposition. The Ge films were then sputter-deposited using a 4 inch intrinsic Ge target (99.999% purity) without further intentional substrate heating at 5 nm/min. The Al heating temperatures were varied at 50°C (Sample ID: 60-50-12), 100°C (Sample ID: 60-100-12), and 150°C (Sample ID: 60-150-12) with Al thickness of 60 nm and Ge deposition time of 12 min to investigate the effect of heating temperature. One control sample (Sample ID: 60-NA-12) did not undergo this heat treatment. Shorter Ge deposition of 1 minute (Sample ID: 60-100-1) and 3 minutes (Sample ID: 60-100-3) were experimented on substrates with 60 nm Al deposition and 100°C heat treatment as well with the aim to investigate the mechanism of Ge epitaxial growth on Si by one-step aluminum-assisted crystallization. The Ge samples were analyzed by XRD, TEM and EDS (Phillips CM200 microscope equipped with an EDAX energy dispersive X-ray spectroscopy system) measurements.

5.1. Effects of heating temperature

Figure 11(a) shows the XRD 2θ-Ω diffraction patterns of samples 60-NA-12, 60-50-12, 60-00-12 and 60-150-12. For sample 60-NA-12, the XRD pattern shows a peak at 65.2° corresponding to Al (220) and a strong Si (400) peak at 69.2° from the Si substrate. The Ge film on 60-NA-12 is amorphous due to the absence of a Ge peak. For samples 60-50-12, 60-100-12 and 60-150-12, apart from the Al peak and Si peak, a peak located at 66° is present which corresponds to Ge (400). The absence of any other Ge peaks and the results of X-ray Phi scans indicate the Ge films are single-crystalline Ge (100). **Figure 11(b)** shows the Si (220) and Ge (220) Phi scan patterns collected from sample 60-100-12 by rotating the specimen with respect to the [110] axis. The four (220) reflections are observed in the Ge Phi scan pattern which suggests the film is with fourfold symmetry about an axis normal to the Si substrate [33]. In addition, the Ge (220) reflections align with the Si substrate (220) reflections indicating the Ge is single-crystalline epitaxy film.

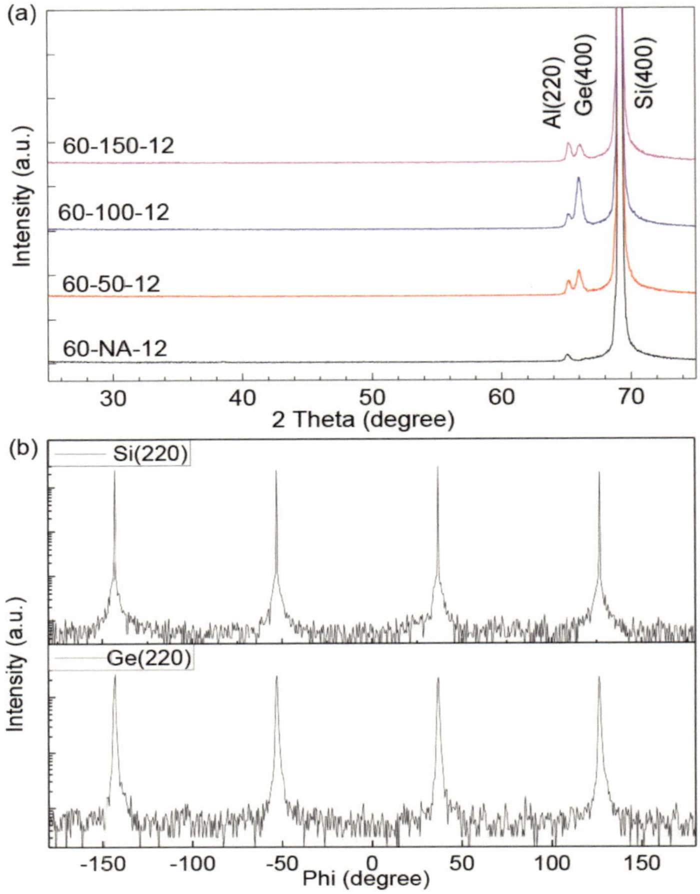

Figure 11. (a) XRD 2θ-Ω diffraction patterns of samples 60-NA-12, 60-50-12, 60-100-12 and 60-150-12, (b) Si (220) and Ge (220) phi scan patterns collected from the sample 60-100-12 showing the epitaxial relationship between the Ge film and Si substrate. (reprinted from Liu et al. [47], with the permission of AIP publishing).

The Ge peak intensity of sample 60-50-12 is lower than that of sample 60-100-12 as shown in **Figure 11(a)** due to incomplete Ge crystallization. The drop in Ge peak intensity for sample 60-150-12 might be due to the lower number of Al grain boundaries compared with that of sample 60-100-12. The Al grain boundaries play an important role in the Al-assisted crystallization process as they supply pathways for the Ge atoms to be epitaxially grown from the Si surface [46]. With increasing heating temperature, Al grain size is enlarged and therefore the density of grain boundaries is reduced [48] verified by the following TEM measurements.

Figure 12 shows the cross-sectional TEM images of samples 60-NA-12, 60-50-12, 60-100-12, and 60-150-12. **Figure 12(a)** shows the absence of Ge crystallization as the Al and amorphous Ge layers on the Si substrate. **Figure 12(b)–(d)** show the Ge epitaxial growth at selected sites on samples 60-50-12, 60-100-12, and 60-150-12. With increasing heating temperature, the crystallization sites decrease probably due to the decrease in density of Al grain boundaries [48], which

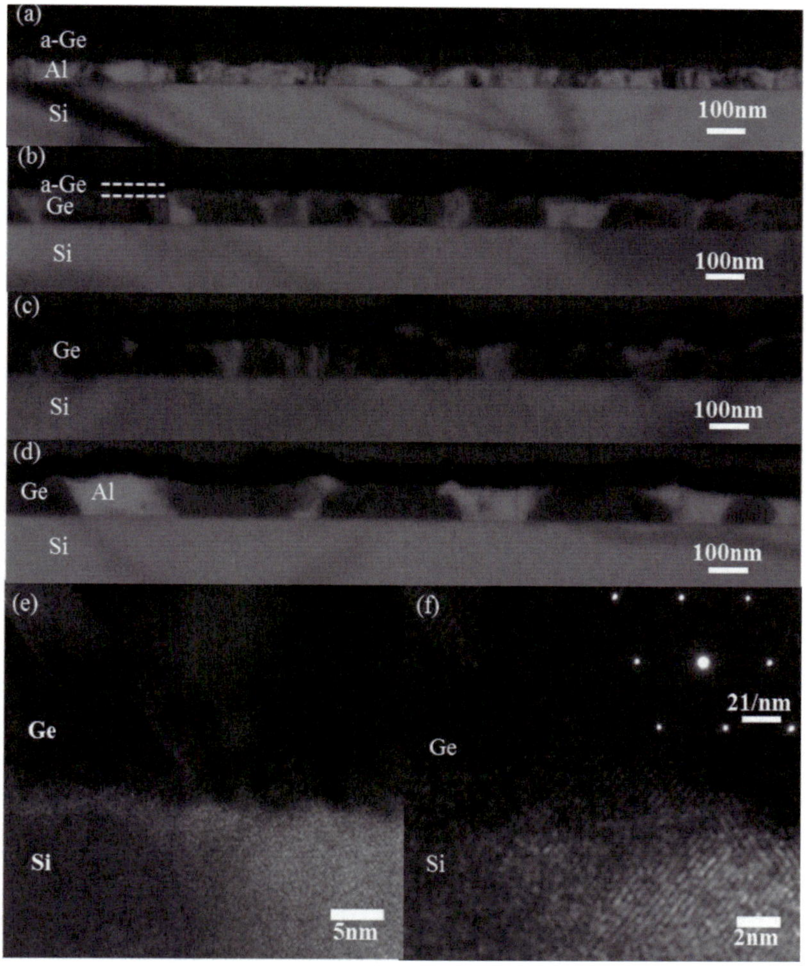

Figure 12. Cross-sectional TEM images of samples (a) 60-NA-12, (b) 60-50-12, (c) 60-100-12, and (d) 60-150-12. (e) Higher magnification and (f) atomic resolution (with SAED pattern in the insert) images of Ge/Si interface on sample 60-100-12. (reprinted from Liu et al. [47], with the permission of AIP Publishing).

are responsible for supplying pathways for the nucleation of the Ge from the Si substrate [46]. An amorphous Ge layer is shown in **Figure 12(b)** verifying the previous observations that Ge crystallization is in-complete on sample 60-50-12. Although sample 60-50-12 has more nucleation sites, it exhibits more discontinuous Ge layer compared with sample 60-100-12 owing to the incomplete crystallization.

The Ge/Si interface of sample 60-100-12 is magnified and shown in **Figure 12(e)**. The epitaxial growth of the Ge layer on Si is revealed and planar defects are observed at the interface in **Figure 12(e)**. **Figure 12(f)** shows the continuous alignment of the atoms from the Si substrate to the grown Ge film suggesting successful epitaxy. Furthermore, the electron diffraction pattern taken from the Ge layer shown in the insert to **Figure 12(f)** indicates the Ge is single crystal [49]. This result is in good agreement with the XRD measurements.

5.2. Mechanism of one-step aluminum-assisted crystallization

To better understand the mechanism of the epitaxial growth of Ge on Si through one-step aluminum-assisted crystallization, EDS mapping of samples that underwent different Ge deposition times (1 minute, 3 minutes and 12 minutes) were carried out. The cross-sectional TEM images (top row), EDS maps of Ge (middle row) and EDS maps of Al (bottom row) of samples 60-100-1, 60-100-3, and 60-100-12 are shown in **Figure 13(a)–(c)**, respectively. They reveal the Ge and Al distributions at different stages of the crystallization process. As shown, the process begins with the Ge nucleating at selected sites at the Si and Al interface. With increasing deposition time, the Ge tends to grow upwards at the initial stage and then grow laterally.

The mechanism of the epitaxial growth of Ge on Si by one-step aluminum-assisted crystallization is discussed as follows. The covalent bonds of Ge are weakened at the interface with the Al layer as a consequence of a screening effect of the free electrons in the Al layer [50]. These Ge atoms have relatively high mobility and may provide the agent for initiating the crystallization process. These mobile atoms tend to lower the Gibbs energy of the system by diffusing to sites of low energy such as the Al grain boundaries. This diffusion is sometimes called grain boundary wetting that reduces total interface energy by replacing the grain

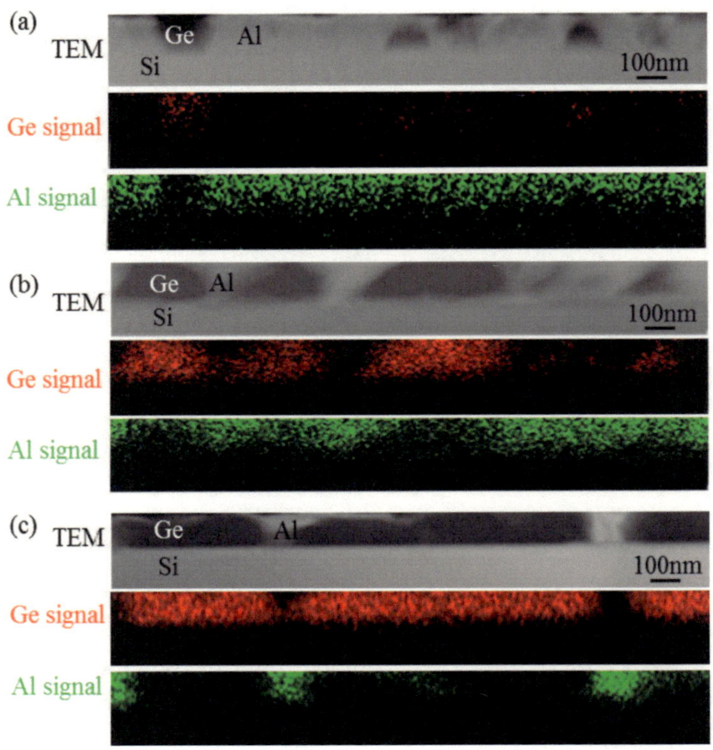

Figure 13. Cross-sectional TEM images (top), EDS maps of Ge (middle) and al (bottom) of samples (a) 60-100-1, (b) 60-100-3, and (c) 60-100-12. (reprinted from Liu et al. [47], with the permission of AIP publishing.

boundary with two interphase boundaries [51]. The diffusion of Ge into the Al grain boundaries forms a pathway to supply the material for crystallization. The crystallization is then driven by the reduction in bulk Gibbs energy when the material changes from amorphous to crystalline [52]. However, this can be counteracted by the increase in interface energy as crystallization proceeds [53]. When a heat treatment is applied, the interface energy at the crystalline-amorphous interface increases, while the interface energy at the crystalline-crystalline decreases [43]. This effectively reduces the energy difference between the crystalline-amorphous and the crystalline-crystalline interfaces favoring the crystallization process. This explains the observed Ge crystallization in the heat treated samples. As studied and discussed in previous work [46], the Al/Si is the preferred interface to Ge/Al for nucleation, as observed on sample 60-100-1 in this work. After the nucleation on Si substrate, the epitaxial growth of Ge continues with further incorporation of Ge atoms though the Al grain boundaries.

6. Conclusions

Epitaxial growth of Ge films on Si has been achieved using magnetron sputtering which is low cost, safe and scalable. The effects of substrate temperature on the properties of the Ge films have been investigated. The surface roughness of the Ge films increases with substrate temperature. Smooth surface with RMS roughness of 0.48 nm can be obtained at 300°C owing to the reduced diffusion length of Ge atoms at low temperature. On the other hand, the crystallinity of the Ge films could be improved by increasing substrate temperature as revealed by XRD and TEM measurements. In addition, the compressive strain in the Ge films decreases with increasing substrate temperature owing to the difference in the thermal expansion coefficients between Si and Ge.

Epitaxial growth of Ge films on Si by magnetron sputtering at low temperature has been achieved through one-step aluminum-assisted crystallization. By applying an *in-situ* low temperature (50–150°C) heat treatment in between Al and Ge sputter depositions, the epitaxial growth of Ge on Si can be achieved as verified by high resolution TEM and XRD analyses. The mechanism of epitaxial growth of Ge on Si substrate by one-step aluminum-assisted crystallization is discussed based on observations on samples with various Ge deposition times. This method significantly lowers the required temperature for and therefore the cost of epitaxial growth of Ge on Si.

Acknowledgements

This work has been supported by the Australian Government through the Australian Research Council (ARC, grant number DP160103433, LP110201112) and the Australian Renewable Energy Agency (ARENA) and by Epistar Corporation and Shin Shin Natural Gas Co., Ltd., Taiwan. Responsibility for the views, information or advice expressed herein is not accepted by the Australian Government.

Author details

Ziheng Liu*, Xiaojing Hao, Anita Ho-Baillie and Martin A. Green

*Address all correspondence to: ziheng.liu@unsw.edu.au

School of Photovoltaic and Renewable Energy Engineering, University of New South Wales, Sydney, Australia

References

[1] Hussain AM et al. Thermal recrystallization of physical vapor deposition based germanium thin films on bulk silicon (100) physica status solidi (RRL). Rapid Research Letters. 2013;**7**(11):966-970

[2] Luryi S, Kastalsky A, Bean JC. New infrared detector on a silicon chip. Electron Devices, IEEE Transactions on. 1984;**31**(9):1135-1139

[3] Beeler R et al. Comparative study of InGaAs integration on bulk Ge and virtual Ge/Si substrates for low-cost photovoltaic applications. Solar Energy Materials and Solar Cells. 2010;**94**(12):2362-2370

[4] King RR et al. 40% efficient metamorphic GaInP/GaInAs/Ge multijunction solar cells. Applied Physics Letters. 2007;**90**(18)

[5] Guter W et al. Current-matched triple-junction solar cell reaching 41.1% conversion efficiency under concentrated sunlight. Applied Physics Letters. 2009;**94**(22)

[6] Tanabe K. A review of ultrahigh efficiency III-V semiconductor compound solar cells: Multijunction tandem, lower dimensional, photonic up/down conversion and plasmonic nanometallic structures. Energies. 2009;**2**(3):504

[7] Green MA. Silicon wafer-based tandem cells: The ultimate photovoltaic solution? 2014

[8] Ringel SA et al. Single-junction InGaP/GaAs solar cells grown on Si substrates with SiGe buffer layers. Progress in Photovoltaics: Research and Applications. 2002;**10**(6):417-426

[9] Roman JM. State-of-the-art of III-V solar cell fabrication technologies, device designs and applications. Advanced Photovoltaic Cell Design. EN548; April 27 2004

[10] Luan H-C et al. High-quality Ge epilayers on Si with low threading-dislocation densities. Applied Physics Letters. 1999;**75**(19):2909-2911

[11] Fukuda Y, Kohama Y. High quality heteroepitaxial Ge growth on (100) Si by MBE. Journal of Crystal Growth. 1987;**81**(1-4):451-457

[12] Currie MT et al. Controlling threading dislocation densities in Ge on Si using graded SiGe layers and chemical-mechanical polishing. Applied Physics Letters. 1998;**72**(14):1718-1720

[13] Choi D et al. Low surface roughness and threading dislocation density Ge growth on Si (001). Journal of Crystal Growth. 2008;**310**(18):4273-4279

[14] Liu Z. Virtual Ge substrates for high efficiency III-V solar cells. 2014. The University of New South Wales

[15] Frank FC, Merwe JHvd. One-dimensional dislocations. I. Static theory. Proceedings of the Royal Society of London. Series A, Mathematical and Physical Sciences. 1949;**198**(1053): 205-216

[16] Volmer M, Weber A. Keimbildung in übersättigten Gebilden. Physical Chemistry. 1926; **119U**(1):277-301

[17] Stranski IN, Krastanow VL, Akad. Wiss. Lit. Mainz Math.-Natur. Kl. IIb 1939. 146

[18] Eaglesham DJ, Cerullo M. Dislocation-free Stranski-Krastanow growth of Ge on Si(100). Physical Review Letters. 1990;**64**(16):1943-1946

[19] Mo YW et al. Kinetic pathway in Stranski-Krastanov growth of Ge on Si(001). Physical Review Letters. 1990;**65**(8):1020-1023

[20] Sakai A, Tatsumi T. Defect-mediated island formation in Stranski-Krastanov growth of Ge on Si(001). Physical Review Letters. 1993;**71**(24):4007-4010

[21] Sieg RM et al. Anti-phase domain-free growth of GaAs on offcut (001) Ge wafers by molecular beam epitaxy with suppressed Ge outdiffusion. Journal of Electronic Materials. 1998;**27**(7):900-907

[22] Ringel SM et al. Toward achieving efficient III-V space cells on Ge/GeSi/Si wafers. Proceedings of the second world conference and exhibition on photovoltaic solar. Energy Conversion. 1998

[23] Voigtländer B et al. Modification of growth kinetics in surfactant-mediated epitaxy. Physical Review B. 1995;**51**(12):7583-7591

[24] Sakai A, Tatsumi T. Ge growth on Si using atomic hydrogen as a surfactant. Applied Physics Letters. 1994;**64**(1):52-54

[25] Copel M et al. Influence of surfactants in Ge and Si epitaxy on Si(001). Physical Review B. 1990;**42**(18):11682-11689

[26] Bean JC et al. Pseudomorphic growth of Ge_xSi_{1-x} on silicon by molecular beam epitaxy. Applied Physics Letters. 1984;**44**(1):102-104

[27] Eaglesham DJ, Cerullo M. Low-temperature growth of Ge on Si(100). Applied Physics Letters. 1991;**58**(20):2276-2278

[28] Wietler TF, Bugiel E, Hofmann KR. Surfactant-mediated epitaxy of relaxed low-doped Ge films on Si(001) with low defect densities. Applied Physics Letters. 2005;**87**(18) 182102

[29] People R, Bean JC. Calculation of critical layer thickness versus lattice mismatch for GexSi1−x/Si strained-layer heterostructures. Applied Physics Letters. 1985;**47**(3):322-324

[30] Nayfeh, A. Heteroepitaxial Growth of Relaxed Germanium on Silicon. PhD thesis Stanford University, 2006

[31] Ferrari C, Rossetto G, Fitzgerald EA. Misfit dislocation and threading dislocation distributions in InGaAs and GeSi/Si partially relaxed heterostructures. Materials Science and Engineering: B. 2002;**91-92**(0):437-440

[32] Kern W. The evolution of silicon wafer cleaning technology. Journal of the Electrochemical Society. 1990;**137**(6):1887-1892

[33] Gilbert SR et al. Epitaxial growth of SrTiO[sub 3] thin films by metalorganic chemical vapor deposition. Applied Physics Letters. 1995;**66**(24):3298-3300

[34] Greene JE. Epitaxial crystal growth by sputter deposition: Applications to semiconductors. Part 2. Critical Reviews in Solid State and Materials Sciences. 1983;**11**(3):189-227

[35] Yamamoto Y et al. Low threading dislocation density Ge deposited on Si (100) using RPCVD. Solid-State Electronics. 2011;**60**(1):2-6

[36] Wolf ID. Micro-Raman spectroscopy to study local mechanical stress in silicon integrated circuits. Semiconductor Science and Technology. 1996;**11**(2):139

[37] Loh TH et al. Ultrathin low temperature SiGe buffer for the growth of high quality Ge epilayer on Si(100) by ultrahigh vacuum chemical vapor deposition. Applied Physics Letters. 2007;**90**(9)

[38] Oda K et al. Improvement of crystallinity by post-annealing and regrowth of Ge layers on Si substrates. Thin Solid Films. 2014;**550**(0):509-514

[39] Sze SM. Physics of Semiconductor Devices. 2nd ed. New York: John Wiley and Sons; 1981

[40] Hartmann JM et al. Reduced pressure–chemical vapor deposition of Ge thick layers on Si(001) for 1.3-1.55-μm photodetection. Journal of Applied Physics. 2004;**95**(10):5905-5913

[41] Ernst F, Pirouz P. Formation of planar defects in the epitaxial growth of GaP on Si substrate by metal organic chemical-vapor deposition. Journal of Applied Physics. 1988;**64**(9):4526-4530

[42] Nast O et al. Aluminum-induced crystallization of amorphous silicon on glass substrates above and below the eutectic temperature. Applied Physics Letters. 1998;**73**(22):3214-3216

[43] Wang ZM et al. Thermodynamics and mechanism of metal-induced crystallization in immiscible alloy systems: Experiments and calculations on Al/a-Ge and Al/a-Si bilayers. Physical Review B. 2008;**77**(4):045424

[44] Zhang T-W et al. Diffusion-controlled formation mechanism of dual-phase structure during al induced crystallization of SiGe. Applied Physics Letters. 2012;**100**(7):071908

[45] Gjukic M et al. Aluminum-induced crystallization of amorphous silicon--germanium thin films. Applied Physics Letters. 2004;**85**(11):2134-2136

[46] Liu Z et al. Epitaxial growth of single-crystalline silicon–germanium on silicon by aluminium-assisted crystallization. Scripta Materialia. 2014;**71**(0):25-28

[47] Liu Z et al. One-step aluminium-assisted crystallization of Ge epitaxy on Si by magnetron sputtering. Applied Physics Letters. 2014;**104**(5) 052107

[48] Tsuji N et al. Strength and ductility of ultrafine grained aluminum and iron produced by ARB and annealing. Scripta Materialia. 2002;**47**(12):893-899

[49] Wei S-Y et al. Epitaxial growth of heavily boron-doped Si by al(B)-induced crystallisation at low temperature for back surface field manufacturing. CrystEngComm. 2013; **15**(9):1680-1684

[50] Hiraki A. Low temperature reactions at Si/metal interfaces; what is going on at the interfaces? Surface Science Reports. 1983;**3**(7):357-412

[51] Wang ZM et al. "Explosive" crystallisation of amorphous germanium in Ge/Al layer systems; comparison with Si/Al layer systems. Scripta Materialia. 2006;**55**(11):987-990

[52] Wang Z et al. Fundamentals of metal-induced crystallization of amorphous semiconductors. Advanced Engineering Materials. 2009;**11**(3):131-135

[53] Benedictus R, Böttger A, Mittemeijer EJ. Thermodynamic model for solid-state amorphization in binary systems at interfaces and grain boundaries. Physical Review B. 1996;**54**(13):9109-9125

Chapter 3

Electrical Properties of Epitaxial Ferroelectric Heterostructures

Andra Georgia Boni, Cristina Florentina Chirila, Raluca Negrea, Corneliu Ghica, Iuliana Pasuk, Ioana Pintilie and Lucian Pintilie

Additional information is available at the end of the chapter

http://dx.doi.org/10.5772/intechopen.70133

Abstract

In the context of miniaturization of devices, ferroelectric materials are used as multifunctional materials for their well-known intrinsic properties, especially for the switching of polarization in an applied electric field. The high-quality epitaxial thin film structures are used for the possibility to study different effects as low dimensions, interface, strain and strain gradients on ferroelectric materials and other electric characteristics, also representing a possibility to obtain new phenomena and properties that can be used for development of new devices with different functionalities. This chapter is a summary of the ferroelectric and dielectric behaviour of epitaxial thin films of $Pb(Zr,Ti)O_3$ (PZT) and $BaTiO_3$ (BTO) obtained by pulsed laser deposition and the correlation with structural quality of the layers and with different electrostatic conditions induced either by electrodes or by the different interlayers. For this purpose in the first part, studies regarding the influence of the substrates and of different top electrodes are performed for $Pb(Zr,Ti)O_3$ (PZT) 52/48. In the second part, we focused on artificial multiferroic structures from alternating layers of PZT 20/80 or $BaTiO_3$ (BTO) as ferroelectric phase and $CoFe_2O_4$ (CFO) as magnetic material. We found that interface configuration and strain engineering could control ferroelectric hysteresis, the capacitance or the leakage current magnitude.

Keywords: ferroelectric thin films, electrical properties, multilayered structures, electrostatic boundary conditions, interfaces

1. Introduction

Ferroelectrics are multifunctional materials possessing special properties derived from the presence of the spontaneous polarization in the absence of an applied electric field. Ferroelectricity is electrical analogue of ferromagnetism, the distinguishing property of ferroelectricity being the possibility of reversing the spontaneous polarization when an external electric field is applied in the opposite direction. As a consequence, the polarization describes a hysteresis loop as magnetization does in ferromagnetic materials. Ferroelectric materials also possess piezoelectric and pyroelectric properties which are used in many electronic applications, such as tunable capacitors, ferroelectric nonvolatile memories, ultrasound sensors or generators and infrared sensors [1–4]. Another interesting topic is related to multiferroic materials—single phase or heterostructures—which possess more than one order parameter (usually magnetic and ferroelectric ordering) and which can lead to new applications if there is a coupling between the order parameters [5–9].

For many applications, the ferroelectrics are used either as bulk ceramics/single crystals or as thin films with different structural qualities, from polycrystalline to epitaxial. The necessary electrical properties for different applications of ferroelectric materials are strongly influenced by the structural quality. For instance, the existence of the grains and grain boundaries in nanostructured thin films/polycrystalline thin films can induce modification on the magnitude of polarization, dielectric constant and so forth. As for the standard semiconductors, the studies of intrinsic electrical properties should be performed on high-quality single-crystal samples. One method to obtain such samples is to deposit thin films of epitaxial quality. The obtaining of epitaxy for ferroelectric materials often involves the deposition of the ferroelectric thin films on single-crystal substrates, with different buffer or electrode layers, resulting in a heteroepitaxial growth of the film. Therefore, the use of materials with different values for the lattice parameters can generate mechanical tensions/deformations and strain in the lattice of the ferroelectric film [10–13].

It is known that, by changing the pressure on ferroelectric bulk ceramics or single crystals, the transition temperature, piezoelectric and dielectric constants can be modified. In the case of thin films, applying significant hydrostatic pressure to induce modification of ferroelectric properties leads to physical cracks of the samples. Thus, the epitaxy offers the possibility to induce strain and strain-polarization coupling for the enhancement of ferroelectric properties. Examples of the influence of the strain in epitaxial heterostructures are enhancement of polarization in $BaTiO_3$ (BTO) [14–16], the shift of the transition temperature for $PbTiO_3$ and $BaTiO_3$ films towards higher values [17], room temperature-induced ferroelectricity in $SrTiO_3$ thin films [14–16] and in un-doped HfO_2 layers [18] or in artificial superlattices [19] formed from non-ferroelectric materials, etc. These phenomena are specific for fully coherent thin films with low density of dislocations. Nevertheless, by growing the thickness of the deposited ferroelectric layer, many dislocations and other defects appear in order to minimize the free energy of the systems. Different lattice relaxation processes could generate strain gradients in the films, which could imply a flexoelectric field. The effects of these supplementary fields lead to different behaviours in polarization switching, shift of the hysteresis loop along voltage axis, modification of remnant polarization and occurrence of diode-like current characteristics with dependence on polarization direction [20–22].

As complex equivalent circuits are used for many applications involving ferroelectric materials, it ensures that good knowledge of the electric properties of these materials is a very important topic besides the deposition method and the structural quality. The ferroelectric materials/thin films should be integrated in a capacitor-like structure for the study of the electrical properties. Therefore, a common way to build such a structure is to deposit a continuous conductor thin film as bottom electrode, on which ferroelectric thin film is further deposited. On top of this structure, the top electrode is deposited using a shadow mask, which delimitates the active area of the capacitor. The specific measurements can be realized by connecting the measurement circuit on the upper and bottom electrodes.

The most used characterization techniques for investigation of ferroelectric properties are hysteresis loops of polarization versus the applied electric field, small signal capacitance measurements (as dependence of capacitance on voltage, frequency or temperature) and leakage current. These characteristics offer information about the ferroelectric character of the structure (e.g. the measurement of a rectangular hysteresis loop or a butterfly shape of the capacitance-voltage characteristics) or the values for different parameters of interest (dielectric constant, coercive field, magnitude of polarization and transition temperature).

One of the most studied classes of ferroelectric materials is oxide ferroelectrics, especially the subclass of the materials with perovskite structure. $BaTiO_3$ and $Pb(Zr,Ti)O_3$ (PZT) are the most investigated materials from this category from the point of view of applications. In this chapter, we will present the electrical and ferroelectric properties for this type of epitaxial ferroelectric thin films, obtained by pulsed laser deposition (PLD), and their dependence on the type of the substrate used for deposition or on the material used for the top electrode. Further on, we will show that, by constructing artificially layered structures from thin films of ferroelectric materials and materials having different electric/dielectric properties, the ferroelectric/electric properties can be modified and engineered to obtain enhanced or even new properties. For example, due to either electromechanical or electrostatic interactions, a tuning of capacitance, switching behaviour or leakage current magnitude can be realized.

2. Epitaxy

The term epitaxy refers to a film growth on a substrate with crystallographic structure close to that of the deposited layer. Epitaxial growth is one of the most important techniques in the present microelectronic industry, allowing a better correlation between structure and the macroscopic properties of thin films. Important problems can be studied in this way, related to physics of surfaces, interfaces and strain engineering. There are a wide variety of growth techniques that can be used to obtain epitaxial thin films including sputtering, metal-organic chemical vapour deposition, pulsed laser deposition, molecular beam epitaxy, physical vapour deposition, etc. In this chapter, we will discuss the heteroepitaxial growth, by pulsed laser deposition, of oxide thin films with ferroelectric/multiferroic properties. There are three known growth modes: (1) Frank-Van der Merwe, layer-by-layer growth; (2) Volmer-Weber, island growth and (3) Stranski-Krastanov, a combination of layer-by-layer and island growth. Thermodynamic approach was used in order to explain these growth modes in close to equilibrium conditions [23]. The balance between free energies from the film surface (γ_F), the substrate surface (γ_S) and the interface between them (γ_I) is the key factor that determines the thin film morphology in this approach. Frank-Van der Merwe growth mode is characterized by the fact that the deposited atoms or molecules are more strongly bonded to the substrate than in between them, thus forming a continuous layer on the substrate. In the case of the Volmer-Weber growth mode, the atoms and molecules are more strongly bonded in between them, forming island on the substrate, thus the wetting of the substrate is poor. Stranski-Krastanov mode is characterized by the fact that, at the beginning of the growth, the atoms and molecules form complete monolayers on the substrate (layer-by-layer growth, usually 1–5 monolayers), and then islands start to develop (growth mode changes to island formation). The transition from monolayers to islands is caused by increased tensions (strain) from increasing the layer thickness. Experimentally, the growing of thin films is not an equilibrium process; therefore kinetic effects have to be taken into account, leading to the occurrence of different growth modes. The above-mentioned growth modes and their schematic representations are shown in **Figure 1**.

A preliminary substrate preparation has to be achieved in order to obtain heteroepitaxial thin films on single-crystal $SrTiO_3$ (STO) substrates. The substrate preparation consists in transforming an optically polished surface into a step-and-terrace surface that is well ordered even on an atomic scale. For this purpose, the $SrTiO_3$ substrates are etched in NH_4-HF solution to remove Sr residue and to obtain a purely Ti-terminated surface and to get a high-quality step-and-terrace structure on the surface. All step edges should have equal height (single unit cell ~ 0.4 nm), and the steps should be approximately parallel and equidistant. This can be obtained after an annealing process, at elevated temperatures, of the etched substrates. The substrate transformation after each processing step is investigated by atomic force microscopy, and the results are presented in **Figure 2**.

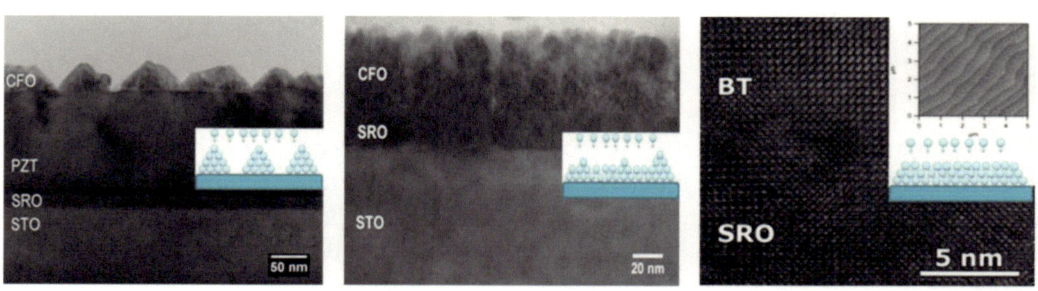

a) b) c)

Figure 1. Growth modes for epitaxial thin films: (a) Frank-Van der Merwe, (b) Stranski-Krastanov and (c) Volmer-Weber.

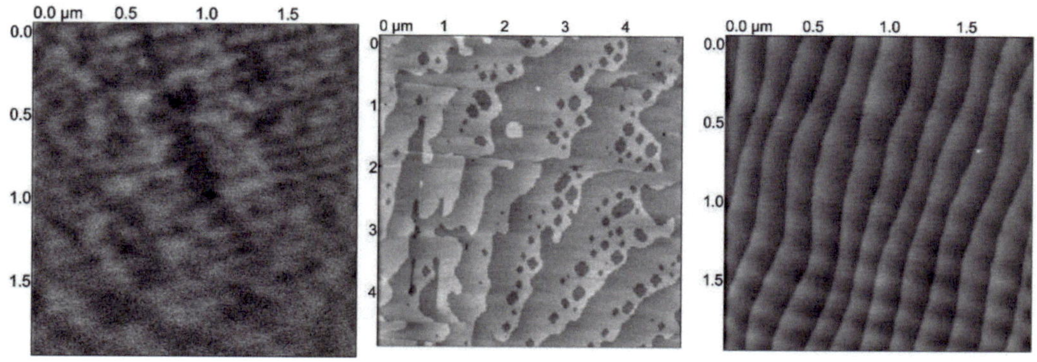

Figure 2. Atomic force microscopy images obtained on STO substrate after each processing step: upper image, fresh substrate; middle image, substrate after etching; and lower image, substrate after thermal annealing.

3. Growth method

Pulsed laser deposition, as the name suggests, is a technique, which uses pulses of laser radiation to remove material from the surface of a solid target. The technique involves complex physical aspects as follows:

1. Interaction between the laser radiation and the target material—high temperature is achieved over a small area (the area of the laser spot) due to the high power of the laser beam in the short period of the laser pulse; this leads to a number of processes occurring at the target surface, such as material decomposition, ionization and evaporation. One has to mention that the target has the same chemical composition as the film intended to be deposited on the substrate.
2. Transfer of the ablated material: the evaporated material from the step 1 forms a plasma plume over the substrate on which the film has to be deposited.
3. Deposition of the ablated material on the substrate—this step consists in the adsorption/chemisorption of the ablated material on the surface of the substrate. One has to mention that the substrate is, usually, heated.
4. Nucleation and growth of the film on the surface of the substrate—the main condition for this process to occur is represented by the balance between free energies from the film surface, the substrate surface and the interface between them.

Each step involved in the pulsed laser deposition process is crucial for obtaining the hetero-epitaxial growth of the thin films.

3.1. Experimental setup

The experimental setup is designed for the deposition of thin films and multilayers from oxide materials and consists of an excimer laser source (KrF, λ = 248 nm) with a pulse duration of

20 ns; a target carrousel with four targets of 2″ diameter, allowing permanent rotation of each target; a substrate holder with controlled motion on five axes and possibility to heat the substrate up to 1000°C; a deposition chamber allowing base vacuum down to 10^{-7} mbar; and high-pressure reflection high-energy electron diffraction (RHEED) system for in situ characterization. The entire system is controlled by PC and is presented in **Figure 3**.

The energy density of the laser pulse (fluence) can reach 5 J/cm^2, and the repetition rate is in the 1–10 Hz range. The laser beam is delivered on the target surface at an angle 45° to the normal. The pressure inside the chamber can be varied during the deposition by changing the flow rate of deposition gas (O_2 or Ar) using mass flow controllers. The substrate temperature is controlled with a proportional–integral–derivative controller (PID controller) temperature controller from room temperature (RT) up to 1000°C. The distance between target and substrate can be adjusted from 40 to 80 mm. All these parameters have to be optimized for each material in order to get the desired heteroepitaxial thin films.

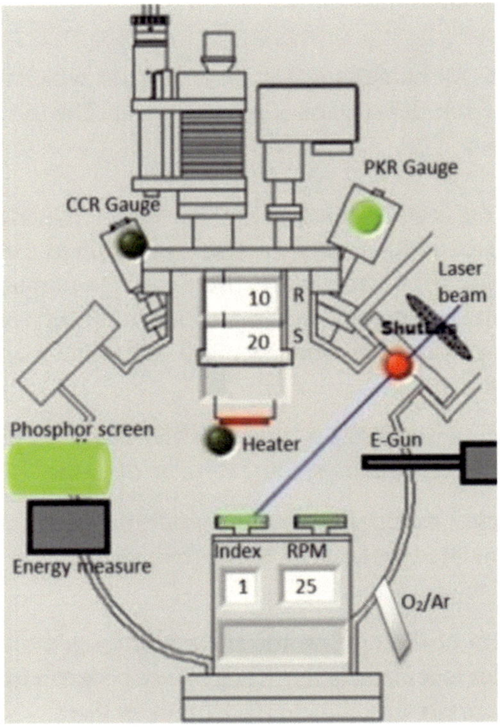

Figure 3. Schematic view of the pulsed laser deposition system.

4. Epitaxial ferroelectric thin films: structural and electrical properties

One representative and widely studied material of the class of perovskite ferroelectrics is lead titanate-zirconate Pb(Zr,Ti)O_3 (PZT). Its phase diagram is quite complex, and its structural, ferroelectric, pyroelectric and piezoelectric properties depend on temperature, strain and Zr/Ti ratio. Around Zr/Ti ratio of 52/48 this material presents a morphotropic phase boundary, and this composition is often preferred due to enhanced dielectric constant and/or piezoelectric coefficient [24–27].

In the last years, great efforts have been dedicated to the epitaxial growth of ferroelectric thin films with the purpose of obtaining enhanced properties compared to that of the polycrystalline ones. There are many reports on obtaining high-quality epitaxial PZT thin films, deposited by various methods on single-crystal substrates with perovskite structures such as $SrTiO_3$ (STO), but there is an increased need to obtain the same material performances on substrates (Si) allowing rapid integration of ferroelectric materials in the existing complementary metal–oxide–semiconductor (C-MOS) technology. First attempts to deposit PZT layers on Si were realized in the context of constructing metal-ferroelectric-semiconductor field-effect transistors, and it was found out that silicates or other parasitic phases are formed at the PZT-Si interface [28]. To overcome these problems, insulating buffer layers were used as barriers to avoid diffusion of Pb atoms towards Si interface, but this method leads to high depolarization fields. As a consequence, the polarization magnitude decreases, and the retention properties are deteriorated [29–32].

In this context, our first results presented in this section consist in evidencing epitaxial deposition by PLD of PZT on Si (001) using as interlayer a MBE-deposited thin film of STO which acts as a barrier for Pb diffusion and as a template for the growth of the subsequent layers. A 20 nm film of $SrRuO_3$ (SRO) has been deposited by PLD to serve as bottom electrode, and then the PZT film was grown. Even if an epitaxial structure is obtained using this configuration of deposited layers, many structural and electrical differences are observed compared to the same ferroelectric capacitor structure deposited on single-crystal STO substrate [33].

XRD 2θ–ω diagrams are presented in **Figure 4** for both types of samples, PZT/SRO/STO/Si and PZT/SRO/STO, showing only (00l) (l = 1,2,3,4) maximas for PZT, SRO and STO layers, indicating an out-of-plane-oriented pseudocubic structure for both cases. Around SRO and STO peaks, the layered fringes evidenced in the inset figures are specific for epitaxial thin films and indicate very smooth and parallel interfaces. The PZT out-of-plane lattice parameter is calculated from this data, and a significantly larger value is obtained in the case of the single-crystal STO substrate (c_{PZT} = 4.113 Å) compared to the case of Si substrate with STO buffer layer (c_{PZT} = 4.055 Å). In addition, the rocking curves recorded around 002 lines suggest a better alignment of the crystal planes for PZT films deposited on STO substrate, with a full width at half maximum of 0.2°, compared to 0.6° in the case of the PZT films deposited on STO-buffered Si substrate. Phi scans were performed on tilted crystalline planes to evidence the cube-on-cube growing relation between PZT, SRO and STO and to determine the (001) orientation relation between the planes of the Si substrate and those of the oxide layers (see **Figure 4**). The epitaxial growth is confirmed by obtaining the four peaks, related to the four-fold rotation axis of the pseudocubic symmetry. The in-plane orientation relations are such that PZT[100]//SRO[100]//STO[100]//Si[110] and are schematically represented in inset figures of Phi scans.

Transmission electron microscopy (TEM) investigations are performed for both structures for a complete structural characterization. It can be easily observed in both cases that the

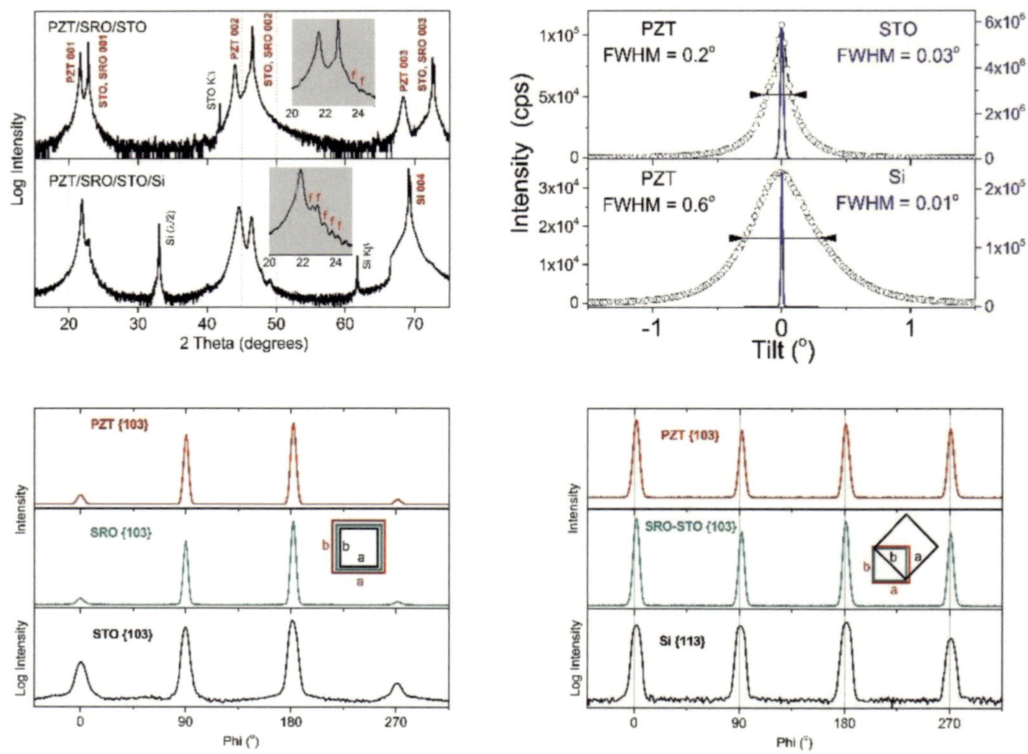

Figure 4. (a) XRD 2Theta-Omega scans (the insets are details around the 001 lines, showing the layer fringes of SRO, or of both SRO and STO thin films, respectively), (b) rocking curves taken at PZT 002, STO 002 and Si 004 lines and (c) phi scans obtained in asymmetric geometry by locating the {103} planes of STO and PZT and the {113} planes of Si (insets, sketches of the in-plane relationships deduced from the phi scans).

SRO and PZT layer thickness is of about 100 and 20 nm, respectively. In the case of the structure deposited on Si substrate, a bright thin layer with thickness of 4 nm is detected at the interface between Si and STO layers, and it is attributed to a native layer of SiO_2. Epitaxial relation between the constituent layers and the orientation relationship between crystallographic planes are revealed from the selected area electron diffraction patterns (SAED). The results obtained for the lattice constant of PZT from SAED images are in accordance with those obtained by XRD, again indicating a more relaxed PZT in the case of Si substrate compared to a more elongated unit cell in the case of PZT deposited on STO substrate. High-resolution TEM (HRTEM) images are acquired in order to observe the quality of the interfaces between the constituent layers. A strain contrast is revealed in the PZT layer in the case of the sample deposited on Si substrate, with a distorted region at the interface with the bottom SRO electrode, region-containing clusters of dislocations (**Figures 5** and **6**).

All the differences observed in the structural quality of the deposited layers and interfaces are having a significant impact on the macroscopic electrical properties of the two structures, as will be presented further on.

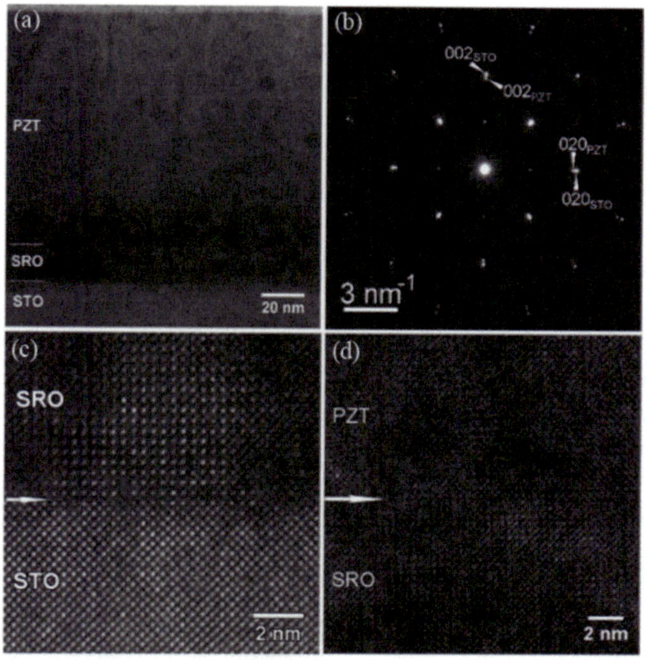

Figure 5. (a) TEM image at low magnification of PZT/SRO/STO heterostructure, (b) SAED pattern corresponding to TEM image (a), (c) HRTEM image of the STO-SRO interface and (d) HRTEM image of the SRO-PZT interface.

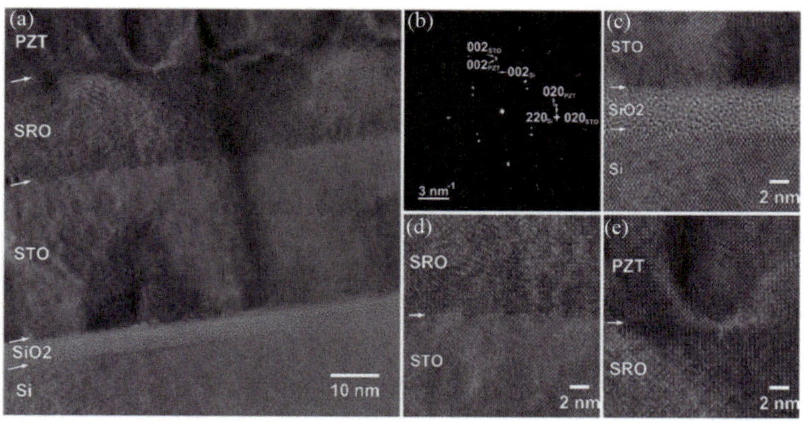

Figure 6. (a) HRTEM image at low magnification of PZT/SRO/STO/Si heterostructures, (b) SAED pattern corresponding to TEM image (a), (c) HRTEM image of the Si-SiO_2-STO interfaces, (d) HRTEM image of the STO-SRO interface and (e) HRTEM image of the SRO-PZT interface.

The presence of the hysteresis loop, describing the change in spontaneous polarization when an external electric field is applied on the ferroelectric capacitor, is the most important property of a ferroelectric material. The defining parameters of the hysteresis loop are saturated polarization (maximum value of polarization), remnant polarization (the value of polarization at zero applied field) and coercive field (the required electric field to have zero polarization). The classic circuit to record a hysteresis loop is based on the Sawyer-Tower experiment [34]. A similar principle used nowadays, computer controlled equipment, is able to record at the same time both current and charge (polarization) hystereses. The current loop recorded for a ferroelectric capacitor should present the two peaks, one for each polarity of the applied voltage. The peaks are attributed to polarization switching from one direction to the other. The polarization-voltage loop is obtained by integration of the current loop.

The ferroelectric character is checked for the two structures by recording the hysteresis loops, and the obtained results are presented in **Figure 7**. Even if both samples present an epitaxial relation between deposited layers and ferroelectricity is evidenced in both cases by the presence of the switching peaks in the current hysteresis, many relevant properties and values are different. For instance, in the case of the structure deposited on STO substrate, the switching peaks are sharper, and the polarization loop is more rectangular than the case of the structure deposited on Si substrate. In addition, the values of the remnant polarization are different: 26 μC/cm² for PZT on Si compared to 48 μC/cm² for PZT on STO substrate. The coercive voltage is lower in the case of PZT deposited on Si, 1.3 V compared to 2.5 V for the other structure.

The presence of ferroelectricity is also revealed by the butterfly shape of the capacitance-voltage (C-V) characteristics, which is also related to the switching phenomena by the relation$\varepsilon = \frac{1}{\varepsilon_0}\frac{\partial E}{\partial P}$. The differences between the C-V measurement and the previously described P-V measurement consist in the fact that the former is performed by superimposing a small amplitude AC voltage (to measure the capacitance) over a DC voltage (setting the polarization value), while the latter is performed by applying a variable voltage (sin or triangle waveform) on the sample. Therefore, the C-V measurement is quasi-static, while P-V is dynamic. The C-V characteristics

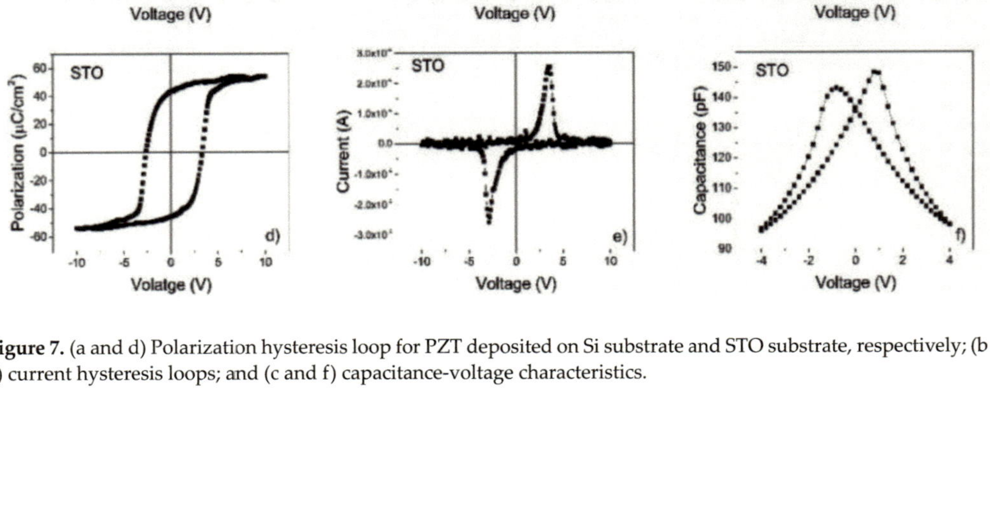

Figure 7. (a and d) Polarization hysteresis loop for PZT deposited on Si substrate and STO substrate, respectively; (b and e) current hysteresis loops; and (c and f) capacitance-voltage characteristics.

are also presented in **Figure 7**. It can be observed that the structures present different shapes of the characteristic and that the values for capacitance and coercive voltages are dependent on the used substrate. In the case of the PZT film deposited on the Si substrate, the value of the dielectric constant at 0 V is 650, much higher than the value of 240 obtained for the PZT film grown on STO substrate. Furthermore, an asymmetry can be observed in the case of PZT/SRO/STO structure. The two capacitance peaks have different values for positive and negative voltage polarities, with a shaper maxima and higher value for positive voltages.

The differences observed between the two structures, and mentioned above, are related to the previously described differences in the structural quality of the ferroelectric films deposited on different substrates. The higher polarization value obtained for the PZT film deposited on the STO substrate is correlated to a higher strain in this case, while the lower value for the film deposited on Si substrate is explained by the increased density of defects in the ferroelectric layer. These defects may suppress the switching of ferroelectric domains, determining a lower polarization value and a slower reversal of polarization, with a larger width of the switching current peaks.

An imprint voltage around 1 V is observed in the hysteresis loop of the structure deposited on the STO substrate, suggesting the presence of an internal electric field oriented towards the top electrode. This internal field cannot be assigned to different work functions of the bottom SRO electrode (4.6–4.9 eV) [35] and top Pt electrode (5.65 eV) [36] as the same electrodes are used for the PZT film deposited on Si substrate. However, the imprint is very much reduced in this case, leading to a more symmetric hysteresis loop, as shown in **Figure 7 (d)**. This is an indication that the internal electric field has a different origin, such as a gradient of the strain distribution or non-homogenous spatial distribution of defects like oxygen vacancies.

Despite these differences, the values of the most important parameters, such as remnant polarization and dielectric constant, belong to the same order of magnitude. This proves that high-quality epitaxial PZT films can be grown by PLD on Si substrates with STO buffer layer, allowing their rapid integration with semiconductor technology. The electric properties can be further tuned by changing the electrode material of the ferroelectric capacitor [37, 38]. The influence of the electrode-ferroelectric interface on the macroscopic electrical properties of epitaxial PZT films has been previously studied [39] for films with tetragonal structure deposited on single-crystal STO substrate, but it was less studied for epitaxial PZT films deposited on Si substrate and with composition near the morphotropic phase boundary [38]. Therefore, different materials were used as top electrodes (Pt, Ir, Ru and an oxide with metallic conduction, SRO) on PZT layer deposited on Si substrate with STO buffer layer.

The hysteresis loops and C-V characteristics obtained at room temperature for all four types of electrodes are presented in **Figure 8**. A first observation is that the shape of the hysteresis loops and the values of the remnant polarizations and coercive fields depend on the material used for the top electrode. For example, in the case of SRO top electrode, the remnant polarization has the highest value around 20 $\mu C/cm^2$ and the highest coercive voltage of 3.7 V. Moreover, both hysteresis and C-V loop are almost symmetric, which is expected due to symmetry of the electrode configuration for this structure (both top and bottom electrodes are SRO). The small shift of the hysteresis, of 0.2 V towards positive voltages, observed in this case can be explained

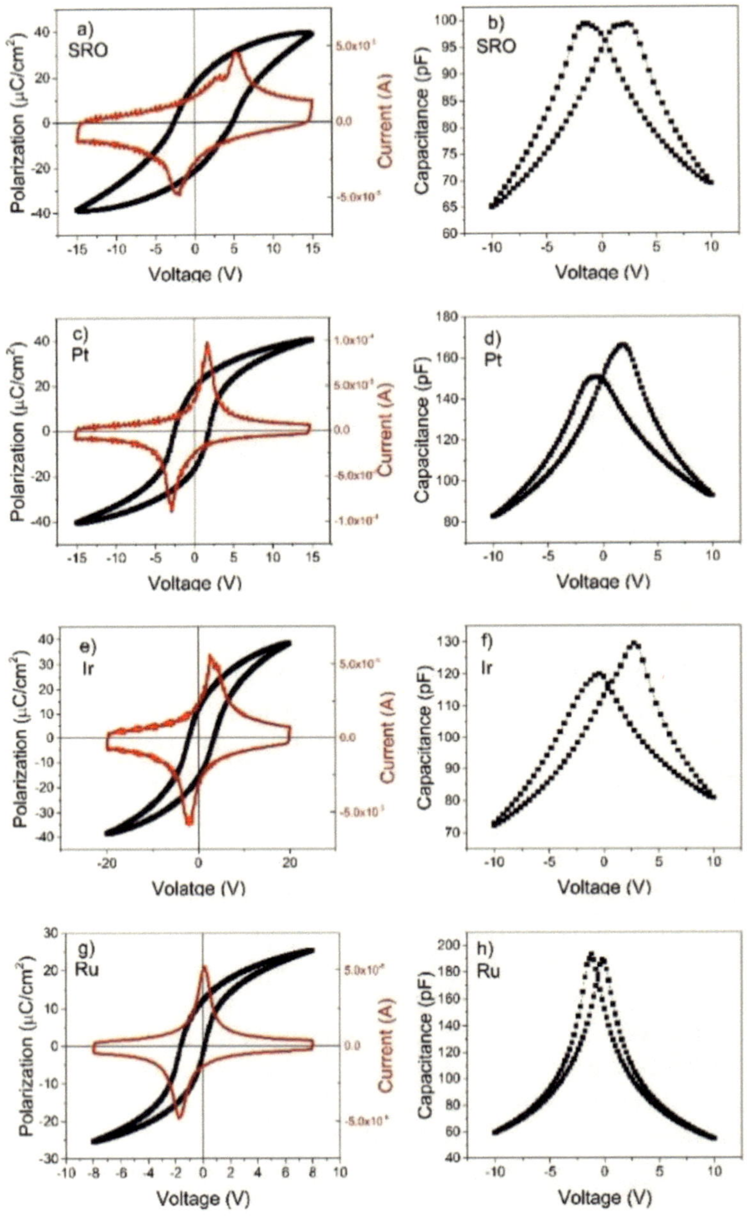

Figure 8. Hysteresis loops and capacitance-voltage characteristics obtained for different top electrodes: (a,b) for SRO, (c,d) for Pt, (e,f) for Ir and (g,h) for Ru, respectively.

by different deposition sequences (PZT deposited on SRO at bottom interface compared to SRO deposited on PZT at top interface), leading to small differences in the electronic properties of the two PZT/SRO interfaces.

The structures with Pt and Ir top electrodes have similar characteristics. This is an expected result considering that Pt and Ir are in the same group of precious metals, with the same structure of the electron shells (same period of the Mendeleev's table). For these two structures, polarization is 12 μC/cm², and the coercive voltage is 2.5 V. A shift of the hysteresis loop towards positive voltages is expected in the case of Pt, due to the higher work function for Pt than SRO bottom electrode, but the magnitude of the internal field is much lower than the difference between work functions. In addition, in the case of Ir, there is no shift of the hysteresis loop, even if the difference between work functions is almost 1 eV. These results confirm again that the origin of imprints is related to the different defect distributions at the top and bottom electrode interfaces and not related to the differences between the work functions of the top and bottom electrodes.

The polarization and coercive voltage have much lower values in the case of Ru top electrode—8 μC/cm² and 1.2 V, respectively—with an imprint of 0.6 V towards negative voltages. The significant differences obtained in the case of Ru top electrode can be assigned to a Ru oxidation, forming RuO_2, which also has a high conductivity. This process leads to an increased density of oxygen vacancy (which acts as a donor-type defect) at the top interface and, in consequence, to an electric field oriented towards the bottom contact.

The dielectric constant obtained from C-V measurements at maximum applied voltage, where contribution from polarization switching is reduced, is dependent on the material used as top electrode: 392, 523, 443 and 309 for SRO, Pt, Ir and Ru, respectively. One can observe from **Figure 8** that the value of measured capacitance is dependent on the value of the applied voltage. The explanation for this behaviour is that the ferroelectric-electrode interface behaves like a Schottky contact, with an associated Schottky capacitance dependent on interface properties as in the following equation: $C = \sqrt{\frac{q\,\varepsilon_0\,\varepsilon_{st}\,N_{eff}}{2(V+V_{bi})}}$ [40]. Therefore, the values of N_{eff} and V_{bi} can change when the material for the top electrode is changed, affecting in this manner the capacitance of the entire metal-ferroelectric-metal structure and leading to different capacitance values for negative and positive voltages as observed in the C-V characteristics presented in **Figure 8**.

5. Ferroelectric multilayered thin films: structural and electrical properties

One way to obtain new electrical properties/new phenomena, of interest for new applications, is to deposit multilayered structures by combining ferroelectric thin films with thin films from materials having different properties, for example, ferroelectric/paraelectric, ferroelectric/dielectric and ferroelectric/ferromagnetic, or by introducing composition gradients. Examples of new phenomena experimentally evidenced in multilayered heterostructures are negative capacitance in ferroelectric superlattices, enhancement of the electro-resistance or multiple ferroelectric states. One of the most known categories of multilayered structures is the one of artificial multiferroics, obtained by combining ferroelectric thin films (PZT, $BaTiO_3$-(BTO)) with thin films having magnetic properties ($CoFe_2O_4$-CFO, $La_xSr_{1-x}MnO_3$-LSMO). Besides at least two order parameters (ferroelectric polarization and magnetization), these materials also can present magneto-electric coupling mediated by interfacial strain or charge, making them very suitable for future applications and devices.

The electrical and ferroelectric characteristics of multiferroic heterostructures will be presented in this section. The structures were obtained by combining PZT or BTO ferroelectric thin films with CFO layers. The first part of the study consists in analysing the influence of the PZT (20/80)-CFO or BTO-CFO interfaces on the structural, ferroelectric and dielectric properties of the multilayer. Two different configurations, symmetrical (PZT-CFO-PZT or BTO-CFO-BTO) and asymmetrical (PZT-CFO or BTO-CFO), have been selected and deposited on (100) STO single crystal with SRO bottom electrode [41].

The XRD 2θ–ω scans reveal pseudocubic structures of the deposited layers for all cases of symmetric and asymmetric structures and for both ferroelectric layers: the full scan from 10 to 110° presents only 00l peaks for constituent layers: SRO, PZT or BTO and CFO. To prove the epitaxial relation between the deposited layers, azimuth phi scan is performed on {103} skew planes for STO, SRO and PZT and on {115} planes of CFO. The results are shown in **Figure 9** for PZT-based structures which mention that the same results are obtained for BTO-based structures. These results indicate a cube-on-cube epitaxial relation for all four structures, and the in-plane orientation is CFO[100]| |PZT[100]| |SRO[100]| |STO[100].

Figures 10 and **11** present TEM images obtained for multilayered structures. The TEM images at low magnifications reveal the constituent layers as well as their thickness. It can be noticed that the CFO layer has a pyramidal growth with a roughness surface, determined by a Volmer-Weber growth mechanisms determined by the lattice mismatch between PZT and CFO. The first layer of ferroelectric materials (PZT or BTO) is of high quality, as is it expected due to small lattice mismatch between ferroelectric layers and SRO bottom electrode and substrate. The second layer of PZT or BTO in symmetric structures presents an increased density of defects induced by the CFO layers, although the hetero-epitaxy is preserved.

In the experimental study of the multilayered structures, it is very important to decide if their electrical properties are a simple superposition of the bulk properties of the constituent materials or are a result of interface phenomena. For example, when a ferroelectric layer is combined with a paraelectric layer, two competing phenomena are determined by the presence of the interface:

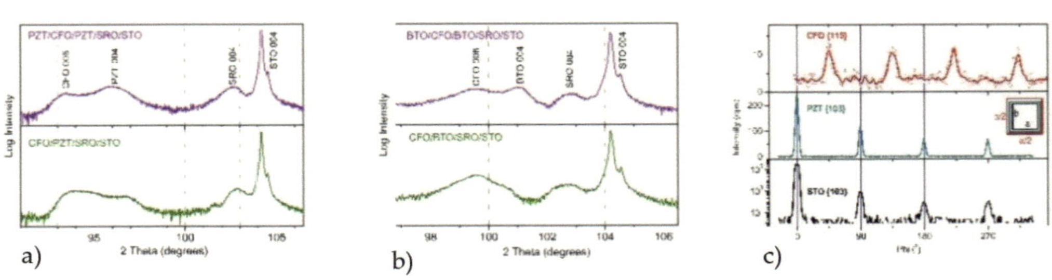

Figure 9. (a) XRD 2θ–ω scans zoomed near STO 004 for PZT-based multilayers structures, (b) XRD 2θ–ω scans zoomed near STO 004 for BTO-based multilayers structures and (c) Phi scans obtained in asymmetric geometry by location the {103} planes of STO and PZT and the {115} planes of CFO.

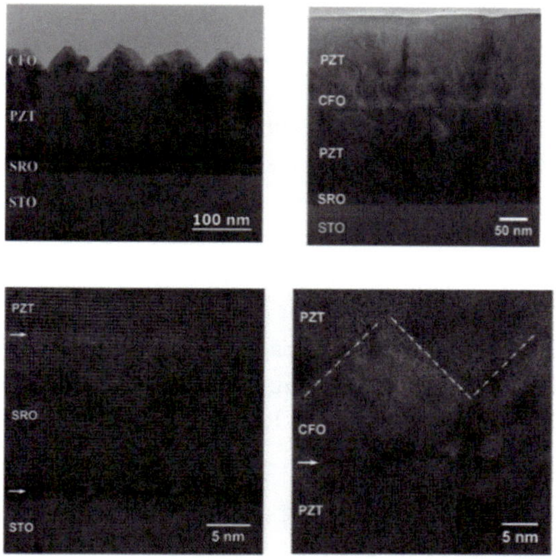

Figure 10. TEM image at low magnification of PZT-CFO and PZT-CFO-PZT heterostructure (first line) and HRTEM images of the STO-SRO, SRO-PZT and PZT-CFO-PZT interfaces (second line).

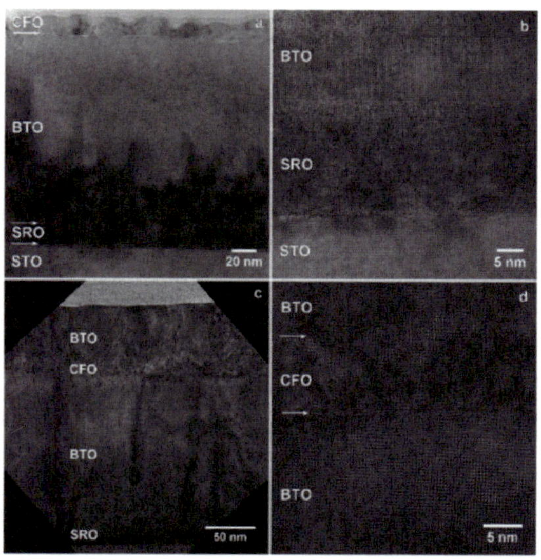

Figure 11. (a)TEM image at low magnification of BTO-CFO heterostructure, (b) HRTEM images of the STO-SRO and SRO-BTO interfaces, (c) TEM image at low magnification of BTO-CFO-BTO heterostructure and (d) HRTEM images of the BTO-CFO-BTO interfaces.

strain fields or strain gradients coming from lattice mismatch between the layers and the associated lattice relaxation mechanisms to reduce the total free energy and depolarization fields with origin in the discontinuity of the polarization charges which will determine different ways of response of the ferroelectric layer such as formation of polydomain structure or transition to a paraelectric state.

The following results show how different electrostatic boundary conditions modify ferroelectric and dielectric properties of multilayered structures. Typical hysteresis loops obtained for epitaxial PZT 20/80 and BTO layers grown on STO substrate with SRO top and bottom electrodes are presented in **Figure 12 (a, d)**. The PZT20/80-based capacitor presents a rectangular shape of the polarization loop, with two sharp current peaks associated to polarization switching from one direction to the other. The remnant and saturated polarization have similar values, around 85 μC/cm^2,, and the coercive field is around 100 kV/cm^2. For the BTO-based capacitor, the remnant polarization is around 15 μC/cm^2, the saturated polarization around 25 μC/cm^2 and the coercive field around 50 kV/cm^2. The polarization hysteresis loop is elongated, and the switching current peaks are broader compared to the PZT case. Those are typical characteristics for these two ferroelectric materials and are further used as references to be compared with the ferroelectric hysteresis loops obtained for asymmetric and symmetric multilayered structures mentioned above.

In what concerns the asymmetric structures, as PZT-CFO or BTO-CFO, the modifications of the hysteresis characteristics are similar for both cases, meaning decrease of the remnant polarization and increase of the coercive field, increase in the width of the switching current peaks and increase of the polarization back switching. These changes are due to imperfect screening of polarization charges at the top interfaces, due to the presence of less conductive CFO layer compared with metallic SRO electrodes. The high depolarization field existing in the system is compensated on the expense of polarization value, which decreases due to the electrostatic coupling between the component layers. Another interesting observation is that, even if the structures are totally asymmetric, with completely different electrostatic boundary conditions at the top and bottom interfaces of the

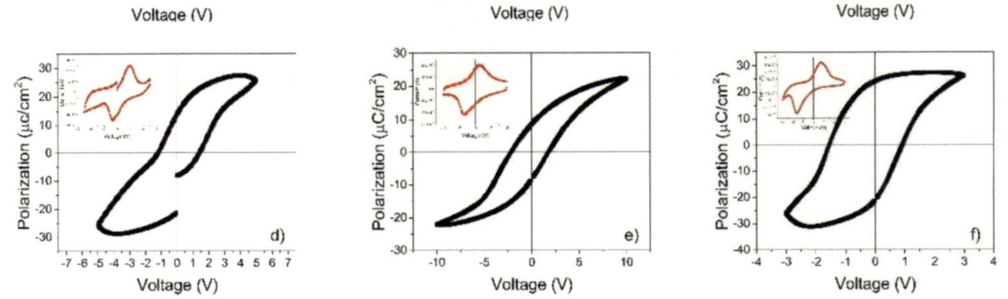

Figure 12. Hysteresis loop for (a) PZT20/80-based capacitor, (b) PZT-CFO asymmetric structure, (c) PZT-CFO-PZT symmetric structure, (d) BTO-based capacitor, (e) BTO-CFO asymmetric structure and (f) BTO-CFO-BTO symmetric structure.

ferroelectric layers, the hysteresis characteristics are approximately symmetric: similar coercive voltages and polarization values for both voltage polarities, with similar amplitude of the switching currents.

The ferroelectric characteristics of symmetric PZT-CFO-PZT and BTO-CFO-BTO structures become similar with the ones of simple metal-ferroelectric-metal structures, with similar values for the polarization and coercive field. The presence of the CFO layer between the ferroelectric films induces a higher leakage current in the case of PZT-based multilayered structure. In the case of BTO-based structure, the effect is opposite, the presence of the CFO layer leading to a lower leakage current and lower back-switching effects, determining in this way a more rectangular shape of the polarization hysteresis loop.

The capacitance-voltage measurements performed at 100 kHz frequency, for the two multilayered structures based on PZT, are depicted in **Figure 13**. The ferroelectric behaviour is also confirmed, for both types of structures, by the butterfly shape of the characteristics. In the case of the asymmetric structure, the capacitance is much lower than in the case of the symmetric structure even if the thickness is lower. Furthermore, the tunability is much lower; the variation of the capacitance value between the peak and the maximum applied voltage is 6 pF, compared to 20 pF in the case of the symmetric structure. In addition, the asymmetry between the values of the capacitance maxima, in the case of the asymmetric structure, can be explained by the presence of asymmetric electrode interfaces, leading to different values of the associated capacitances (different interface properties). The dielectric behaviour is further examined by measuring the dependence on frequency of the capacitance and dielectric losses for the two types of the structures. The results are compared with those obtained on metal-ferroelectric-metal capacitors and on metal-CFO-metal structures. As regards the SRO-PZT-SRO structure, the dielectric behaviour is typical for a ferroelectric/isolator material: there is a small decrease of the capacitance with increasing the frequency, with low values of the dielectric losses. Also, the results obtained on SRO-CFO-SRO structure, presented in **Figure 14 (c–d)** are specific for ferrite-based capacitors [42]:

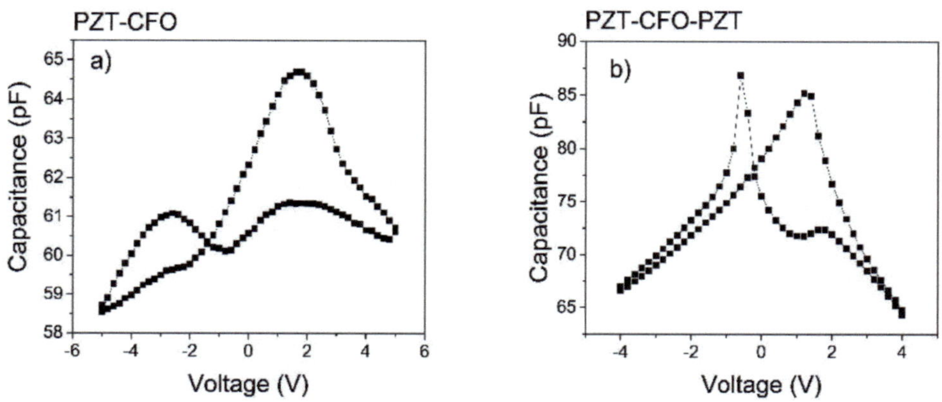

Figure 13. Capacitance voltage characteristics for PZT-based multilayered structures for (a) asymmetric configuration and (b) symmetric configuration.

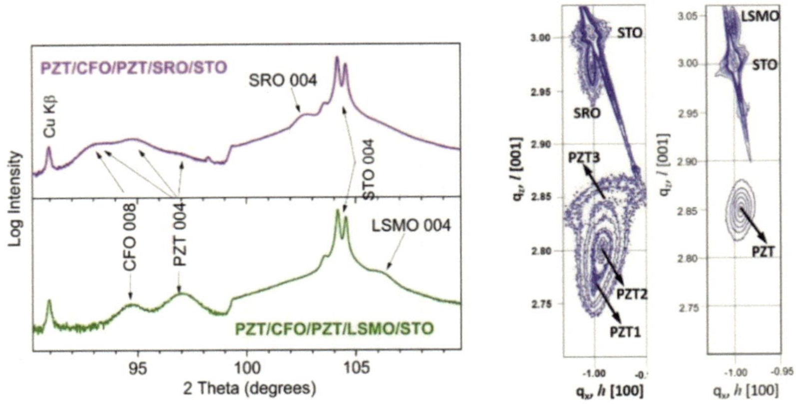

Figure 14. (a) XRD diagrams zoomed near the 004 line of STO and (b) RSM images for PZT-CFO-PZT deposited on SRO bottom electrode (left) and of PZT-CFO-PZT deposited on LSMO bottom electrode (right).

the values of the capacitance present a steplike decrease with frequency, with one order of magnitude difference between low and high frequencies; dielectric losses present a fast decreasing up to 1 kHz, followed by a peak in the frequency range where the capacitance decreases.

The dielectric behaviour for the two types of multilayered structures is different, compared to the behaviour of the single-phase PZT or CFO-based capacitors, and is strongly dependent on the symmetry of the structure. Even if the multilayered structures display steplike decrease of the capacitance with the increase of frequency and peaks in dielectric losses, the values for capacitance and dielectric losses are lower than the case of the simple CFO capacitors. The dependence of the capacitance and dielectric losses on frequency for multilayered structures are determined especially by an additional interface polarization/charge, due to the presence of interfaces between layers of materials with different permittivity and resistivity values. At high frequencies, the value of the capacitance for multilayered structures is closer to that estimated from the serial connection of the capacitors associated to component layers (PZT or CFO). For lower frequencies, the capacitance value is significantly dependent on the number of interfaces, increasing as the number of interfaces in the structure increases. In addition, the position and magnitude of the relaxation peak are strongly related to the configurations of the multilayer structure. Further results will show how different strains and strain gradients influence the ferroelectric and dielectric properties. As an example, we consider two symmetric structures of PZT/CFO/PZT, with thinner PZT layer (50 nm) than previous examples, deposited on two different bottom electrodes SRO and LSMO.

The results of XRD investigations performed on these two symmetric structures, deposited on two different bottom electrodes, are presented in **Figure 14 (a)** for 2θ–ω around 004 line of PZT and in **Figure 14 (b)** for reciprocal space mapping (RSM). The structure deposited on LSMO electrode presents only a peak for PZT in XRD pattern, corresponding to a bulk out of plane lattice parameter, and only one node in RSM map is attributed to a fully relaxed PZT. The structure deposited on SRO bottom electrode presents different characteristics: evident splitting of the PZT lines, associated with two nodes in RSM map, having different in-plane and out-of-plane lattice parameters. These values are associated to the two different PZT layers: one

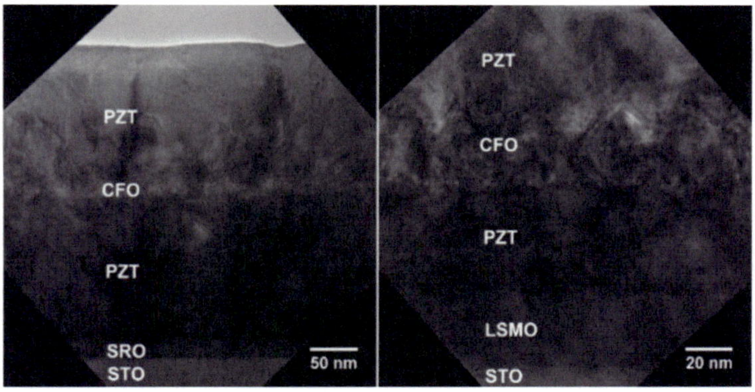

Figure 15. TEM image at low magnification of PZT-CFO-PZT deposited on SRO bottom electrode (left) and of PZT-CFO-PZT deposited on LSMO bottom electrode (right).

fully strained, with in-plane parameter close to STO lattice constant (most probably bottom PZT layer), and one almost fully relaxed, with similar parameters as for the PZT film in structure deposited on LSMO electrode (**Figure 15**).

The ferroelectric behaviour, for both structures, is comparatively presented in **Figure 16** through polarization hysteresis loops and capacitance-voltage characteristics. Both structures present rectangular hysteresis loops and well-evidenced butterfly shape of C-V characteristics. However, a series of differences are easily observed:

- Different values of remnant polarization (90 μC/cm² for structure with SRO bottom electrode and 60 μC/cm² for structure with LSMO bottom electrode)
- Much higher shift of the hysteresis loop towards positive voltages for the structure deposited on SRO

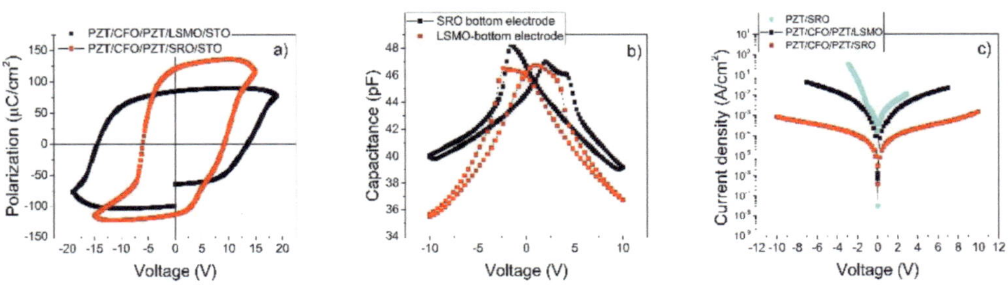

Figure 16. (a) The polarization hysteresis loops; (b) capacitance-voltage characteristics for PZT-CFO-PZT symmetric structure deposited on SRO bottom electrode and LSMO bottom electrode and (c) the dependence of current density on voltage for the two studied structures and compared with a simple thin layer of PZT-based capacitor.

- Higher tunability in the case of LSMO bottom electrode
- A small asymmetry between capacitance maxima in the case of SRO bottom electrode

These differences could be correlated with different structural characteristics determined by XRD and TEM investigations. A totally relaxed structure in the case of the LSMO bottom electrode implies a lower tetragonality and explains lower polarization values. The structural defects, observed in TEM imagines for both PZT layers of this structure, could act as polarization domain pinning centres which determines a slower reversal of polarization from one direction towards the other. The structure deposited on SRO presents two PZT layers with different structural properties. As a consequence, we assume that there is a strain gradient that could be correlated with the existence of an internal electric field pointing towards the top interface that explains the shift observed in the P-V loop of this structure.

A less discussed topic in this chapter, but very important for the operation of ferroelectric devices, is the leakage current. A higher leakage current is detrimental for long-term operation of ferroelectric-based devices. Thus, a significant research effort is dedicated to the identification of the conduction mechanisms, which control the leakage current in ferroelectric thin films and to find a solution to decrease the value of the leakage current. The dependence of the leakage current on voltage (I-V characteristics) is presented in **Figure 17 (c)** for the cases of the two PZT-CFO-PZT structures deposited in SRO and LSMO electrodes. The results are compared with those obtained for a simple PZT capacitor with similar thickness. It is clear that, even if these structures operate at higher voltages, the leakage current is much lower than for a simple PZT layer, the differences being around two orders of magnitude for the structure deposited on SRO.

It was shown in the previous section that the number of interfaces in this multilayered heterostructures determines the dielectric behaviour. Further, **Figure 17** presents the results of the measurements regarding the dependence of the capacitance and dielectric losses, on frequency and temperature, for these two symmetrical structures, with different strain conditions and structural quality of the ferroelectric layers. The values of capacitance and dielectric losses at temperatures below 250 K are similar to those obtained for single-phase PZT capacitor presented in **Figure 18**. This behaviour is modified towards a Maxwell-Wagner mechanism with an increase in temperature, specific for multilayered structures with interfaces between materials with different electric properties. The transition between low-temperature and high-temperature dielectric behaviour can be correlated with a strong variation of the resistivity of the CFO layer with temperature as is exemplified in **Figure 17 (f)**. Therefore, the difference in resistivity between PZT and CFO layers will increase with temperature, favouring in this way Maxwell-Wagner polarization mechanism. The capacitance at lower frequencies and the relaxation marked by the peak in dielectric losses are strongly dependent on the structure and on the temperature. It can be observed that for the structure deposited on LSMO electrode, the values of the capacitance at lower frequencies are higher, and the frequency where the maximum in dielectric losses occurs is higher than for the structure deposited on SRO, at the same temperature.

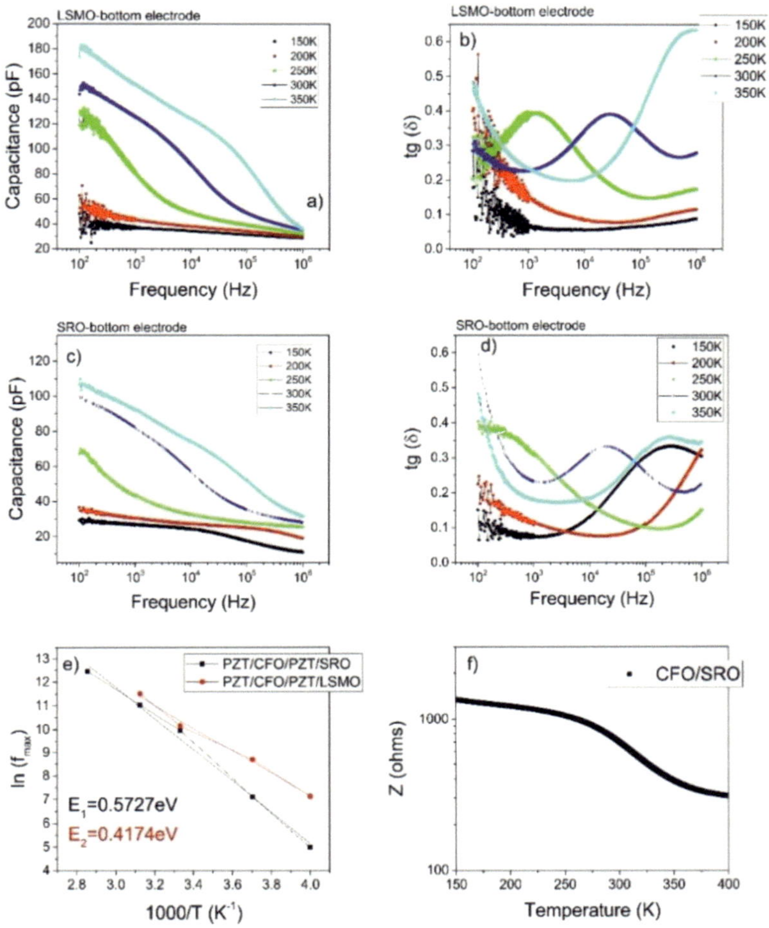

Figure 17. (a and b) The dependence of capacitance and dielectric losses, respectively, on frequency and for different temperatures for LSMO bottom electrode case; (c and d) the dependence of capacitance and dielectric losses, respectively, on frequency and for different temperatures for SRO bottom electrode case; (e) the Arrhenius plot of the maximum dielectric losses frequencies and (f) the variation of impedance of a thin layer CFO-based capacitor on temperature.

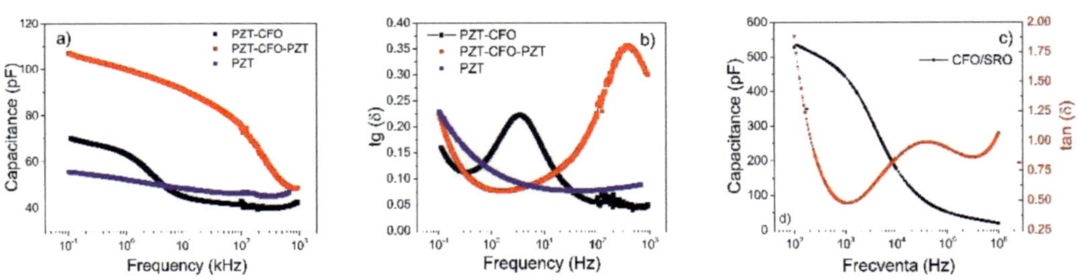

Figure 18. (a) The dependence of capacitance on frequency, (b) the dependence of the dielectric loses on frequency comparatively presented for PZT-based structures and (c) the dependence of capacitance and dielectric losses on frequency for a thin layer of CFO-based capacitor.

6. Conclusions

This chapter presents the electrical properties of epitaxial ferroelectric thin films and multilayers. A short description of the deposition/growing steps used to obtain high-quality epitaxial ferroelectric structures, with sharp interfaces, is presented at the beginning. The main experimental results show how ferroelectric and dielectric properties depend on the structural quality of the ferroelectric layer and on the electrostatic boundary conditions.

Acknowledgements

The authors acknowledge the financial support of the Romanian Ministry of Education-Executive Unit for Funding High Education, Research, Development and Innovation (MEN-UEFISCDI) through the Nucleus Program PN16-4801; the Idea-Complex Research Grant PN-II-ID-PCCE-2011-2-0006 (Contract No. 3/2012); the IFA-CEA (Contract No. C503/2016); and the CNCS-UEFISCDI Project of PN-II-PT-PCCA-2013-4-0470 (Contract No. 238/2014).

Author details

Andra Georgia Boni*, Cristina Florentina Chirila, Raluca Negrea, Corneliu Ghica, Iuliana Pasuk, Ioana Pintilie and Lucian Pintilie

*Address all correspondence to: andra.boni@infim.ro

National Institute of Materials Physics, Bucharest-Magurele, Romania

References

[1] Scott JF. Applications of modern ferroelectrics. Science. 2007 Feb 16;**315**(5814):954-959

[2] Lallart M, editor. Ferroelectrics - Applications [Internet]. InTech; 2011. Available from: http://www.intechopen.com/books/ferroelectrics-applications

[3] Okuyama M, Ishibashi Y, editors. Ferroelectric Thin Films [Internet]. Berlin, Heidelberg: Springer Berlin Heidelberg; 2005. (Ascheron CE, Kölsch HJ, Skolaut W, editors. Topics in Applied Physics; vol. 98). Available from: http://link.springer.com/10.1007/b99517

[4] Scott JF. Ferroelectric Memories [Internet]. Berlin, Heidelberg: Springer Berlin Heidelberg; 2000. (Itoh K, Sakurai T, editors. Springer Series in Advanced Microelectronics; vol. 3). Available from: http://link.springer.com/10.1007/978-3-662-04307-3

[5] Eerenstein W, Mathur ND, Scott JF. Multiferroic and magnetoelectric materials. Nature. 2006;**442**:759-765. doi: 10.1038/nature05023. Available from: http://www.nature.com/nature/journal/v442/n7104/full/nature05023.html

[6] Khomskii D. Classifying multiferroics: Mechanisms and effects. Physics. 2009 Mar 9;**2**:20

[7] Scott JF. Data storage: Multiferroic memories. Nature Materials. 2007 Apr;**6**(4):256-257

[8] Loidl A, von Loehneysen H, Kalvius GM. Multiferroics. Journal of Physics: Condensed Matter. 2008 Oct 7;**20**(43):430301

[9] Martin LW, Crane SP, Chu Y-H, Holcomb MB, Gajek M, Huijben M, et al. Multiferroics and magnetoelectrics: Thin films and nanostructures. Jounal of Physics: Condensed Matter. 2008 Oct 29;**20**(43):434220

[10] Matthews J editor. Epitaxial Growth. USA. Elsevier; 2012. p. 401

[11] Pintilie L, Teodorescu CM, Ghica C, Hrib LM, Chirila C, Trupina L, et al. Interfaces in epitaxial structures based on oxide ferroelectrics. In: Composite, Ceramic, Quasi-Crystals, Nanomaterials, High Temperature Protection Coatings. Quebec/Wilmington, Canada/USA: FLOGEN; 2014

[12] Nguyen MD, Dekkers M, Houwman E, Steenwelle R, Wan X, Roelofs A, et al. Misfit strain dependence of ferroelectric and piezoelectric properties of clamped (001) epitaxial Pb(Zr0.52,Ti0.48)O3 thin films. Applied Physics Letters. 2011 Dec 19;**99**(25):252904

[13] Pertsev NA, Kukhar VG, Kohlstedt H, Waser R. Phase diagrams and physical properties of single-domain epitaxial Pb({Zr_{1-x},Ti_x) O_3 thin films. Physical Review B. 2003 Feb 28;**67**(5):054107

[14] Jang HW, Kumar A, Denev S, Biegalski MD, Maksymovych P, Bark CW, et al. Ferroelectricity in strain-free $SrTiO_3$ thin films. Physical Review Letters. 2010 May 13;**104**(19):197601

[15] He F, Wells BO, Shapiro SM. Strain phase diagram and domain orientation in $SrTiO_3$ thin films. Physical Review Letters. 2005 May 2;**94**(17):176101

[16] Zubko P, Catalan G, Buckley A, Welche PRL, Scott JF. Strain-gradient-induced polarization in $SrTiO_3$ single crystals. Physical Review Letters. 2007 Oct 19;**99**(16):167601

[17] Schlom DG, Chen L-Q, Eom C-B, Rabe KM, Streiffer SK, Triscone J-M. Strain tuning of ferroelectric thin films*. Annual Review of Materials Research. 2007;**37**(1):589-626

[18] Polakowski P, Müller J. Ferroelectricity in un-doped hafnium oxide. Applied Physics Letters. 2015 Jun 8;**106**(23):232905

[19] Bousquet E, Dawber M, Stucki N, Lichtensteiger C, Hermet P, Gariglio S, et al. Improper ferroelectricity in perovskite oxide artificial superlattices. Nature. 2008 Apr 10;**452**(7188):732-736

[20] Jeon BC, Lee D, Lee MH, Yang SM, Chae SC, Song TK, et al. Flexoelectric effect in the reversal of self-polarization and associated changes in the electronic functional properties of BiFeO3 thin films. Advanced Materials. 2013 Oct 1;**25**(39):5643-5649

[21] Lee D, Jeon BC, Yoon A, Shin YJ, Lee MH, Song TK, et al. Flexoelectric control of defect formation in ferroelectric epitaxial thin films. Advanced Materials. 2014 Aug 1;**26**(29):5005-5011

[22] Zhang J, Xu R, Damodaran AR, Chen Z-H, Martin LW. Understanding order in compositionally graded ferroelectrics: Flexoelectricity, gradient, and depolarization field effects. Physical Review B. 2014 Jun 13;**89**(22):224101

[23] Crystal Growth Beginners - AbeBooks [Internet]. Available from: https://www.abebooks.com/book-search/title/crystal-growth-beginners/

[24] Jaffe B, Cook WR, Jaffe H. In: Cook BJR, Jaffe H, editors. Piezoelectric Ceramics [Internet]. Academic Press; 1971. p. v. Available from: http://www.sciencedirect.com/science/article/pii/B9780123795502500041

[25] Kim DM, Eom CB, Nagarajan V, Ouyang J, Ramesh R, Vaithyanathan V, et al. Thickness dependence of structural and piezoelectric properties of epitaxial $Pb(Zr_{0.52,}Ti_{0.48})O_3$ films on Si and $SrTiO_3$ substrates. Applied Physics Letters. 2006 Apr 3;**88**(14):142904

[26] Uchino Kenji. Piezoelectric Actuators and Ultrasonic Motors [Internet]. Springer; 1997. Available from: http://www.springer.com/us/book/9780792398110

[27] Noheda B, Gonzalo JA, Cross LE, Guo R, Park S-E, Cox DE, et al. Tetragonal-to-monoclinic phase transition in a ferroelectric perovskite: The structure of $PbZr_{0.52}Ti_{0.48}O_3$. Physical Review B. 2000 Apr 1;**61**(13):8687-8695

[28] Zhu Y, Yan P, Yi T, Cao L, Li L. Interface diffusion and chemical reaction on the interface of a PZT film/Si(III) sample during annealing treatment in N_2 and vacuum. Surface and Interface Analysis. 1999 Nov 1;**27**(11):972-980

[29] Lin A, Hong X, Wood V, Verevkin AA, Ahn CH, McKee RA, et al. Epitaxial growth of $Pb(Zr_{0.2}Ti_{0.8})O_3$ on Si and its nanoscale piezoelectric properties. Applied Physics Letters. 2001 Apr 2;**78**(14):2034-2036

[30] Basit NA, Kim HK, Blachere J. Growth of highly oriented $Pb(Zr, Ti)O_3$ films on MgO-buffered oxidized Si substrates and its application to ferroelectric nonvolatile memory field-effect transistors. Applied Physics Letters. 1998 Dec 28;**73**(26):3941-3943

[31] Wang Y, Ganpule C, Liu BT, Li H, Mori K, Hill B, et al. Epitaxial ferroelectric $Pb(Zr, Ti)O_3$ thin films on Si using $SrTiO_3$ template layers. Applied Physics Letters. 2002 Jan 7;**80**(1):97-99

[32] Wen-Chieh Shih, Pi-Chun Juan, and Joseph Ya-min Lee. Fabrication and characterization of metal-ferroelectric $(PbZr_{0.53}Ti_{0.47})O_3$-Insulator (Y_2O_3)-ssemiconductor field effect transistors for nonvolatile memory applications. Journal of Applied Physics. 2008 May 1;**103**(9):094110

[33] Chirila C, Boni AG, Pasuk I, Negrea R, Trupina L, Rhun GL, et al. Comparison between the ferroelectric/electric properties of the $PbZr_{0.52}Ti_{0.48}O_3$ films grown on Si (100) and on STO (100) substrates. Journal of Material Science. 2015 Mar 14;**50**(11):3883-3894

[34] Physics of Ferroelectrics [Internet]. Berlin, Heidelberg: Springer Berlin Heidelberg; 2007. (Topics in Applied Physics; vol. 105). Available from: http://link.springer.com/10.1007/978-3-540-34591-6

[35] Hartmann AJ, Neilson M, Lamb RN, Watanabe K, Scott JF. Ruthenium oxide and strontium ruthenate electrodes for ferroelectric thin-films capacitors. Applied Physics A. 2000 Feb;**70**(2):239-242

[36] Schaeffer JK, Fonseca LRC, Samavedam SB, Liang Y, Tobin PJ, White BE. Contributions to the effective work function of platinum on hafnium dioxide. Applied Physics Letters [Internet]. 2004 Sep 6;**85**(10). Available from: http://www.osti.gov/scitech/biblio/20632781

[37] Pintilie L, Alexe M. Metal-ferroelectric-metal heterostructures with Schottky contacts. I. Influence of the ferroelectric properties. Journal of Applied Physics. 2005 Dec 15;**98**(12):124103

[38] Boni AG, Chirila C, Pasuk I, Negrea R, Trupina L, Le Rhun G, et al. Electrode interface controlled electrical properties in epitaxial $Pb(Zr_{0.52}Ti_{0.48})O_3$ films grown on Si substrates with $SrTiO_3$ buffer layer. Thin Solid Films. 2015;**593**:124-130

[39] Pintilie I, Teodorescu CM, Ghica C, Chirila C, Boni AG, Hrib L, et al. Polarization-control of the potential barrier at the electrode interfaces in epitaxial ferroelectric thin films. ACS Applied Material Interfaces. 2014 Feb 26;**6**(4):2929-2939

[40] Sze SM. Physics of Semiconductor Devices [Internet]. 2nd ed. John Wiley & Sons; Available from: https://archive.org/details/PhysicsOfSemiconductorDevices

[41] Chirila C, Ibanescu G, Hrib L, Negrea R, Pasuk I, Kuncser V, et al. Structural, electric and magnetic properties of $Pb(Zr_{0.2}Ti_{0.8})O_3$–$CoFe_2O_4$ heterostructures. Thin Solid Films. 2013 Oct 31;**545**:2-7

[42] Gutiérrez D, Foerster M, Fina I, Fontcuberta J, Fritsch D, Ederer C. Dielectric response of epitaxially strained $CoFe_2O_4$ spinel thin films. Physical Review B. 2012 Sep 10;**86**(12):125309

Chapter 4

Tm on W(110): A Growth Study by Scanning Tunneling Microscopy

David Coffey, José I. Arnaudas, David Serrate and Miguel Ciria

Additional information is available at the end of the chapter

http://dx.doi.org/10.5772/intechopen.70218

Abstract

Exploring bottom-up procedures to achieve island and particles with a defined size can open opportunities in many applications. This contribution focuses on the growth of epitaxial Tm islands, below the monolayer range, on the W(110) surface by studying in situ the diffusion process at high temperature, between 700 and 1200 K, by means of scanning tunnel microscopy (STM) to determine the topography of the Tm deposits as a function of the coverage and thermal treatments of an initial room temperature deposit. Samples subject to a prolonged heating process, spending several hours at temperatures below 700 K, show that the average Tm islands size remains constant at higher temperatures, in contrast with samples subject to a faster heating. It is observed that the presence of carbon strongly limits the diffusion of Tm, thus leading to the formation of pseudomorphic nanometric islands instead of a full monolayer.

Keywords: surface diffusion, scanning tunneling microscopy, rare earth nanostructures, nucleation, self-organization

1. Introduction

Growing rare earth (RE) on the (110) surface of bcc metals (Mo, W, or Nb) has been widely used to obtain layers with the basal plane of the rare earth on this film plane [1–5]. Although the (110)bcc and (0001)hcp surfaces do not match with each other and therefore, the concept of epitaxial growth is not applicable; it has proven to still be possible to prepare quite perfect superlattices and films. Historically, the preparation of rare earth artificial superlattices dates from 1985 [1], with the Gd-Y system showing a novel magnetic behavior that has been relevant in the history of nanostructured magnetic materials, as it was the first observation of antiferromagnetic coupling between ferromagnetic blocks occurring through a nonmagnetic spacer [4].

The (110) surface of W offers the opportunity of studying the magnetic behavior of ordered RE overlayers on top of a conductive but nonmagnetic substrate. In general, the body-centered cubic crystals of refractory metals such as W and Mo promote the two-dimensional layer growth of RE with no intermixing [6–8] with the (110) surface, yielding morphologies with low corrugation.

Thulium represents an interesting case, since it is the only heavy RE which orders magnetically along the *c* axis of the bulk hcp crystal structure. Between the Néel temperature, $T_N = 58$, and 40 K, the magnetic moments are ferromagnetically ordered within the hcp basal-plane layers and have an incommensurate sinusoidal modulation along the *c* axis. The easy axis is parallel to the *c* axis, owing to the strong crystal-field anisotropy. Upon decreasing the temperature, the propagation wave number along the *c* axis increases, and below ~30 K, a ferrimagnetic structure with a seven-layer repeat distance develops: the magnetic moments point up along the *c* axis in three layers and down in the consecutive four layers [9, 10]. As the reduction of thickness along the *c* axis could drastically influence the magnetic behavior of Tm, it is interesting to investigate the limit case of one atomic layer. Thulium grows heteroepitaxially on the W(110) surface, with the [1010] direction of the Tm layer along the [110] direction of the (110) substrate, similar to the Nishiyama-Wasserman orientation (see **Figure 1**) observed in other RE/W(110) systems [11, 12].

In nanostructured systems, there exists a strong relation between the magnetic properties of the nanostructure and its crystalline configuration and size, both being controllable by the preparation procedure. Tm on W(110) presents a large variety of morphologies depending on the coverage and thermal treatment after deposition, ranging from single atoms and clusters, to nanowires, to crystalline hexagonal monolayer islands for coverages below the monolayer and a hexagonal pyramidal Stranski-Krastanov growth for a multilayer coverage, where a

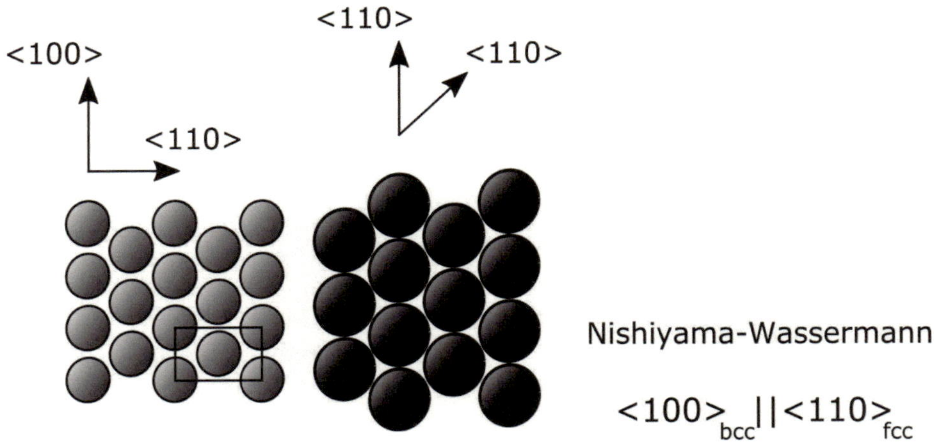

Figure 1. Nishiyama-Waserman orientation relationship for a closed-packed fcc or hcp overlayer (right) on a bcc (110) surface (left). The Miller indices of the directions in the overlayer are those appropriate for an fcc structure. Adapted from [13].

full monolayer is first formed on which subsequent pyramidal multilayer islands are able to grow. The study presented here is focused on the sub-monolayer regime and its evolution with temperature treatments. It is shown that the size of the islands can be controlled by the density of impurities on the W(110) surface, which diffuses from the bulk to the surface over time at high temperatures.

2. Experimental details

2.1. Scanning tunneling microscopy

The scanning tunneling microscope (STM) is a lensless microscope, able to bypass the diffraction limits imposed by the finite numerical aperture and wavelength, where the image is reconstructed by measuring a pointwise matrix of the interaction between the sample and a probe; the probe consists of a very sharp wire, which, when brought close enough to a metallic surface, conducts via quantum tunneling [14, 15]. By moving this probe along the surface, the variation of the tunnel current provides information of the local density of states (LDOS) at the surface of the sample, therefore making STM strictly a surface technique. Due to the exponential dependence of quantum tunneling with the width of the barrier, STM is extremely sensitive to features at the atomic scale, which goes hand in hand with the weakness of being very vulnerable to contamination at said scale. To reduce the exposure of samples to unwanted molecules, experiments are typically performed in ultra high vacuum conditions (UHV, $P \leq 10^{-10}$ mbar), maintaining vacuum unbroken along the whole experimental process, from sample preparation to STM measurements.

The STM used in this work is a SPECS Aarhus HT-STM [16], which can operate over a wide range of temperatures, 90–1300 K. A large copper block surrounds the sample acting as both temperature reservoir and damping mass to reduce vibrations. High temperatures are obtained by thermal radiation from a W filament that can be approached to the back of the sample. The STM systems are equipped with a Nanonis SPM control system for data acquisition and interfacing with the microscopes. Data treatment and analysis are performed using WSxM [17] and Gwyddion [18].

STM tips are prepared by electrochemically etching a W wire in a NaOH solution. The wire is positioned through a small flat ring where a drip of the solution is placed, forming a thin layer by surface tension, minimizing the length of the wire exposed to etching which provides for a more robust tip.

2.2. Preparation of Tm/W(110)

2.2.1. Cleaning and carbon contamination of the W(110)

The preparation of a clean W(110) surface is procedurally less complicated than other surfaces, but more demanding in the design of the heating stage and sample holder, as they involve materials such as W or Mo which are difficult to machine, but able to withstand temperatures above 2300 K.

The main contaminant in a tungsten single crystal is carbon, which is dissolved in the bulk and diffuses to the surface over time. The procedure to remove said C, described in [19], consists in annealing cycles in an oxygen-rich atmosphere ($T \sim 1500$ K, $P_{02} \sim 5 \times 10^{-7}$ to 1×10^{-8} mbar) which causes the C adsorbates to react with the oxygen, forming CO molecules that can desorb from the surface. The second step in the procedure is to perform a short, high temperature flash of the sample ($T > 2300$ K) which removes atomic oxygen adsorbates and other remaining impurities from the sample. By performing several annealing-flash cycles, until the flashes do not produce a significant pressure spike ($P_{\mathrm{Flash}} \leq 1 \times 10^{-10}$ mbar), a high quality clean surface is obtained. An STM image of a clean W(110) surface is shown in **Figure 2a**, with 34 adsorbates in a 40×40 nm^2 area, a ~0.3% of a monolayer. Carbon adsorbates appear as depressions on the surface, as they present a lower LDOS. To the right, **Figure 2b**, the LEED pattern corresponding to the clean surface is shown, where the rectangular lattice of the W(110) surface results in clear and well-defined spots; the [110] and [001] directions are indicated by lines.

As with any diffusion process, the rate at which C reaches the surface is strongly related to the temperature of the crystal, which during the experiments performed in this chapter is in the 700 to 1200 K range for hours at a time, strongly enhancing the C segregation. This produces a change in the quality of the surface over time, affecting the growth of Tm structures. Understanding how the surface evolves as C contamination grows is therefore important for high temperature experiments. Locally, carbon reconstruction patches will start to form where the adsorbate density is high enough; over time, the carbon reconstruction will cover the whole surface. An STM image of said reconstruction is shown in **Figure 3** accompanied by the corresponding LEED pattern showing the 15×3 relation [20].

While LEED measurements are able to distinguish between the two situations presented in **Figures 2** and **3**, quantifying the non-reconstructed adsorbate density is beyond the

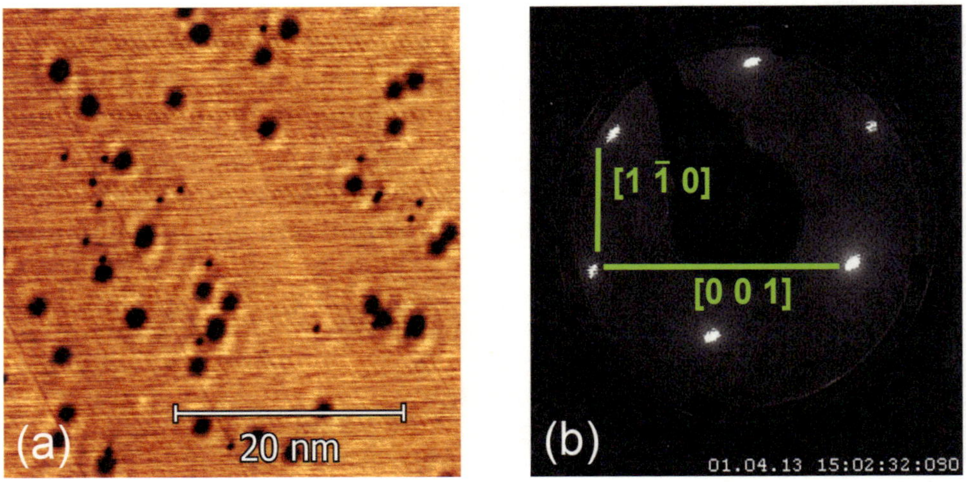

Figure 2. (a) STM image of a clean W(110) surface taken at $V_{\mathrm{Bias}} = 100$ mV, $I_t = 1$ nA. Carbon adsorbates appear as dark spots around and serve as scatter centers for the surface state, which is clearly visible around them. (b) LEED pattern corresponding to the clean W(110) surface, with the [110] and [001] directions indicated by lines.

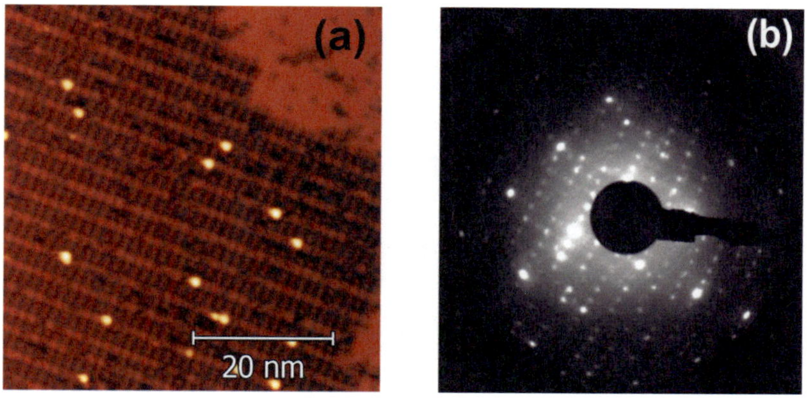

Figure 3. (a) STM image of the Carbon reconstruction on W(110) taken at V_{Bias} = 100 mV, I_t = 1 nA. (b) Corresponding LEED pattern illustrating the 15 × 3 reconstruction.

capabilities of this and other area analysis techniques (such as Auger), and STM characterization is required to ensure an adequate surface for experiments sensitive to even low densities of impurities.

2.2.2. Evaporation of Tm

Evaporation of Tm has been performed by using an electron beam evaporator. The e-beam heats a molybdenum crucible where the evaporant material is located in the form of flakes, obtaining a rate near 1 ML/min, as measured by a fluxmeter at the exit of the evaporator, where the current is proportional to the number of ions crossing the fluxmeter section per unit of time.

2.3. Tm monolayer: structural characterization

Figure 4a presents an STM image, taken at 4.2 K in a low temperature instrument, of the Tm ML along with the corresponding LEED pattern in **Figure 4d**, illustrating the situation. The STM image on the left, **Figure 4a**, shows a quasi-hexagonal symmetry on the Tm layer which should be noted is not atomic resolution, but rather a Moiré pattern caused by the overlap of the triangular Tm lattice and the rectangular W(110) surface, shown in greater detail in **Figure 4b**; the actual atomic resolution of the Tm layer can be seen in **Figure 4c**. The LEED pattern in **Figure 4d** presents the spots due the W(110) seen in **Figure 2b** and an added quasi-hexagonal structure corresponding to the Tm layer; satellite points forming an hexagon surrounding the Tm spots are due to the Moiré pattern.

The lattice parameter measured along [001] is 4.08 Å, while the two other sides of the isosceles triangle measure 3.92 Å. This means that the Tm ML displays a fairly distorted hexagonal structure, forming in fact a rhombic or isosceles triangular lattice. Thus, the lattice mismatch between Tm and W is large enough to produce in the first monolayer of thulium, an asymmetric distortion of the hcp structure that, with respect to the bulk lattice parameters, is compressed along the [110] W by 1% and expanded along [001] W by about 15%; the W lattice

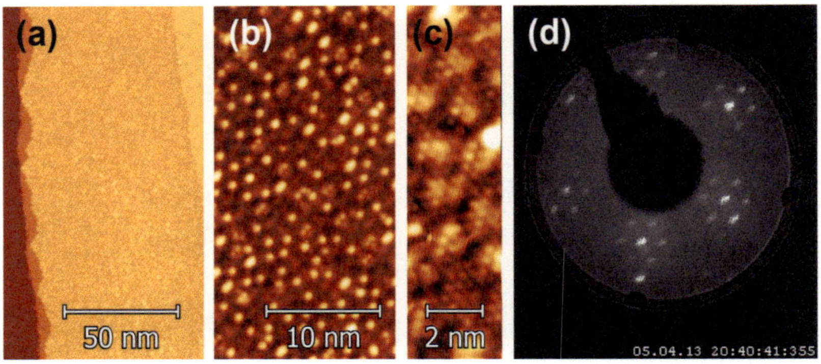

Figure 4. (a) STM image of a Tm/W(110) sample showing the Tm monolayer on W(110); the clean W surface can be seen to the sides of the Tm layer. (b) Detail of the Moiré pattern of the Tm monolayer and (c) detail of the underlying atomic resolution. (d) LEED pattern corresponding to the Tm monolayer on W(110) showing the spots corresponding to the crystalline structure surrounded by satellites corresponding to the Moiré pattern.

parameter is slightly enlarged respect to the bulk, with a_W = 3.173 Å. This results in a Tm: W coincidence match of 2:3 and 7:9 along the [110] and [001] W directions, respectively.

3. Experimental results

3.1. Low coverage: single atoms and clusters

One of the peculiarities of this system is that for low coverages, even at room temperature, single atoms and small clusters present a very low diffusion and do not aggregate into larger objects until higher temperatures, as illustrated by **Figure 5a**.

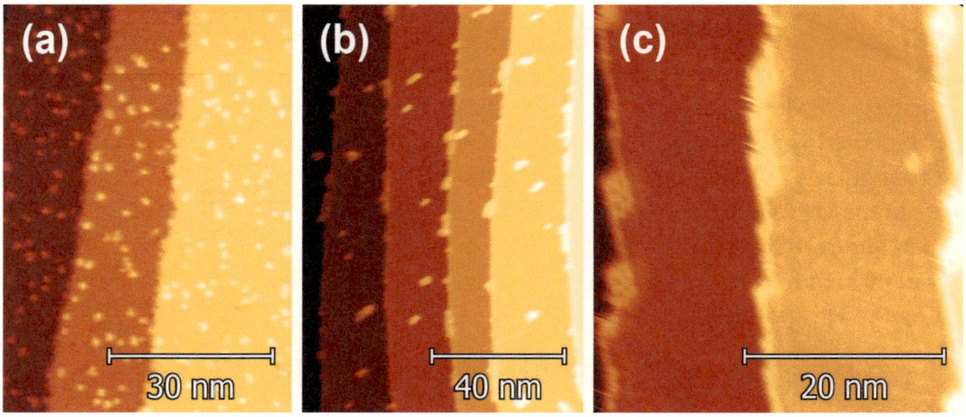

Figure 5. Tm/W(110) sample with a low Tm coverage (a) at RT, as deposited, (b) at 700 K after 4 h in the 550–700 K range, causing a high density of C adsorbates on the tungsten surface and (c) at 900 K, after 3 additional hours in the 700–900 K range, showing the surface covered in carbon reconstruction patches.

Increasing the temperature of the sample to $T = 700$ K facilitates diffusion, resulting in migration of Tm to the step edges, as well as aggregation into small islands (**Figure 5b**). It should be noted that changes in the configuration of the sample occur in a time window smaller than the thermalization time required by the system, as no strong evolution is observed at this temperature from $t_{700\,K} = 10$ to 30 min (where t_T is the time t that the sample has been at a given temperature T). Increasing the temperature above 900 K produces a notable change in the configuration, as Tm structures acquire visible hexagonal characteristics. **Figure 5c** shows the resulting sample at this temperature, where all Tm have migrated to the step edges forming a wire along the edge that expands into islands with sharp hexagonal angles.

In contrast to the previous example, a strikingly different behavior is observed for a similar starting coverage if the heating process is altered. **Figure 6a** shows the sample as deposited at room temperature, and while the coverage is roughly 1.5–2.0 times that of **Figure 5a**, it is still low enough for it to consist in small clusters as well as some atom-like objects.

In order to try and capture the diffusion process, longer time periods are spent measuring this sample at lower temperatures (600, 650, and 700 K), without observing any significant change at any given temperature, pointing to an equilibrium for the island size that is reached in the first few minutes after increasing the temperature, before the drift is low enough to allow STM measurements. Increasing the time spent at lower temperatures, however, does have a significant effect regarding the presence of C on the surface, which is much more evident in **Figure 6b** than for an equivalent temperature (730 K) reached in a shorter time, as in **Figure 5b**; the other difference between the two images is that in this case, Tm does not migrate to the step edge or form larger islands, as the diffusion process is inhibited by the presence of a large quantity of carbon adsorbates.

The difference in behavior is stronger as the temperature is increased as the large, regular islands with hexagonal characteristics, where all the Tm accumulates leaving free the majority of the tungsten surface are nowhere to be seen; instead, small irregular islands are distributed

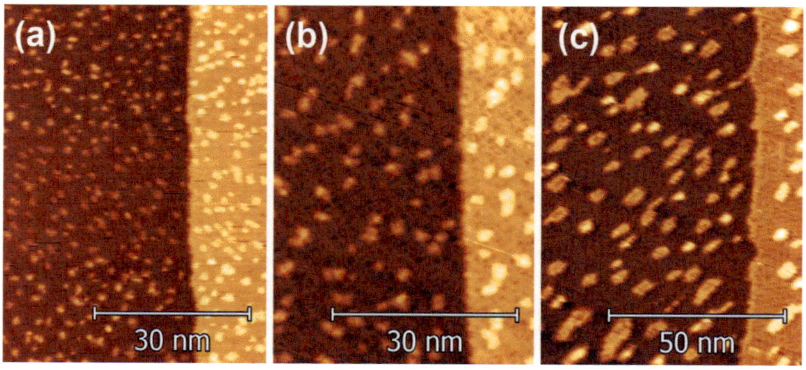

Figure 6. Tm/W(110) sample with a low Tm coverage (a) at RT, as deposited, (b) at 730 K after 4 h in the 600 to 730 K range, causing a high density of C adsorbates on the tungsten surface and (c) at 1100 K, after 3 h in the 750–1100 K range, showing the surface covered in carbon reconstruction patches.

along the surface with multiple carbon reconstruction patches covering the rest of the tungsten. **Figure 6c** shows this behavior, including double-layer islands due to the constraints imposed by the growing carbon reconstruction, even after increasing the temperature up to 1100 K.

3.2. Mid coverage: 0.5 monolayers

At higher coverages, Tm still does not tend to aggregate into a single continuous layer but rather tends to form small objects a few nanometers wide (**Figure 7a**) which, when annealed at temperatures around 600 K, evolve into small flat islands (**Figure 7b**). Following a slow heating process, as done for lower coverages, produces a similar behavior: the segregation of carbon is strongly enhanced which favors the formation of large patches of carbon reconstruction, inhibiting the diffusion of Tm and thus limiting the size of the islands; with the island size constrained by carbon, increasing the temperature up to 950 K induces a structural change where the islands become regular in shape, with hexagonal features, but without merging into large patches, but rather staying mainly as 10–15 nm wide individual islands.

It is not until the coverage is near the full monolayer that Tm does, in fact, aggregate rather than forming smaller individual objects. In this state, Tm presents a high corrugation and an irregular appearance, as seen in **Figure 8a**. Increasing the temperature to 780 K favors the mobility of Tm, allowing for the formation of a smoother but still irregular layer with a more compact structure, as shown by the fact that the tungsten surface is clearly visible in **Figure 8b** and that maintaining the sample at this temperature does not increase the proportion of visible tungsten, therefore discarding reevaporation of Tm as the cause of the observed decrease in coverage. As for the case of low coverage shown in **Figure 5**, at temperatures around 950 K, the Tm layer forms a hexagonal structure a single atomic layer in height, although in the form of large patches in this case (pictured in **Figure 8c**), as performing a faster annealing process conserves a surface clean enough for the Tm to diffuse all along the step edges, leaving the rest of the W step free of smaller Tm objects.

The evolution of the system as seen in **Figure 8** can actually be monitored by STM, as it changes in a scale of minutes. **Figure 9** shows the same area of the course of 30 min at $T = 780$ K, showing a dynamic behavior. The evolution pictured occurs 40 min after setting the temperature

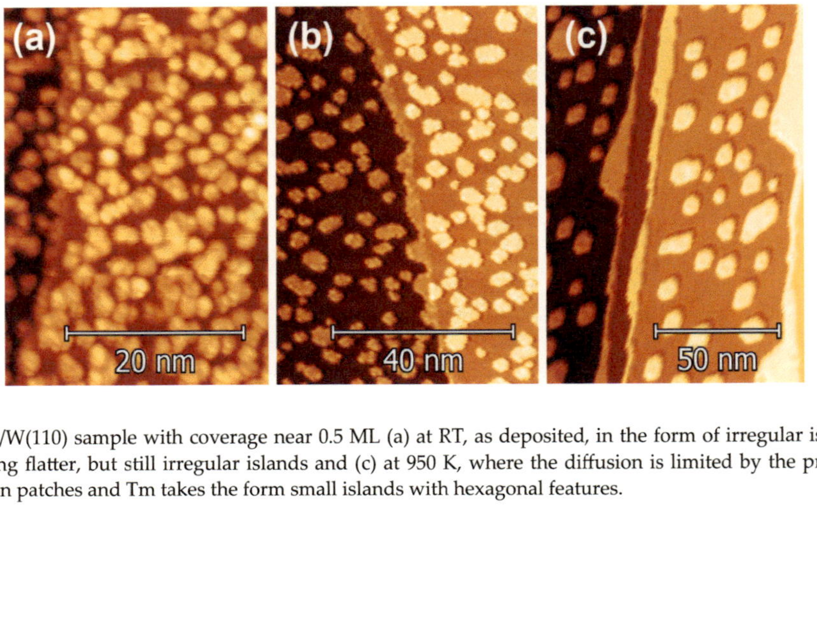

Figure 7. Tm/W(110) sample with coverage near 0.5 ML (a) at RT, as deposited, in the form of irregular islands (b) at 750 K showing flatter, but still irregular islands and (c) at 950 K, where the diffusion is limited by the presence of C reconstruction patches and Tm takes the form small islands with hexagonal features.

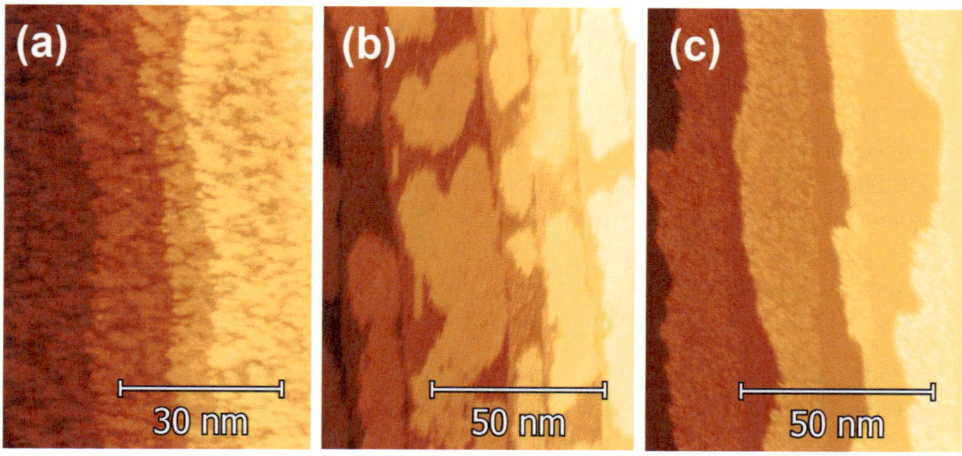

Figure 8. Tm/W(110) sample with a coverage around 0.5 ML coverage (a) as deposited at RT (b) at 750 K and (c) at 950 K. Islands change from irregular and rough in (b) to the crystalline with a Moiré pattern (c) for higher temperatures.

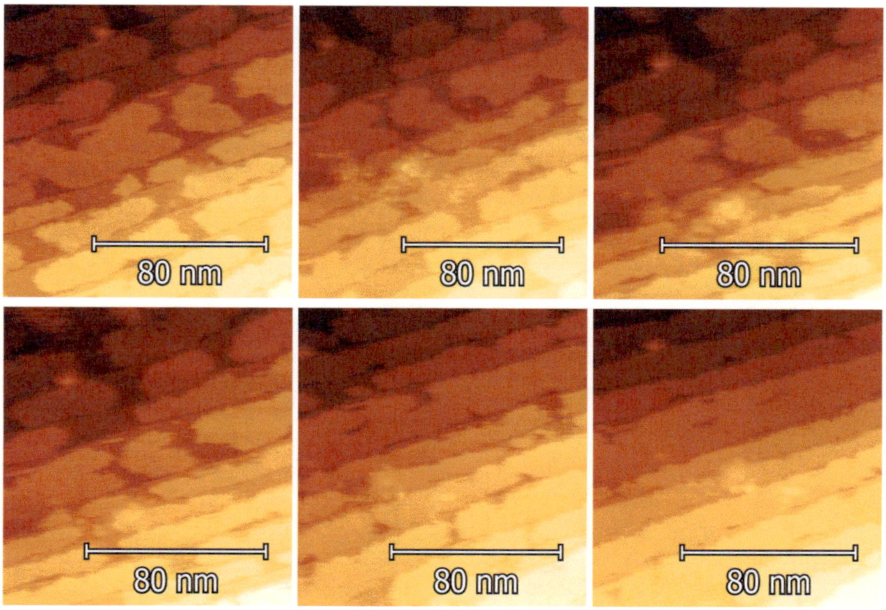

Figure 9. From left to right, top to bottom: Evolution over 30 min of the sample at 750 K. Thulium islands expand, covering the whole surface.

set point; some self-assembled nanowires can be seen in the first images, although the overwhelming majority of Tm is forming irregular-shaped islands. Over time, the islands expand, approaching the full monolayer and covering the wire structures. As no new material is added during this time, the increase in surface covered by the Tm islands is due to the system evolving to a new, less compact configuration. At higher temperatures, this is then followed by a structural change into a more compact hexagonal lattice, as seen from **Figure 8b** and **c**.

3.3. High coverage: above 1 monolayer

For higher initial coverages, above the monolayer, the Stranski-Krastanov growth mode is evident, with Tm forming multiple-layer high hexagonal pyramids in **Figure 10a**. Annealing this sample illustrates the origin of this growth mode: the W-Tm interface is energetically more favorable than the Tm-Tm interface; therefore, completion of the initial wetting layer takes priority over subsequent layers, as illustrated by the fact that the said higher Tm layers are reevaporated at lower temperatures than the wetting layer. Thus, annealing at high temperature, a multilayer sample, such as **Figure 10a**, leads to a similar result as for submonolayer samples, with the already discussed heteroepitaxial monolayer with a Moiré pattern, pictured in **Figure 10b**.

At high temperatures, carbon segregation to the surface is fast enough to observe its effect over consecutive scans. **Figure 11** shows the same area of the Tm monolayer at 1000 K over the course of 5 min. The small dark patch in **Figure 11a** is an area with carbon reconstruction which grows over time, while other small patch appears (**Figure 11b**), eventually merging with each other, forming a larger area of carbon reconstruction along the step edge. The growth of the carbon reconstruction patches displaces Tm from the layer, rather than growing below it, as it can be seen in **Figure 11b** and **c**: small areas on the Tm layer with a greyer color appear to have carbon beneath, while a darker area, similar in hue to the reconstruction along the step edge, seems to have displaced the Tm layer.

This effect is also seen on other samples, as shown in **Figure 12**, while the time evolution is not captured, it can be seen that Tm is displaced by the carbon reconstruction onto the second layer in **Figure 12a**, and 30 min later in **Figure 12b**, the Tm layer is flat once again, with Tm organized along to the step edge and carbon adsorbates and reconstruction patches on the outer side of the step. As the coverage is lower than the initial, it suggests a reevaporation at this temperature of the material that moves onto the second layer.

In this case, the carbon reconstruction patches within the layer have grown to the point where it is favorable for Tm to move to the second layer. Over time, this Tm is reevaporated reducing the total coverage, while the remaining material reorganizes forming a continuous layer with the carbon reconstruction patches along the step edges.

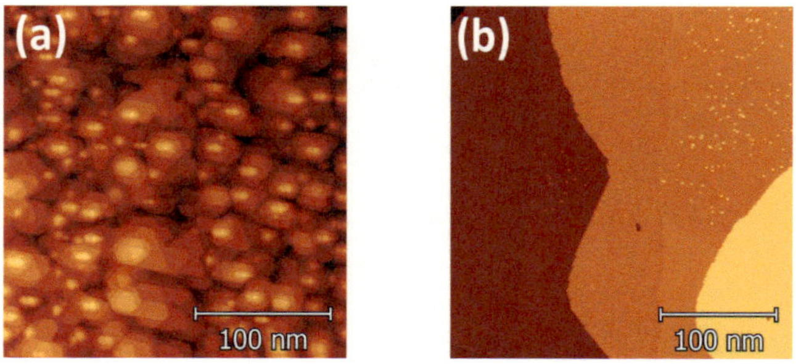

Figure 10. (a) Multilayer Thulium sample as deposited at room temperature showing hexagonal pyramidal islands. (b) After annealing at 1100 K, leaving only monolayer Tm islands on the W(110) surface; carbon reconstruction patches can also be seen.

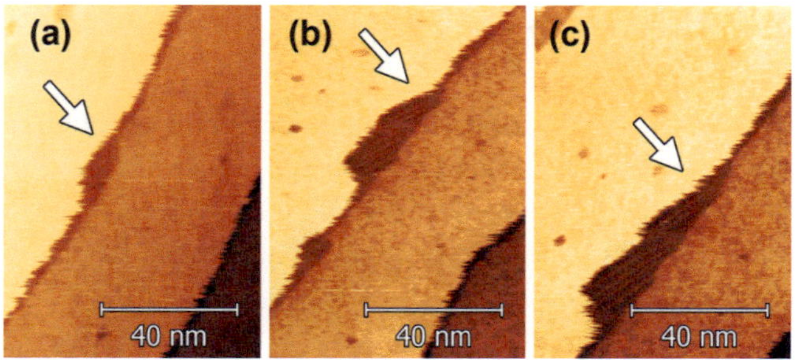

Figure 11. Evolution over 5 min of the sample at 1000 K, showing the growth of a carbon reconstruction patch, displacing the Tm layer: (a) At eh beginning, (b) the time between and (c) at the end of the sequence.

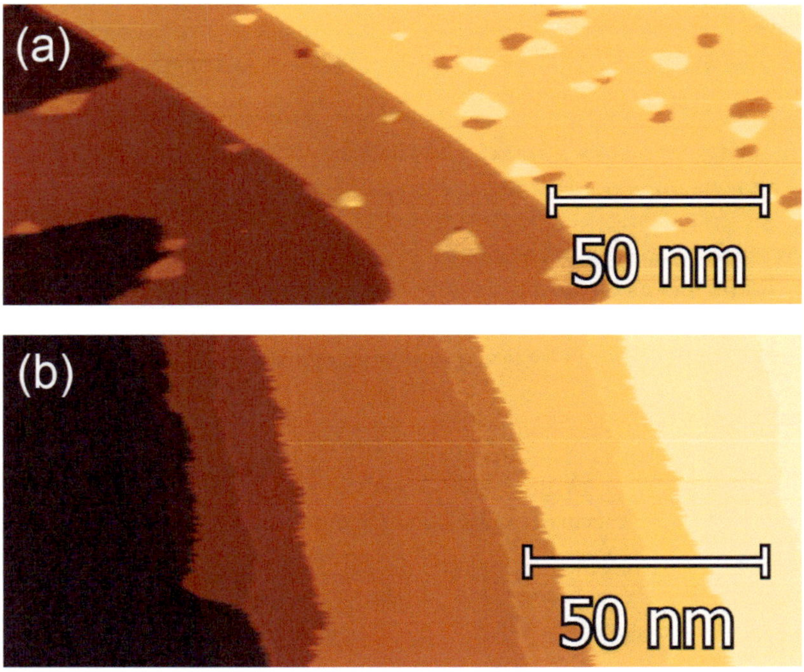

Figure 12. (a) Tm being displaced by carbon reconstruction patches at 1050 K. (b) The same sample 30 min later while keeping the temperature at 1050 K, showing that Tm has compensated for the displaced material by migrating to the step edge.

4. Analysis

4.1. Size distribution of Tm islands

To study the effect of carbon adsorbates on the diffusion of Tm on the W(110) surface and thus on the resulting island size, it is interesting to first introduce the basic ideas involved in the description of nucleation on ideal surfaces, omitting the effect of impurities, anisotropies, steps, and other deviations from a perfect surface.

Once an atom or molecule is absorbed onto a surface, it can be reevaporated or diffused along the surface. Adatoms diffusing on the surface can encounter each other leading to the formation of dimers, clusters, and larger two-dimensional islands. These islands, as a whole, are stable and do not diffuse further; their shape and size, however, can change by dissociation and diffusion of the adatoms on the edges.

Island nucleation and growth (precoalescence stage) continues until islands grow to the point of merging with each other. At this point, continued growth of the layer consists in the filling of the remaining holes (post-coalescence stage). Thus, at constant temperature, the number of islands increases over time until coalescence between them begins, reducing the total number of individual islands. For the case of three dimensional islands, if no additional material is deposited on top of the island, but rather there is a transference between layers, overcoming the Ehrlich-Schwoebel energy barrier associated with the interlayer jumping is required.

The behavior of single-component metallic systems is well understood, based on scaling properties of measured island densities and shapes as a function or temperature and covering. Several reviews present a thorough presentation of the state of the art [21, 22]. Usually, the diffusion process is presented in the context of a clean surface: only the adatoms and the surface are considered to obtain the relevant parameters [23]. In some cases of practical interest, the diffusion happens on surfaces partly covered by other impurities, and the diffusion is modified with respect to that observed in a clean surface, e.g., adsorbed hydrogen atoms enhances the self-diffusion of Pt by two orders of magnitude [24].

4.1.1. Island density

Two factors influence the nucleation and growth of islands: Deposition of atoms onto the surface with a flux F and thermally activated diffusion of adatoms along the surface with a diffusion coefficient D:

$$D = D_0 \exp(-U/k_B T) \tag{1}$$

where U is the diffusion barrier and k_B the Boltzmann constant. For atomic systems, $D_0 = \nu a^2$, where ν is an attempt frequency, and a is the lattice constant of the substrate. Thus, with the definitions of F and D, the mean time for a unit cell to be hit by an atom is $1/Fa^2$, and the mean time after which the atom leaves the cell by diffusion is a^2/D. Some conclusions can be obtained observing Eq. (1): adatom diffusion is thermally activated, thus increasing temperature produces an increment of D, and adatoms can diffuse over longer distances. As a result, the density N of stable islands becomes smaller with increasing T, and the number of atoms forming the island increases.

In order to describe and quantify the structures observed by STM or AFM as function of coverage Θ or temperature, a complete set of equations has been obtained in the literature [21]. A fundamental concept is the critical island size or critical nucleus i. This variable denotes the critical cluster size, which becomes stable on adding an atom. Clusters with $i+1$ atoms are more likely to grow than to dissociate. i depends on the substrate temperature, increasing

http://dx.doi.org/10.5772/intechopen.70218

when T increases, and other factors as the crystal symmetry of the substrate and the structure of a film, either amorphous or crystalline.

4.1.2. Mean field calculation

The mean-field nucleation theory has been used to obtain a simple expression for the concentration n_x of 2-D islands of size i for the case of complete condensation [21, 22, 25]:

$$n_x \propto \left(\frac{D}{F}\right)^{-\chi} \exp\left[\frac{E_i}{(i+2)\,k_B\,T}\right] \tag{2}$$

where $\chi = i/(i+2)$, and E_i is its bonding energy. A stable cluster means that it grows more rapidly than it decays, for instance by a dissociation process, during the course of deposition.

4.2. Influence of carbon adsorbates in island size

While the discussion above pertains to the case of clusters with a small number of atoms, the dependence of the equilibrium island size with temperature is a general conclusion. **Figure 13** shows the island size evolution with temperature of a sample, where little carbon has segregated over the whole sequence, leading to a final island size that is actually limited by the step size, as seen in the lower section of **Figure 13**, illustrating the sample at different stages.

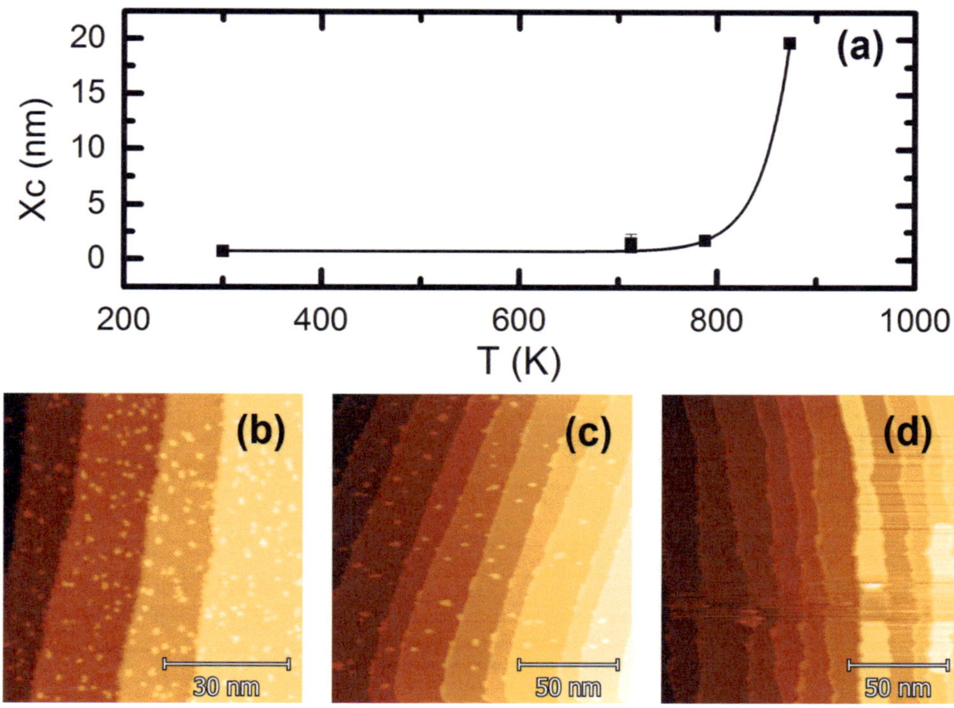

Figure 13. (a) Island size evolution with temperature. Below, from left to right, the evolution of the island size on the same sample at (b) room temperature, (c) 730 K and (d) 950 K. Note that the island size at high T is limited by the step size.

This is in stark contrast with the behavior described for samples, where the annealing process is performed over a longer time, allowing for a greater density of carbon adsorbates on the surface, as seen previously in **Figures 6** and 7. Plotting the island size as a function of temperature for several samples illustrates how different the behavior is between a clean surface and one saturated with the 15 × 3 carbon reconstruction.

It is not trivial to quantify the number of impurities, especially with experiments at high temperature where resolution might be compromised to avoid risking tip crash, further complicated by drift issues. **Figure 14** shows the island size evolution for several sets of samples, divided roughly in three categories depending on the carbon density on the surface: Low Carbon (blue hues in the graph), where the sample remains reasonably clean after the annealing process and island size is limited mainly by the size of the steps and the amount of Tm available; Mid Carbon, where carbon reconstruction patches and a high density of carbon adsorbates strongly limit the mean island size; High Carbon, with carbon reconstruction patches covering the whole surface.

A careful control over the carbon density on the W(110) surface can thus provide a reliable method for obtaining self-assembled islands with a definite size, ranging from being limited by the atomic step size to islands under 5 nm in diameter.

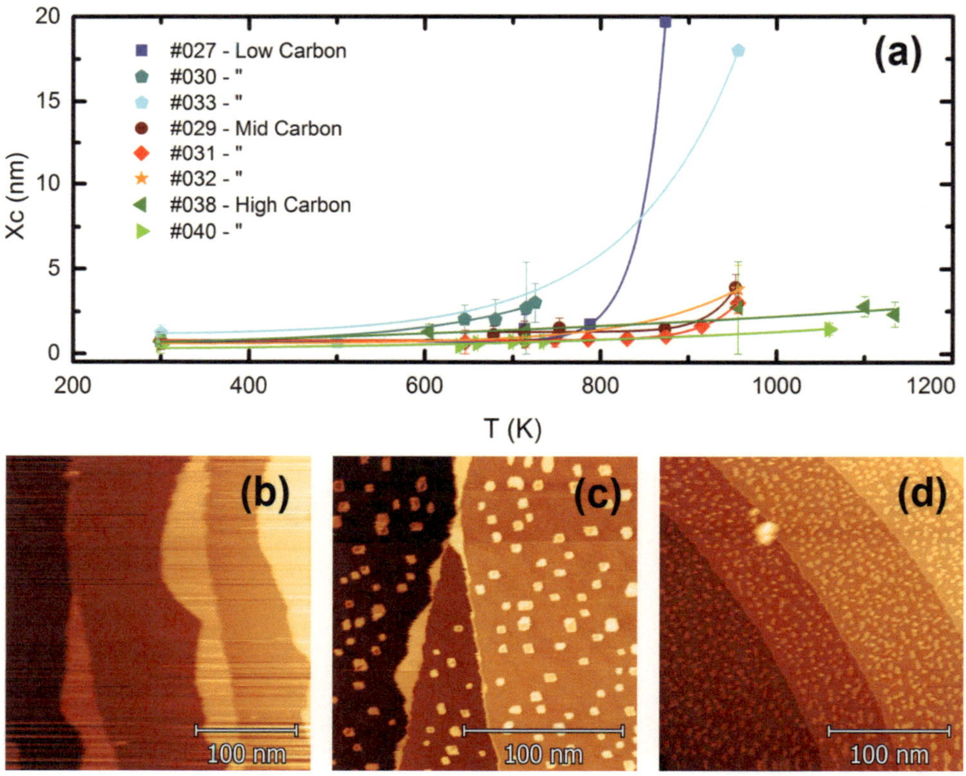

Figure 14. (a) Island size evolution with temperature for a set of samples. Below, representative final states for different samples, corresponding to the curves marked as (b) sample 030 at a final temperature of 950 K, with low carbon presence resulting in large islands; (c) sample 031 at 950 K, with a strong presence of carbon reconstruction, but segregated after the islands already increased in size; and (d) sample 040 at 1050 K, where very small regular islands can be observed all over the surface, due to the inhibited diffusion due to the high density of carbon on the surface.

5. Conclusion

In this chapter, the preparation of single atomic layer Tm films on a W(110) substrate is explored. Sub-monoloyer Tm deposits can present a varied morphology depending on initial coverage, substrate temperature and adsorbate density. The key to achieving high-quality pseudomorphic Tm films resides in an annealing process after evaporation, which motivates the study presented here, where STM experiments are performed at high temperature to observe in situ the evolution of the Tm film. Measurements over time evidence the diffusion of carbon adsorbates from the bulk of the W crystal onto the surface. The effect of carbon impurity density on the diffusion process of Tm atoms is studied by observing the evolution of multiple samples at different initial coverages following different thermal processes, with longer times leading to a higher adsorbate density. It is observed that the presence of carbon strongly limits the diffusion of Tm, thus leading to the formation of pseudomorphic nanometric islands instead of the full monolayer. The ultimate result of the procedure described in this chapter indicates that with careful control of the impurity density, structures with mean Tm island size down to 5 nm in diameter can be obtained, an interesting achievement considering that the magnetic anisotropy of Tm foresees perpendicular magnetization in these islands. This work highlights the role of surface adsorbates on the diffusion process of single atoms, and how it governs the nucleation of islands; experimental evidence is presented of control over the mean island size by means of inducing a change in the diffusion parameters on the surface in a controlled fashion by using impurities existing in the substrate, a procedure that could be extrapolated to other magnetic materials to create self-ensembled nanosystems.

Acknowledgements

This work was supported by the Spanish MICINN (Grants MAT2015-66726-R and MAT2013-46593-C6-3-P), Gobierno de Aragón (Grant E81), and Fondo Social Europeo. We thank Dr. José Luis Diez-Ferrer for providing the images of **Figure 3**.

Author details

David Coffey[1,2,3], José I. Arnaudas[2,3], David Serrate[2,3] and Miguel Ciria[1,3]*

*Address all correspondence to: miguel.ciria@csic.es

1 Instituto de Ciencia de Materiales de Aragón, Consejo Superior de Investigaciones Científicas, Zaragoza, Spain

2 Laboratorio de Microscopías Avanzadas, Fundación Instituto de Nanociencia de Aragón, Universidad de Zaragoza, Zaragoza, Spain

3 Departamento de Física de la Materia Condensada, Universidad de Zaragoza, Zaragoza, Spain

References

[1] Kwo J, Gyorgy EM, McWhan DB, et al. Magnetic and structural properties of single-crystal rare-earth Gd-Y superlattices. Physical Review Letters. 1985;**55**:1402-1405

[2] Weller D, Alvarado SF, Gudat W, et al. Observation of surface-enhanced magnetic order and magnetic surface reconstruction on Gd(0001). Physical Review Letters. 1985; **54**:1555-1558

[3] Kwo J, Hong M, Nakahara S. Growth of rare-earth single crystals by molecular beam epitaxy: The epitaxial relationship between hcp rare earth and bcc niobium. Applied Physics Letters. 1986;**49**:319-321

[4] Majkrzak CF, Cable JW, Kwo J, et al. Observation of a magnetic antiphase domain structure with long-range order in a synthetic Gd-Y superlattice. Physical Review Letters. 1986;**56**:2700-2703

[5] Salamon MB, Sinha S, Rhyne JJ, et al. Long-range incommensurate magnetic order in a Dy-Y multilayer. Physical Review Letters. 1986;**56**:259-262

[6] Kolaczkiewicz J, Bauer E. The adsorption of Eu, Gd and Tb on the W(110) surface. Surface Science. 1986;**175**:487-507

[7] Nicklin CL, Binns C, Norris C, et al. Valence state of low-dimensional thulium structures grown on molybdenum (110). Surface Science. 1994;**307-309**:858-862

[8] Bodenbach M, Höhr A, Laubschat C, et al. Surface electronic structure of Tm(0001) and Yb(111). Physical Review B. 1994;**50**:14446-14451

[9] Koehler WC, Cable JW, Wollan EO, et al. Magnetic structures of thulium. Physical Review. 1962;**126**:1672-1678

[10] Brun TO, Sinha SK, Wakabayashi N, et al. Temperature dependence of the periodicity of the magnetic structure of thulium metal. Physical Review B. 1970;**1**:1251-1253

[11] Li H, Tian D, Quinn J, et al. Structural and electronic properties of ultrathin films of Gd, Tb, Dy, Ho, and Er. Physical Review B. 1992;**45**:3853-3856

[12] Tober ED, Ynzunza RX, Westphal C, et al. Relationship between morphology and magnetic behavior for Gd thin films on W(110). Physical Review B. 1996;**53**:5444-5448

[13] Barret S, Dhesi S. The Structure of Rare-Earth Metal Surfaces. London: Imperial C; 2001

[14] Binnig G, Rohrer H, Gerber C, et al. Surface studies by scanning tunneling microscopy. Physical Review Letters. 1982;**49**:57-61

[15] Binnig G, Rohrer H. Scanning tunneling microscopy. Helvetica Physica Acta. 1982;**55**:726

[16] Petersen L, Schunack M, Schaefer B, et al. A fast-scanning, low- and variable-temperature scanning tunneling microscope. Review of Scientific Instruments. 2001;**72**:1438

[17] Horcas I, Fernández R, Gómez-Rodríguez JM, et al. WSXM: A software for scanning probe microscopy and a tool for nanotechnology. Review of Scientific Instruments. 2007;**78**:13705

[18] Nečas D, Klapetek P. Gwyddion: An open-source software for SPM data analysis. Central European Journal of Physics. 2012;**10**:181

[19] Bode M, Krause S, Berbil-Bautista L, et al. On the preparation and electronic properties of clean W(110) surfaces. Surface Science. 2007;**601**:3308

[20] Bode M, Pascal R, Wiesendanger R. STM study of carbon-induced reconstructions on W(110): Strong evidence for a surface lattice deformation. Surface Science. 1995;**344**:185-191

[21] Brune H. Microscopic view of epitaxial metal growth: Nucleation and aggregation. Surface Science Reports. 1998;**31**:121-229

[22] Einax M, Dieterich W, Maass P. *Colloquium* : Cluster growth on surfaces: Densities, size distributions, and morphologies. Reviews of Modern Physics. 2013;**85**:921-939

[23] Brune H, Bales GS, Jacobsen J, et al. Measuring surface diffusion from nucleation island densities. Physical Review B. 1999;**60**:5991-6006

[24] Horch S, Lorensen HT. Enhancement of surface self-diffusion of platinum atoms by adsorbed hydrogen. Nature. 1999;**398**:134

[25] Venables JA. Nucleation calculations in a pair-binding model. Physical Review B. 1987;**36**:4153-4162

Section 2

Epitaxy Monte Carlo Simulation and Reactor Design

Chapter 5

Monte Carlo Simulation of Epitaxial Growth

Celso I. Fornari, Gabriel Fornari,
Paulo H. de O. Rappl, Eduardo Abramof and
Jerônimo dos S. Travelho

Additional information is available at the end of the chapter

http://dx.doi.org/10.5772/intechopen.70220

Abstract

A numerical Monte Carlo (MC) model is described in detail to simulate epitaxial growth. This model allows the formation of structural defects, like substitutional defects and vacancies, and desorption of adsorbed atoms on the surface. The latter feature supports the study of epitaxial growth at very high kinetic regime. The model proposed here is applied to simulate the homoepitaxial growth of Si. The results obtained fit well to the experimental reports on (0 0 1) silicon homoepitaxy. The easy implementation of a large number of microscopic processes and the three-dimensional spatial information during the film growth suggests that the model can be applied to simulate the growth of binary, ternary, or more compounds and even the growth of superlattices and heterostructures.

Keywords: Monte Carlo simulation, molecular beam epitaxy, epitaxial growth, lattice-matched substrates

1. Introduction

Computer simulation has been successfully applied to the study of surface growth by molecular beam epitaxy (MBE) [1]. These simulation results play a key role in understanding and elucidating various aspects of MBE growth. These simulations provide an atomistic interpretation of the changes in characteristic reflection high energy electron diffraction (RHEED) patterns, observed in real-time, that are related to the different growth modes, i.e., from island nucleation to layer-by-layer growth mode. Additionally, by simulation, the dependency of substrate temperature and the growth rate in the vacancy formation and structural defects can be evaluated [2–6]. However, the majority of these computational models for simulating neglects desorption effect and, consequently, do not explain growth at high substrate temperatures, where the growth rate is decreased due to desorption of atoms adsorbed on the surface. Besides, the models often use structural approximations, in which the formation of defects is not allowed. In this work, a numerical Monte Carlo (MC) model is presented, in which desorption processes and structural defects are allowed, being possible to study limit cases, i.e., low kinetic energy regime, where structural defects are more likely, and high kinetic energy regime, where the desorption rate competes with the deposition rate.

This chapter is organized as follows: Section 2 presents a review on molecular beam epitaxy, focusing on the growth modes and the most common experimental techniques used to characterize thin epitaxial films. Section 3 presents a brief review of computational models of epitaxial growth simulation, and the model developed in this work. Section 4 presents the simulation results of silicon (Si) homoepitaxial growth on Si (0 0 1). Finally, Section 5 presents the main conclusions and new applications of the proposed model.

2. Molecular beam epitaxy

Molecular beam epitaxy (MBE) is a state-of-the-art ultra-high vacuum (UHV) thin film growth technique. The materials that will compose the films are sublimated from highly stable effusion cells, forming the molecular beam, which is deposited on an independently heated substrate. The substrate is a monocrystalline material and the impinging atoms, from the solid source, follow the substrate crystalline orientation. The physicochemical interaction mechanisms between two phases, in this case, vapor solid, with the growing solid, is called epitaxy, which has a Greek root: *epi* means "above" and *taxis* means "an ordered manner" [7].

Since the system operates in a UHV environment, the mean free path of the vapor species is much larger than the distance between the solid sources and the substrate. The interactions of the molecular beam, before colliding with the substrate surface, can be neglected. The substrate temperature is controlled independently of the effusion cells and is kept low relative to the temperatures of the cells. Therefore, the growth far from the thermodynamic equilibrium makes it possible to compensate distinct thermodynamic properties of the different materials, besides allowing the growth of high-quality monocrystalline films.

The UHV conditions in the MBE growth chamber result in an extremely low background impurity level, which allows the growth of samples with high doping control. Besides, several materials can be sublimated simultaneously and, by means of independent shutter control, multi-layer systems, with very sharp interfaces, can be obtained. Among that, a wide range of alloy composition and different doping levels can be achieved. The growth conditions are reproducible, highly stable and can vary over a wide range, which is crucial for optimizing the growth for specific materials. Additionally, compatibility with *in situ* analysis tools provides insights into the microscopic processes involved in the growth. The reflection high-energy electron diffraction (RHEED), for example, probes the film surface and, due to the small angle of the incident electron beam, provides information on surface morphology, in real time, during the growth [8].

In the past decades, the interest in MBE was promoted mainly by the exciting properties of semiconductor structures due to two characteristics: high control in the atomic level and reproducibility. Nowadays, after the theoretical prediction of the topological phase of the matter, the MBE technique has shown to be a promising way to obtain high-quality samples, without using counter doping to suppress free carriers due to the structural defects [9].

2.1. Growth mechanisms

Growth of MBE can take place on a substrate composed of the same material, e.g., silicon epitaxy on Si substrates [10], or on different materials, e.g., Ge epitaxy on Si [11]. The first is called homoepitaxy and the second is called heteroepitaxy. These growth conditions may lead to different growth mechanisms, due to the differences in the lattice parameters. Basically, there are three different modes of growth: Volmer-Weber, Stranski-Krastanov, and Frank-van der Merwe, which will occur depending on the experimental parameters and the lattice mismatches.

In Volmer-Weber's mode, the interaction of the adsorbed atoms is much stronger among them than with the substrate surface, which leads to the formation of clusters or three dimensional islands. As growth proceeds, these islands expand, widening their volumes, whose height greatly exceeds the thickness of a monolayer (ML), leading to the simultaneous growth of atomic layers with rough surface, as shown in **Figure 1(a)** [8].

In Frank-van der Merwe's mode, the atoms adsorbed on the surface have a stronger interaction with the surface, leading to the formation of a complete ML before another starts to grow. This layer-by-layer growth mode is often referred as a bi-dimensional growth and is shown in **Figure 1(b)** [8].

The last mode of growth, called Stranski-Krastanov, is characterized by mixing both previously mentioned. In this mode, the adsorbed atoms begin to grow layer-by-layer. When the critical layer thickness is reached, which value depends on the specific physical properties of each compound, the elastic energy accumulated in the growth relaxes, resulting in the formation of clusters or islands. This mode of growth is shown in **Figure 1(c)** [8].

Of course, all these growth modes are important, since each has a particular application. For the growth of topological insulators, such as bismuth chalcogenide compounds, the Frank-van der Merwe mode is necessary, because the aim here is to minimize structural defects during the formation of the epitaxial layers [12]. However, to manufacture low-dimensional

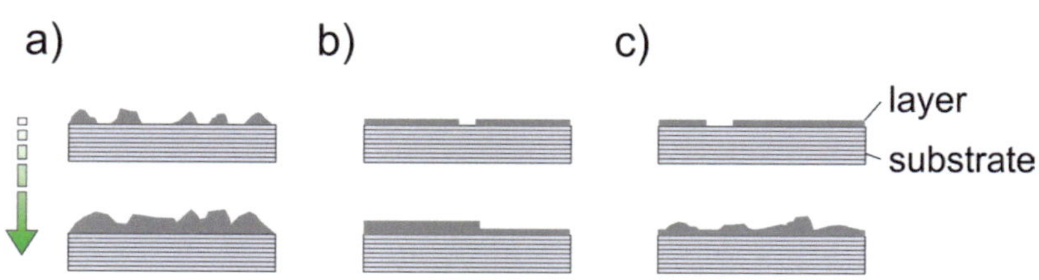

Figure 1. Modes of growth: (a) Volmer-Weber; (b) Frank-van der Merwe; and (c) Stranski-Krastanov.

structures, such as quantum dots, when island morphology is required, the growth parameters of the Volmer-Weber mode are determined to control the density, size, and distribution of the islands.

2.2. Characterization techniques

RHEED and atomic force microscope (AFM) are the most common experimental techniques applied to characterize the surface of epitaxial films. The first, as mentioned, provides information *in situ* and in real time on surface reconstruction, growth rate, and growth mode. The second allows probing the surface morphology of the film.

The RHEED equipment, basically, consists of an electron gun, in the energy range of 10–50 keV, and a phosphorescent screen. The electron beam is directed to the sample surface, at low angle (<5°). The de Broglie wavelength of electrons for this energy range is 0.17–0.06 Å, which corresponds to the interatomic distances in the crystalline lattice. Therefore, whenever the difference between the incident beam and the diffracted one is equal to a vector of the reciprocal lattice, there will be a maximum of diffraction.

The AFM technique provides high-resolution images of the surface morphology. A piezoelectric is used to scan the sample surface, and a very sharp tip is used to probe the surface. The tip deflection is measured at each point, providing spatial information in the real space of the surface. This information can be directly compared to the simulation results, in order to validate the growth models.

3. Molecular beam epitaxial models

3.1. Epitaxial growth models

The most common nucleation models used in the simulation of growth by MBE are either completely deterministic or totally stochastic. The deterministic models are based on the temporal evolution of differential equations and study microscopic parameters or stability properties of the growing surface. The deterministic models do not contain explicit spatial information of the growing surface [13–15].

Alternatively, to the analytic simulations, there are models that consider the stochastic nature of the microscopic processes. These stochastic calculations are typically implemented in the form of molecular dynamics (MD) or kinetic Monte Carlo (MC) simulations [16–18]. The MC simulation allows the easy implementation of a large number of microscopic processes. The rate of each microscopic process is obtained from first principle calculations or from experiments. These models are often in agreement with experimental works and provide spatial information of the growing surface [19].

The solid-on-solid (SOS) approach is very common in Monte Carlo simulation models [20]. In this approximation, one atom can only be accommodated on another atom, and therefore, structural defects, such as vacancies, are not allowed. Since bulk defects are not allowed in the

SOS model, the growth quality can only be evaluated by the growing surface. This approach has been widely used in growth simulations of MBE due to the successful interpretations of RHEED intensity oscillations and the surface atomistic processes.

3.2. The Monte Carlo epitaxial model

This MC computational growth model was implemented using the nearest-neighbor and lattice-gas approximation. The first approximation, as mentioned, considers the interaction only with the in-plane nearest neighbor. **Figure 2(a)** shows an example for a cubic lattice structure, which has a maximum of four in-plane neighbors. Therefore, the potential energy of each lattice point is determined by analyzing the number of bonds between the closest atoms. This approximation has been widely used in MBE simulations [21]. The second approximation considers a fixed crystalline structure, with sites at which the atoms from the solid source can be accommodated. In this case, the possible crystalline positions can be either occupied or empty. Once a fixed crystalline structure is considered, no strain is accounted in this model. **Figure 2(b)** illustrates the lattice gas approximation. It is good to emphasize that these approximations do not allow studying surface reconstruction, since the atomic positions are predetermined and the atoms can fill or not each one. Besides, these approximations allow only the study of homoepitaxy or, in some cases, heteroepitaxy of lattice-matched substrates, e.g., the growth of bismuth telluride on BaF_2 (1 1 1), where the lattice mismatch is of only 0.04% [12]. In similar cases, the lattice gas approximation can be employed in the numerical models.

The processes considered, in the model presented in this chapter, are shown in **Figure 3**. After deposition (a), surface atoms are allowed either to migrate to another position (b–e) or to desorb (g), and structural defects can be formed (f). The surface is defined as an occupied lattice position with dangling bonds. This statement is also valid for atoms near a hole, since these atoms have one or more dangling bonds due to the presence of the hole itself. This means that the neighboring atoms can migrate and fill the hole, turning on the bulk diffusion process.

Models in which bulk defects are considered allow for the study of structural defect formation at low kinetic energy regimes, e.g., growth by MBE at low substrate temperatures. In addition,

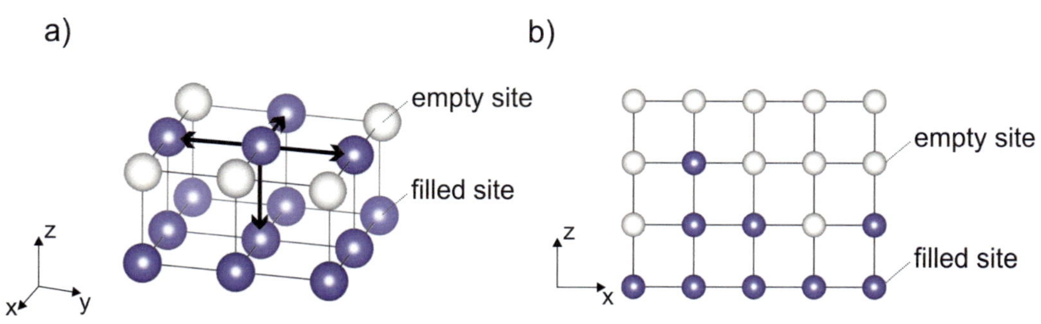

Figure 2. (a) Determination of the number of nearest neighbor to a chosen position of a cubic lattice. In this example, the atom has three in-plane bonds and one out-of-plane bond. (b) The lattice gas approximation considers a fixed empty crystalline structure, in which the atoms from the solid source are accommodated, initiating epitaxial growth.

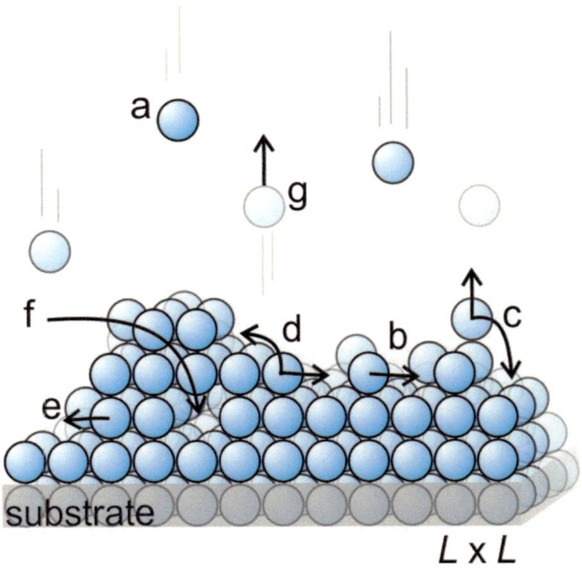

Figure 3. Allowed processes during MC simulation of thin film growth by MBE: (a) deposition of atoms, arriving from the solid sources, on the surface of a growing film; (b) surface diffusion to a more stable position; (c) atom in an unstable position, without lateral bonds, which can either desorb to the vacuum volume or diffuse to a more stable position; (d) surface diffusion or migration to a higher kink lattice; (e) possible surface diffusion forming overhang; (f) hole formed in the structure volume; and (g) desorption.

the desorption process of atoms at the surface is critical for investigating high-energy growth regimes, since desorption of adsorbed atoms (adatoms) significantly affects the growth rate and morphology of films at very high substrate temperatures.

The implementation of this model can be divided into two steps: deposition and surface analysis. The first step is responsible for the seating of the atoms on the growing surface, which, in a computational view, can be seen as a square matrix being filled with numbers. In sequence, the second step starts, and each position of the matrix is analyzed. At this point, for each matrix position, a calculation is made and, depending on the result, one of the possibilities described in **Figure 3** can occur, except for the process shown in **Figure 3(a)**.

These steps are repeated until a certain growth time, as shown in **Figure 4**. For each loop interaction, a time unit is added to the deposition time. By using deposition rates equivalent to the experimental ones and increasing the number of site-analysis during the second step, i.e., the number of times each position is analyzed, it is possible to use an approximately equivalent experimental time unit.

3.2.1. Deposition rate

Depending on the temperature of the effusion cells, different growth rates are achieved. These rates can be experimentally determined, *in situ*, by measuring RHEED intensity oscillations or using a quartz crystal microbalance. In addition, these rates can be determined by directly measuring the film thickness after growth.

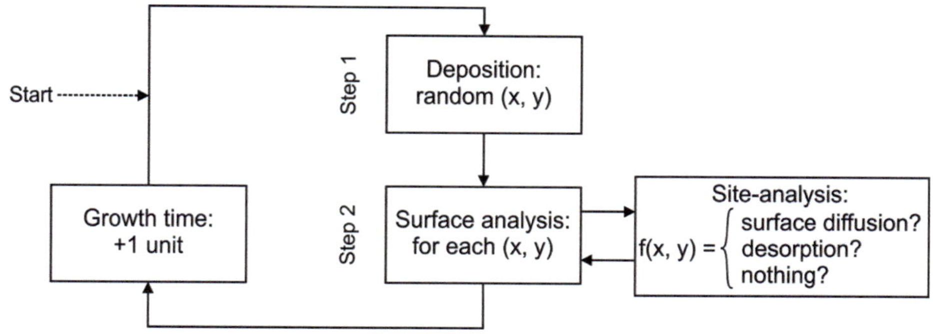

Figure 4. Flow chart showing the two-step model. During one time unit, a number of atoms are randomly deposited on the growing surface. After deposition, all surface sites are randomly analyzed. Surface diffusion or desorption can occur during this process. The deposition time is increased by one unit, and the process is repeated until the total growth time is achieved.

In the model, a certain amount of atoms is deposited on the first step of each round of the simulation. This amount depends directly on the substrate area, as given by Eq. (1),

$$N_X = L^2 . D \tag{1}$$

In this equation, N_X is the number of atoms from specie X that will be deposited per second, and L is the lateral dimension of the substrate (the size of one side of the square matrix). The constant D is the reciprocal of the experimental deposition rate, given in seconds per mono layers (s/ML), i.e., $D = 1/GR$, where GR is the experimental growth rate in ML/s. In this model, it is possible to simulate the growth of materials composed by one or more atomic species. In that case, the first step of the model will be repeated N_{tot} times, with $N_{tot} = N_X + N_Y + \ldots$, where X, Y indicate each atomic species. To fit the simulation to the experimental data, D must be carefully determined for each atomic species in the model.

3.2.2. *Surface analysis*

In the second step, each position of the surface is analyzed randomly. The surface control is managed by a linear dynamic data structure (implemented through a linked list), containing all spatial coordinates and the atomic species.

In a certain drawn position, the probability of surface diffusion or desorption is calculated. These probabilities are given as the product of an observation time, τ, and a hopping rate, $R_{D,E}$, for diffusion (D) or desorption (E), as shown in Eq. (2),

$$P_{D,E} = \tau . R_{D,E} \tag{2}$$

Theoretically, the observation time should be as low as possible, $\tau \to 0$. However, a very low observation time leads to a high computational time, since the second step is repeated until the integrated observation time is equal to one unit, as shown in Eq. (3). In practice, a good value is $\tau \approx 0.01$, which means that the second step of the model is repeated 100 times,

$$\sum_{i=1} \tau_i = 1. \tag{3}$$

The hopping rate is determined by an Arrhenius equation, as shown in Eq. (4),

$$R_{D,E} = R_0 . e^{-\frac{E_{D,E}}{k_B T}} \tag{4}$$

In this equation, $E_{D,E}$ is the energy for diffusion (D) or desorption (E), k_B is the Boltzmann constant, and T is the substrate temperature. The typical vibration frequency of the atom,R_0, is function of Planck constant, h, and is given by Eq. (5), typically around 10^{13} Hz,

$$R_0 = \frac{2 k_B T}{h}. \tag{5}$$

The energies for diffusion or desorption can be obtained from *ab initio* calculations or by experimental results. These values can be fitted to the experimental data to reproduce the experimental results.

The energy depends on the number of bonds that each atom possesses. Equation (6) presents the energy required for an atom to diffuse (E_D) or desorb (E_E),

$$E_{D,E} = mE_{D0,E0} + nE_{DL,EL}, \tag{6}$$

where m is the number of out-of-plane bonds, n is the number of in-plane bonds (see **Figure 2(a)**), and the sub-indices $D0$ and $E0$ indicate the energy required to diffuse or desorb, respectively, for one out-of-plane bond.

The sub-indices DL and EL indicate the energy required to diffuse or desorb, respectively, for one in-plane bond. In practice, in a cubic lattice, $m = 0$ or 1 and $n = 0, 1, 2, 3$ or 4. When an atom is located on a vacancy, which is an unoccupied site in the lattice, $m = 0$.

3.2.3. Probabilities

At each round, in the second step, all surface positions are analyzed once. For each position, the number of in-plane and out-of-plane bonds is computed, and then, the probabilities for diffusion and desorption are calculated. A random number is generated, and if it reaches one of these events, the atom will migrate to another available position or desorb. **Figure 5(a)**

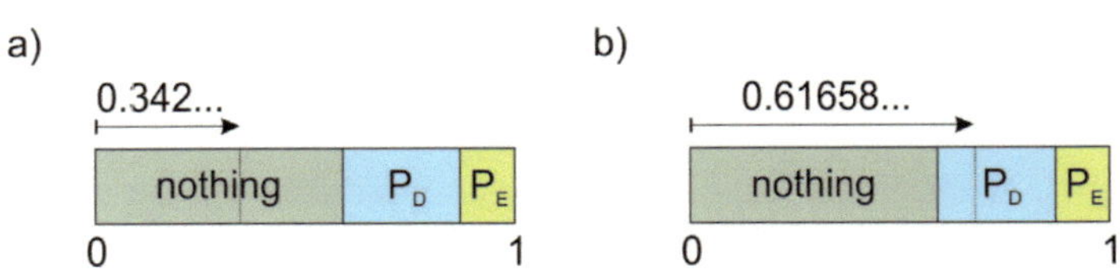

Figure 5. (a) Example of a draw number where none of the events is reached and (b) when a surface diffusion event is achieved.

shows a draw number, where none of the events was reached, and **Figure 5(b)** shows the activation of a surface diffusion event.

When a diffusion event is reached, the atom will migrate to any random neighboring position in the lattice. For a desorption event, the atom is simply removed from the surface. Since the MBE equipment operates under UHV conditions, the mean free path inside the growth chamber is very high, and these atoms can stick to the MBE wall chamber without interacting with the molecular beam.

For very high substrate temperatures, some lattice positions can be considered unstable, since the probability of diffusion and/or desorption can surpass one unit. Whenever the sum of the probabilities surpasses one unit, as shown in Eq. (7), a "right event" is achieved,

$$P_D + P_E > 1 . \quad (7)$$

This situation tries to describe limiting conditions, where the growing rate is overlapped by the desorption rate. Since this event occurs at high substrate temperatures, the adatoms have a very high surface mobility. To accomplish this condition, each time a "right event" is reached all available surface positions are analyzed. If one or more of these positions offers enough bonds to avoid another "right event," the atom migrates randomly to one of these positions. Otherwise, if none of the surface positions is stable enough, i.e., if Eq. (6) is true, the atom desorbs to the vacuum.

To avoid border effects, a periodic boundary condition was implemented. Atoms located at the borders of the substrate have only two or three in-plane bonds. By increasing the substrate temperature, these positions can become more unstable. The toroid considers a closed substrate, such as: $x(1) \leftrightarrow x(L)$ and $y(1) \leftrightarrow y(L)$. In this sense, if an equivalent position is occupied, $x(1) \leftrightarrow x(L)$, the atom located at $x(1)$ gets a bond.

3.2.4. Random numbers

To obtain the stochastic nature of the microscopic processes, the Mersenne Twister random number generator was employed, which is a 623 dimensionally equidistributed uniform pseudorandom number generator with an accuracy of up to 32 bits [22].

3.2.5. Quantifying results

Surface roughness is estimated based on the exposed surface, without computing any structural defects. The simulated film roughness is calculated by:

$$\sigma = \frac{1}{L^2} \sum_{x=0}^{L} \sum_{y=0}^{L} \sqrt[2]{\left(h_{(x,y)} - \bar{h}\right)^2}. \quad (8)$$

The term $h_{(x,y)}$ represents the highest position occupied by an atom, and $\bar{h}$ is the average thickness of the entire film. The equation for roughness estimation resembles the R_{sq} coefficient, obtained experimentally from atomic force microscopy (AFM), scanning tunneling microscopy (STM), or X-ray reflectivity measurements, which allows a direct comparison between simulation and experimental results reported in the literature (Section 2.2).

RHEED patterns are very sensitive to the surface roughness and morphology of the outer layers. The growth rate can be estimated *in situ* during MBE growth through the RHEED intensity oscillations, which is dynamically calculated by:

$$I(t) = \left(\sum_n (-1)^n . S_n(t)\right)^2. \tag{9}$$

where the term $S_n(t)$ is the exposed cover of the nth layer at time t. This equation is slightly different from the equations used in the SOS models, since it calculates the reflected intensity using only surface atoms exposed to the beam [23].

4. Simulation results

4.1. Silicon (0 0 1)

For the simulation of a (0 0 1) Si homoepitaxy, the energy value for diffusion to out-of-plane bond is $E_{D0} = 1$ eV and the in-plane energy is $E_{DL} = 0.5$ eV. The energy barrier for the evaporation process was determined by fitting the growth rate curves as a function of substrate temperature with the calculated growth rates. The curves were calculated using the vapor pressure of silicon as a function of temperature and the Knudsen equation [8]. The out-of-plane energy barrier found for evaporation is $E_{E0} = 3.8$ eV and the in-plane energy is $E_{EL} = 0.3$ eV. The out-of-plane barrier energy determines the inflection point of the growth rate curve, and the in-plane energy determines directly the derivative of the curve, which is the rate of change as a function of substrate temperature.

During growth simulation, the exposed surface, surface roughness, RHEED intensity, structural defects density, growth rate, and the partial structure, containing all atoms coordinates, were continuously recorded as a function of time. The substrate temperature was investigated in a range varying from room temperature up to the silicon melting point. The external parameters of the simulation are substrate temperature, deposition rate, lattice size, and deposition time.

Figure 6 presents surface roughness in monolayers (ML) for four different substrate temperatures (T_{SUB}). The deposition rate was kept constant at 0.01 ML/s during 1000 s of deposition, which results in films thicknesses with approximately 10 ML. At room temperature (300 K), the atoms on surface do not have sufficient energy to migrate leading to a limited surface diffusion, which favors the formation of structural defects, like vacancies and overhangs. As the substrate temperature increases to 600 K, the diffusion hopping rate raises and atoms situated on unstable positions, with a few bonds, may migrate to more stable positions on the lattice. This process decreases the surface roughness, as observed in **Figure 6**, since the innermost layers tend to be fulfilled. The growing surface at this condition is dominated by coalescence of small clusters. For substrate temperature of 800 K, a layer-by-layer growth mode is reached.

http://dx.doi.org/10.5772/intechopen.70220

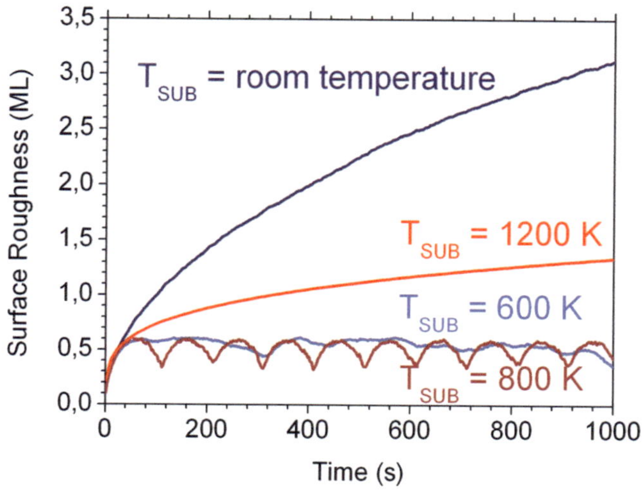

Figure 6. Surface roughness profiles for films grown on different substrate temperatures (T_{SUB}). Due to limited surface diffusion for films grown at room temperature, high roughness is obtained. Raising the substrate temperature, a high surface mobility is achieved, leading to a step-by-step growth mode at 800 K. For even higher substrate temperatures, atoms are allowed to diffuse with elevated rate and the reevaporation becomes significant, enhancing surface roughness.

This condition is evidenced by very clear oscillations on the surface roughness profile, with a period of approximately 100 s. The oscillation ends in a minimum, indicating that the 10th layer is practically completed. Increasing the substrate temperature to 1200 K, the surface roughness is raised to a value around 1 ML, which is a consequence of film thickness inhomogeneity, due to high surface diffusion rates.

The results of RHEED intensity oscillations dynamically calculated for a set of growth conditions are displayed in **Figure 7**. At room temperature, the calculated intensity does not exhibit any oscillations, which indicates that the intensity of the interference pattern is suppressed by the rough surface. The growth at T_{SUB} between 600 and 800 K exhibits well-defined intensity oscillations with a period around 100 s, indicating a growth rate of 1 ML/s. At 600 K, the oscillations are less intense and less symmetric than at 800 K, indicating the growth of more than one layer at the same time. For substrate temperatures around 800 K a layer-by-layer growth mode is achieved, evidenced by the symmetric RHEED intensity oscillation and confirmed by the calculated surface roughness (**Figure 6**). At 1200 K, the intensity oscillations disappear again. The layer-by-layer mode is lost due to higher surface atoms mobility, which leads to rougher surfaces. These results agree very well to experimental work of (0 0 1) Si homoepitaxy [24]. In this report, the RHEED intensity oscillations are weak and vanish after 2 min of deposition at room temperature and at 1270 K, whereas they are intense, well defined, and long-standing for substrate temperatures between 600 K and 900 K. Surface coverage of the first 10 layers was calculated and is presented in **Figure 8(a)** as a function of substrate temperature.

At lower temperatures, more than one atomic layer is filled at the same time, due to low surface diffusion. This is evidenced in the lower panel of **Figure 8**, where the 10th layer starts to be filled around 400 s, whereas the other 9 layers are still been filled. At T_{SUB} = 600 K, the coverage

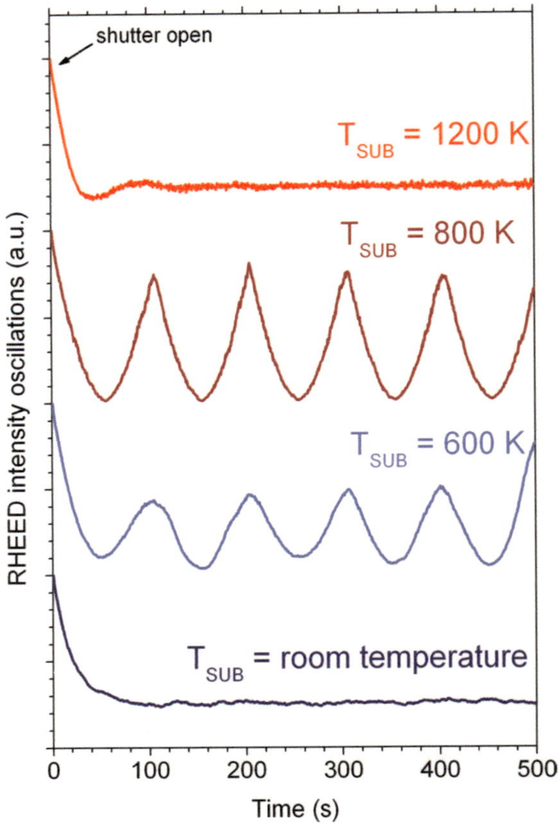

Figure 7. Normalized RHEED intensity oscillations dynamically calculated for several growth substrate temperatures (T_{SUB}). The intensity oscillations are suppressed at room temperature and at 1200 K, indicating rough surfaces at these conditions. For intermediated substrate temperatures, from 600 to 800 K, clearly defined oscillations are observed.

chart shows that simultaneous growth occurs for a maximum of two layers, whereas, at 800 K, the chart evidences the growth of only one layer per time period. Raising T_{SUB} to 1200 K, the high diffusion hopping rate recovers simultaneous layer growth and generates noisy lines on the surface coverage profile. Besides that reevaporation process becomes significant, contributing to an increase of spiked profile due to an abrupt change on the atomic layers coverage.

The exposed surface is presented in **Figure 8(b)** for growth temperatures between 300 K and 800 K.

From room temperature to T_{SUB} = 500 K, the images present a rough surface with cluster of Si atoms close to each other. At 700 K, the surface becomes much more flat, exhibiting large terraces with monolayer islands and a few void lakes. This growth condition is mostly controlled by islands coalescence and is on transition to the layer-by-layer growth regime. Raising the substrate temperature to 800 K, a plain layer-by-layer growth mode is observed, where the lowermost atomic layers are completely fulfilled by a step-flow mechanism before a new layer start to be formed. These results are in agreement with experimental STM images captured in an ultra-high vacuum MBE system right after the (0 0 1) Si homoepitaxy growth [25].

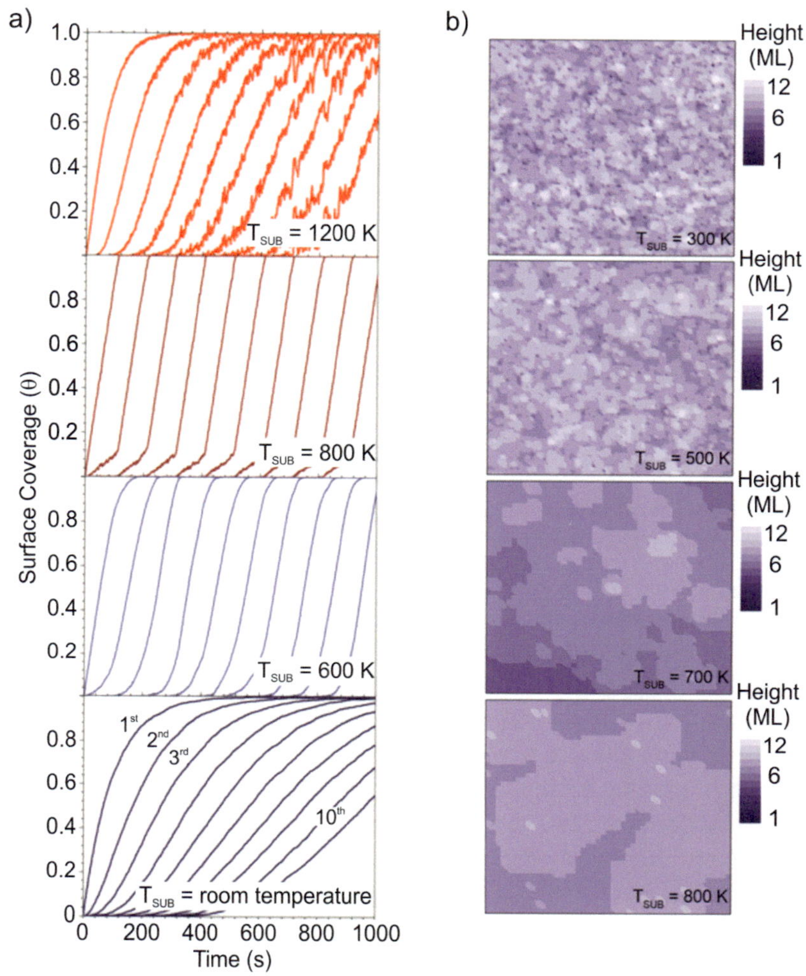

Figure 8. (a) Surface coverage as a function of deposition time of the first ten atomic layers for substrate temperatures from 300 K to 1200 K. (b) Exposed surface for a L = 100 lattice positions and four substrate temperatures. Increasing substrate temperature enhances atoms mobility, lowering surface roughness by widening terraces on the surface (T_{SUB} from 300 to 700 K). At T_{SUB} = 800 K, a transition from cluster to layer-by-layer growth is achieved.

The dependence of the defects density formed during film growth as a function of the substrate temperature is presented in **Figure 9** for different deposition rates. This graph shows that, for each deposition rate, there is a threshold temperature where the defect density starts to decay exponentially. This temperature is higher for higher deposition rates. As the deposition rate increases, higher diffusion rates, i.e., higher temperatures are needed to accommodate the impinging atoms without the formation of holes. The local minima observed in the curves of **Figure 9** correspond to the variation of the diffusion rate due to the average number of bonds of the surface atoms.

At lower substrate temperatures regimes, the hopping rate is decreased, inhibiting the atoms mobility. These growth conditions are favorable to structural defects, which leads to rough

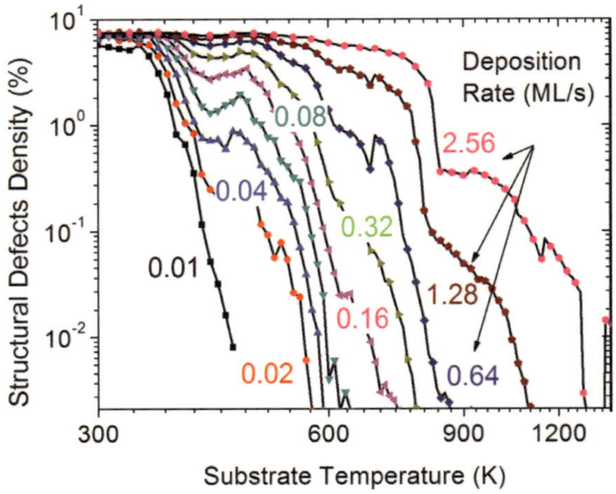

Figure 9. Density of holes formed in the simulated films as a function of substrate temperature for deposition rate varying from 0.01 ML/s up to 2.56 ML/s.

surfaces. For higher temperatures, in an energetically unstable condition with many dangling bonds, the hopping rate for diffusion raises. This condition favors atoms located in unstable positions to search other points in the lattice with lower energy, decreasing the number of structural defects and the surface roughness. With higher substrate temperatures, the atoms located in the surface are allowed to migrate in a short period of time. This condition allows atoms to migrate to a more stable position, like on kinks, giving rise to a step-flow growth mode with flatter surfaces and less structural defects. For even higher substrate temperatures, the hopping rate for diffusion is very high, even for positions with many bonds. In this situation, several diffusion movements are allowed, producing inhomogeneous thicknesses with rougher surfaces. The hopping rate for reevaporation begins to be significant, lowering the growth rate. If substrate temperature continues to increase, the reevaporation rate becomes more and more significant, until growth rate vanishes.

4.2. Applications

This model can be applied to study low and high kinetic growth regimes, since this numerical model of MBE growth considers bulk defects and desorption process and the growth of multiple elements, such as binary and ternary alloys. Besides, it is possible to study thermal treatment effects on crystalline structures, sputtering processes of surfaces, and even the growth on substrates with cleavage steps.

One topic example, which has attracted a lot of attention from experimental scientists, is topological insulator materials. The most common archetypes of this new electronic state of matter are the bismuth chalcogenides compounds, such as Bi_2Te_3 and Bi_2Se_3. This class of matter has a nontrivial topological order, resulting in a metallic behavior attributed to the surface states but insulating behavior in the bulk. However, the presence of spontaneous structural defects in these compounds leads to electric conduction in the bulk, which overlaps with surface states. Of course, in order to achieve practical applications, it is fundamental to have insulating samples in the bulk. The model presented here can be applied to understand the growth dynamics of these compounds, elucidating the mechanisms of structural defect formation.

5. Conclusion

This chapter presented firstly a brief description of the molecular beam epitaxy and the main growth modes achieved using this technique as well as of the most common experimental tools to probe thin film surfaces. The second part described a numerical model based on Monte Carlo method to simulate epitaxial growth.

The model proposed here is applied to simulate the molecular beam epitaxial growth of Si on (0 0 1) Si substrate. The results obtained fit well to the experimental reports on (0 0 1) silicon homoepitaxy. At low substrate temperatures, films exhibit a high density of structural defects and a very rough surface. On the other hand, for high temperatures, large surface terraces are achieved with a low density of bulk defects. At even higher substrate temperatures, the surface roughness increases due to the high mobility of the atoms at the surface, which could lead to non-homogeneity of the film thickness. By further increasing the substrate temperature, the atoms have sufficient energy to desorb, resulting in a reduction of the growth rate.

The large number of microscopic processes and the three-dimensional spatial information during the film growth allow the model to simulate the growth of binary, ternary, or more compounds and even the growth of superlattices and heterostructures. In addition, the model allows an easy implementation to study thermal treatments in crystalline structures and surface sputtering processes.

Author details

Celso I. Fornari[1]*, Gabriel Fornari[2], Paulo H. de O. Rappl[1], Eduardo Abramof[1] and Jerônimo dos S. Travelho[2]

*Address all correspondence to: celso@las.inpe.br

1 Laboratório Associado de Sensores e Materiais—LAS, National Institute for Space Research, São José dos Campos, SP, Brazil

2 Laboratório Associado de Computação e Matemática Aplicada—LAC, National Institute for Space Research, São José dos Campos, SP, Brazil

References

[1] Madhukar A, Ghaisas SV. The nature of molecular beam epitaxial growth examined via computer simulations. Critical Reviews in Solid State and Materials Sciences. 1988;**14**:1-130. DOI: 10.1080/01611598808241266

[2] Family F, Lam PM. Renormalization-group analysis and simulational studies of groove instabilitiy in surface growth. Physica A: Statistical, Mechanics and Its Applications. 1994;**205**:272-283. DOI: 10.1016/0378-4371(94)90504-5

[3] Levi AC, Kotrla M. Theory and simulation of crystal growth. Journal of Physics: Condensed Matter. 1997;**9**:299-344. DOI: 10.1088/0953-8984/9/2/001

[4] Das Sarma S, Tamborenea P. A new universality class for kinetic growth: One-dimensional molecular-beam epitaxy. Physical Review Letters. 1991;**66**:325-328. DOI: 10.1103/PhysRevLett.66.325

[5] Das Sarma S, Lanczycki C, Kotlyar R, Ghaisas S. Scale invariance and dynamical correlations in growth models of molecular beam epitaxy. Physical Review E. 1996;**53**:359-388. DOI: 10.1103/PhysRevE.53.359

[6] Landau DP, Pal S. Monte Carlo simulation of simple models for thin film growth by MBE. Thin Solid Films. 1996;**272**:184-194. DOI: 10.1016/0040-6090(95)06945-3

[7] Bauer G, Springholz G. Molecular beam epitaxy—Aspects and applications. Vacuum. 1992;**43**:357-365. DOI: 10.1016/0042-207X(92)90038-X

[8] Herman MA, Sitter H. Molecular Beam Epitaxy: Fundamentas and Current Status. 2nd ed. Berlin: Springer; 1988. p. 382. DOI: 10.1007/978-3-642-97098-6

[9] Ando Y. Topological insulator materials. Journal of the Physical Society of Japan. 2013;**82**:102001-1-102001-32. DOI: 10.7566/JPSJ.82.102001

[10] Schelling C, Springholz G, Schäffler F. New kinetic growth instabilities in Si(001) homoepitaxy. Thin Solid Films. 2000;**369**:1-4. DOI: 10.1016/S0040-6090(00)00823-3

[11] Eaglesham DJ, Cerullo M. Low-temperature growth of Ge on Si(100). Applied Physics Letters. 1991;**58**:2276-2278. DOI: 10.1063/1.104898

[12] Fornari CI, Rappl PHO, Morelhão SL, Abramof E. Structural properties of Bi_2Te_3 topological insulator thin films grown by molecular beam epitaxy on (111) BaF_2 substrates. Journal of Applied Physics. 2016;**119**:156303-1-156303-9. DOI: 10.1063/1.4947266

[13] Ratsch C, Venables JA. Nucleation theory and the early stages of thin film growth. Journal of Vacuum Science & Technology A: Vacuum, Surfaces, and Films. 2003;**21**:s96-s109. DOI: 10.1116/1.1600454

[14] Kariotis R, Lagally MG. Rate equation modelling of epitaxial growth. Surface Science. 1989;**216**:557-578. DOI: 10.1016/0039-6028(89)90395-6

[15] Aumann CE, Kariotis R, Lagally MG. Rate equation modeling of interface width. Journal of Vacuum Science & Technology A: Vacuum, Surfaces, and Films. 1989;**7**:2180-2185. DOI: 10.1116/1.575953

[16] Liang YY, Yoon SF, Fitzgerald EA. Kinetic Monte Carlo simulation of quantum dot growth on stepped substrates. Journal of Physics D: Applied Physics. 2013;**46**:495102. DOI: 10.1088/0022-3727/46/49/495102

[17] Günther V, Mauß F, Klauer C, Schlawitschek C. Kinetic Monte Carlo simulation of the epitaxial growth of Si(1 0 0). Physica Status Solidi (c). 2012;**9**:1955-1962. DOI: 10.1002/pssc.201200340

[18] Günther V, Mauß F. Si(100)2×1 epitaxy: A kinetic Monte Carlo simulation of the surface growth. Physics Procedia. 2013;**40**:56-64. DOI: 10.1016/j.phpro.2012.12.008

[19] Stumpf R, Scheffler M. Ab initio calculations of energies and self-diffusion on flat and stepped surfaces of Al and their implications on crystal growth. Physical Review B, Condensed Matter. 1996;**53**:4958-4973. DOI: 10.1103/PhysRevB.53.4958

[20] Weeks JD, Gilmer GH, Jackson KA. Analytical theory of crystal growth. The Journal of Chemical Physics. 1976;**65**:712-720. DOI: 10.1063/1.433086

[21] Marmorkos IK, Das Sarma S. Kinetic simulation of molecular beam epitaxial growth dynamics. Surface Science. 1990;**237**:L411-L416. DOI: 10.1016/0039-6028(90)90511-6

[22] Matsumoto M, Nishimura T. Mersenne twister: A 623-dimensionally equidistributed uniform pseudo-random number generator. ACM Transactionson Modeling and Computer Simulation. 1998;**8**:3-30. DOI: 10.1145/272991.272995

[23] Marmorkos IK, Das Sarma S. Atomistic numerical study of molecular-beam-epitaxial growth kinetics. Physical Review B. 1992;**45**:11262-11272. DOI: 10.1103/PhysRevB.45.11262

[24] Sakamoto T, Kawai NJ, Nakagawa T, Ohta K, Kojima T, Hashiguchi G. Rheed intensity oscillations during silicon MBE growth. Surface Science. 1986;**174**:651-657. DOI:10.1016/0039-6028(86)90487-5

[25] Jernigan GG, Thompson PE. Temperature dependence of atomic scale morphology in Si homoepitaxy between 350 and 800°C on Si (100) by molecular beam epitaxy. Journal of Vacuum Science & Technology A: Vacuum, Surfaces, and Films. 2001;**19**:2307-2311. DOI: 10.1116/1.1384559

Chapter 6

Silicon Epitaxial Reactor for Minimal Fab

Ning Li, Hitoshi Habuka, Yuuki Ishida,
Shin-ichi Ikeda and Shiro Hara

Additional information is available at the end of the chapter

http://dx.doi.org/10.5772/intechopen.69986

Abstract

Cost-effective and mass production of size-controlled wafers becomes one of the future trends for electronic devices. Herein, we design a Minimal Fab system for the growth of half-inch-diameter silicon wafer devices. Different from the conventional chemical vapour deposition (CVD) systems, a new-type of CVD reactor was designed and developed for the Minimal Fab. The minimal CVD reactor has a small reaction chamber for rapid growth processes. It employed (i) a vertical gas flow, (ii) heating modules using concentrated infrared light, (iii) chlorine trifluoride gas for quick reactor cleaning and (iv) optimized epitaxial growth conditions so that the reactor cleaning is not necessary. Reducing the total gas flow rate is an effective way to increase the wafer temperature. The heating process was further assisted by the absorption of infrared light by the precursor trichlorosilane. The slimly designed reflector could help in improving the heating speed.

Keywords: chemical vapour deposition, infrared light heating, reflectors, cleaning process

1. Introduction

The electronic device fabrication follows the two major trends, that is, the larger silicon wafer diameter and smaller design rule [1], mainly for economic reasons. A huge number of the device chips are produced in this manufacturing system. These trends require a huge investment for developing and preparing the plant.

However, we have the other technical trend [2] that the highly integrated device chips are customized and applied to various fields including the information technologies. For this purpose, a small amount of various chips are flexibly produced. Here, the Minimal Fab [3–6] is expected to flexibly produce just the right number of electronic device chips, from one to million, on-demand and on-time, consuming less material and power. For this concept, a small silicon wafer having 12.5 mm diameter allows the great flexibility. It enables the very quick processes of the lithography, thin film formation, annealing and others. The instruments have a very small footprint. The chemical vapour deposition (CVD) process and its reactor should be developed as the key technology.

Ordinary CVD reactors used for epitaxial growth of large-diameter wafers consume a large amount of gases and heating power. For fast growth of small-scale devices, such as 12.5-mm wafer, Minimal Fab is designed better because a slow gas flow rate and slim heating modules are used. In addition, the environment inside the growth chamber is easily maintained and therefore the regular chamber cleaning for ordinary CVD reactors becomes less necessary for Minimal Fab. However, some parameters that are ignored in ordinary CVD systems may become effective in Minimal Fab. Special care should be paid to finely tune those important parameters.

To fabricate reactors for small substrates, the thermal condition becomes different when compared to that for the large substrates [7]. The concentrated infrared flux effectively heats the small substrate; the thermal process becomes very rapid. In addition to the heating system, the highly reactive gas of chlorine trifluoride (ClF_3) [8] is chosen for the reactor cleaning because it can easily remove the silicon film, unnecessarily formed in the CVD reactor, at various temperatures even at room temperature.

In Section 2 [9], a small footprint CVD reactor for producing silicon thin films is explained. This employs the technical issues of (i) vertical gas flow, (ii) a concentrated infrared flux and (iii) *in situ* reactor cleaning using chlorine trifluoride gas. Steps (i) and (ii) achieve a less heating energy and a rapid cooling. The cleaning process is rapid by (iii). The heating step is not necessary because the chlorine trifluoride gas is reactive even at room temperature. Section 3 explains the practical thermal condition [10, 11]. For achieving the rapid process, the infrared light absorption by the precursor gas is useful. Simultaneously, the cleaning-free process becomes possible. In Section 4, the heat transport near the wafer is evaluated [12]. The thin plates are recognized to be the suitable reflector material. For the quick temperature up and down, the reactor parts set near the wafer should be small, slim and thin. The wafer rotation and the highly heat-conductive susceptor help in achieving the symmetrical and uniform profile of silicon epitaxial film thickness.

2. Silicon chemical vapour deposition process for minimal fab

In this section, the CVD reactor was designed and developed, taking into account the thermal process using the reflector, which concentrates infrared flux. The reactor cleaning process using the highly reactive chlorine trifluoride gas is also employed.

2.1. Chemical reaction for silicon epitaxy

The chemical reactions at silicon substrate surfaces [13] are briefly shown as follows:

$$SiHCl_3 \rightarrow {}^*SiCl_2 + HCl\uparrow \quad ({}^*\text{: Chemisorbed at surface}) \tag{1}$$

$${}^*SiCl_2 + H_2 \rightarrow {}^*Si + 2HCl\uparrow \tag{2}$$

Similar to the ordinary silicon epitaxial growth, trichlorosilane ($SiHCl_3$) is used in a hydrogen ambient. Firstly, trichlorosilane is chemisorbed at the surface to form the intermediate species, $^*SiCl_2$. Next, it reacts with hydrogen to form silicon. Thus, the overall chemical reaction is written as follows:

$$SiHCl_3 + H_2 \rightarrow Si + 3HCl\uparrow \tag{3}$$

Assuming the Eley-Rideal model, the growth rate is expressed as follows:

$$\text{Growth rate} = \frac{k_{\text{Ad}} k_r [\text{SiHCl}_3][\text{H}_2]}{k_{\text{Ad}}[\text{SiHCl}_3] + k_r[\text{H}_2]}, \tag{4}$$

Where k_{Ad} and k_{r} are the rate constant for Eqs. (1) and (2), respectively. When k_{Ad} [$SiHCl_3$] is significantly larger than k_{r} [H_2], the epitaxial growth rate becomes as follows:

$$\text{Growth rate} = k_r[\text{H}_2] \tag{5}$$

Because the hydrogen concentration in the reactor is nearly constant, the epitaxial growth rate simply depends on the wafer temperature.

Generally, the epitaxial growth rate at 1100°C is about 1–4 μm/min by the ordinary horizontal cold wall reactor and is about 8 μm/min by the high-speed rotation vertical reactor.

In order to avoid the formation of surface defects, such as light point defects, the gas phase chemical reaction is suppressed by maintaining the gas phase temperature low, that is, by employing the cold wall environment.

2.2. Thermal condition

Figure 1 shows the heat transport for the silicon wafer in the minimal CVD reactor. The half-inch silicon wafer is heated by the infrared flux, Q_{IR}, emitted from the halogen lamps. The heat is emitted from the half-inch silicon wafer, Q_{Em}, as radiation heat. The gas flow containing the precursor gas from the gas nozzle takes the heat away from the wafer, Q_{Flow}.

By concentrating the infrared flux, Q_{IR} effectively reaches and locally heats the substrate with minimizing the heat loss. The distance between the silicon wafer and the gas nozzle can be changed by finely adjusting Q_{Flow}. In order to effectively heat the half-inch wafer, the half-inch wafer is placed below the reflector, as shown in **Figure 2(a)**. The half-inch silicon wafer is heated using the infrared flux coming from the upper outside.

By employing this geometry, Q_{Em} is high, because the emission of radiation heat is not disturbed. Although the Q_{Em} value decreases the wafer temperature, the increasing rate of the wafer temperature is high because of the high Q_{IR} value. Additionally, the cooling rate of the wafer becomes high.

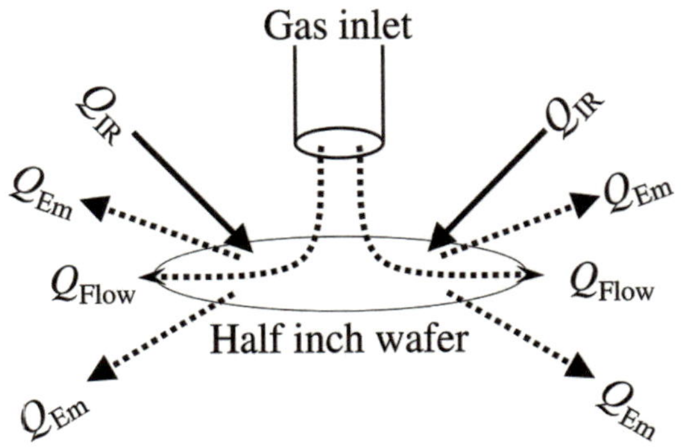

Figure 1. Heat transport for the silicon wafer in the minimal CVD reactor.

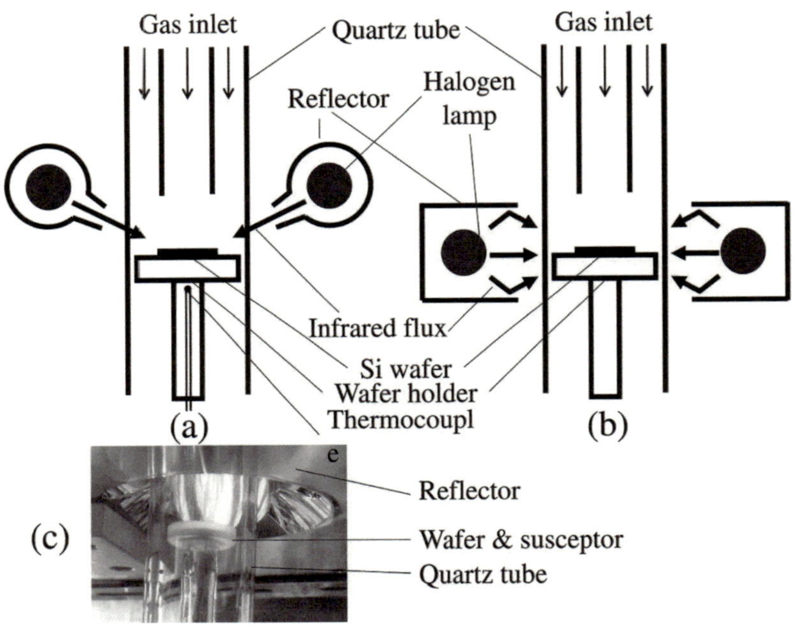

Figure 2. The minimal CVD reactor, (a) using the reflector concentrating the infrared flux to the silicon wafer, (b) using the ordinary type reflector, in which the silicon wafer is enclosed and (c) photograph of the minimal reactor.

2.3. Cleaning process

The reactor cleaning is necessary for removing the film produced around the substrate, such as at the wafer holder and the quartz tube. The typical silicon CVD process [14] utilizes hydrogen chloride gas near 1200°C. Such high-temperature process requires a long period for increasing and decreasing the temperature. By contrast, the chlorine trifluoride gas is useful, because the following chemical reaction occurs at any temperatures, even at room temperature [8],

$$3Si + 4ClF_3 \rightarrow 3SiF_4\uparrow + 2Cl_2\uparrow \tag{6}$$

2.4. Reactor

Figure 2(a) shows the reactor, which consists of a quartz tube (inner diameter of 24 mm), a quartz wafer holder, gas inlets, three halogen lamps, and a reflector. A half-inch silicon wafer is set on the quartz wafer holder. The inner zone and the outer zone gas inlets have diameters of 7 and 24 mm, respectively. A half of the total gas was introduced to each of the inlets. The distance between the silicon wafer and the bottom of the inner zone gas inlet was 3 cm, in this section. **Figure 2(c)** shows a photograph which depicts the half-inch silicon wafer heated by the infrared flux.

Figure 2(b) shows the CVD reactor having an ordinary type reflector. The half-inch silicon wafer is enclosed in the reflector. The infrared flux approaches *via* the various paths experiencing multiple reflections by the reflector. Comparing this with the minimal CVD reactor, the half-inch wafer cooling rate is shown to be influenced by the reflector geometry. This will be explained in detail in Section 2.6.

The wafer temperature is measured in ambient nitrogen by an R-type thermocouple, directly attached to the backside of the half-inch silicon wafer. During the silicon film deposition in ambient hydrogen, the thermocouple is placed in the quartz wafer holder, as shown in **Figure 2(a)**.

2.5. Process

For the silicon film formation, the carrier gas and precursor gas are vertically introduced to the silicon wafer. The carrier gas is hydrogen (H_2) and the precursor gas is trichlorosilane. The cleaning gas is chlorine trifluoride diluted to 5% in ambient nitrogen at atmospheric pressure. The total gas flow rate is 0.2–1.2 slm. The half-inch silicon wafer is heated to 800–1100°C.

The silicon CVD process is shown in **Figure 3**. After the silicon wafer is heated, the trichlorosilane gas is introduced for the silicon deposition. After terminating the trichlorosilane gas supply, the wafer is cooled down and unloaded from the reactor. The chlorine trifluoride gas is introduced into the reactor at atmospheric pressure and at room temperature for

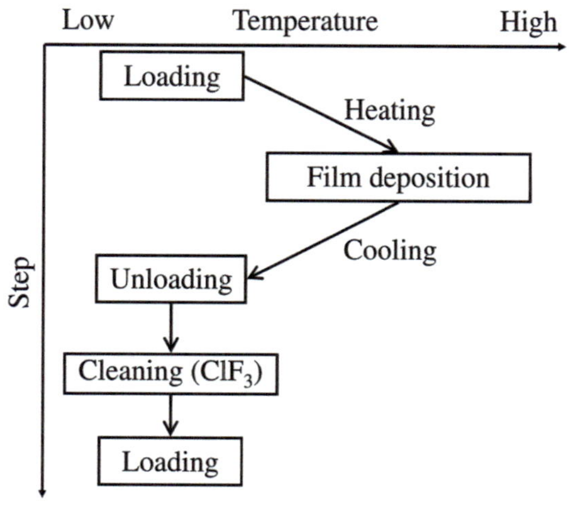

Figure 3. Silicon chemical vapour deposition process for the Minimal Fab.

removing the silicon film formed on various places in the reactor. The flow rate of nitrogen and chlorine trifluoride gas is 1 and 0.05 slm, respectively.

2.6. Cooling rate

The capability of the reflector is first evaluated by the cooling rate of the half-inch silicon wafer. **Figure 4** shows the cooling rate of the minimal CVD reactor at the nitrogen flow rate of 1 slm, using the reflectors of minimal and ordinary type, shown in **Figure 2(a)** and **(b)**, respectively. **Figure 4** shows the decreasing temperatures of the half-inch wafer. The cooling rate of the minimal CVD reactor is −22 K/s, which is 20% greater than that using the ordinary-type reflector.

2.7. Heat balance

The electric power for heating the silicon wafer is evaluated with adjusting the gas flow rate and the trichlorosilane gas concentration. The half-inch wafer temperature is obtained following the relationship between the temperatures on the backside of the half-inch wafer and the temperature below the quartz wafer holder measured by thermocouple. Additionally, based on the silicon film growth rate saturation [13] at temperatures lower than 1000°C and at the high trichlorosilane gas concentrations, the half-inch wafer temperature is obtained.

The half-inch wafer temperature in a steady state is influenced by various conditions, such as the lamp voltage, V (V), the total gas flow rate, F_{TotalGas} (sccm) and the trichlorosilane gas concentration, C_{TCS} (%). Using a least-squares approximation, the half-inch wafer temperature, T_{Wafer} (°C), is obtained, assuming that each parameter linearly influences the T_{Wafer} value.

$$T_{\mathrm{wafer}}(^{\circ}\mathrm{C}) = 90\sqrt{V(V)} - 0.13F_{\mathrm{TotalGas}}(\mathrm{sccm}) - 4.3C_{\mathrm{TCS}}(\%) + 310(^{\circ}\mathrm{C}) \quad (7)$$

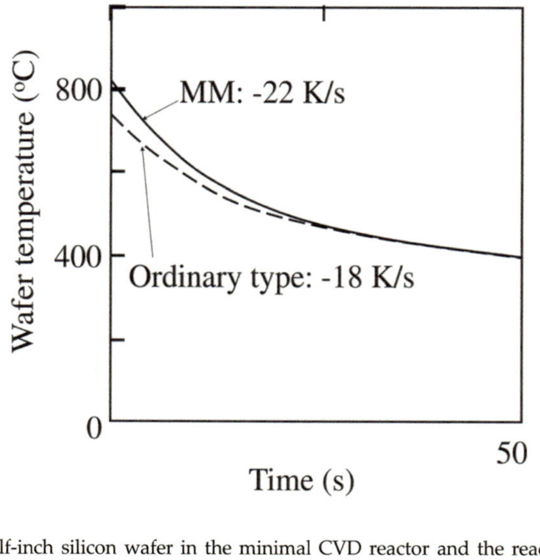

Figure 4. Cooling rate of half-inch silicon wafer in the minimal CVD reactor and the reactor using the ordinary type reflector. (In ambient nitrogen: 1 slm and atmospheric pressure.).

http://dx.doi.org/10.5772/intechopen.69986

The T_{Wafer} value increases by the increasing V value and decreases by the increasing $F_{TotalGas}$ value and C_{TCS} value. Particularly, the increase in $F_{TotalGas}$ of 10 sccm and that of C_{TCS} of 0.25% induces a 1°C decrease in T_{Wafer}.

Following Eq. (7), the half-inch wafer temperature is estimated as a function of the lamp voltage and the total gas flow rate at the trichlorosilane concentration fixed to be 3.5%, as shown in **Figure 5**. Eq (7) shows that the lamp voltage becomes 10–20 V lower by decreasing the total gas flow rate.

2.8. Reactor cleaning

As shown in **Figure 6(a)**, a polysilicon film is formed on the wafer holder and on the inner wall of the quartz tube during the film deposition. The thick silicon film is formed at a height near that of the half-inch silicon wafer, because the reactor inner wall near the substrate is high.

Because the film on the quartz wall might produce particles to cause surface defects, it must be removed. Thus, chlorine trifluoride gas [2] at 5% is introduced into the reactor at room temperature in the ambient nitrogen. **Figure 6(b)** shows a reactor before introducing the chlorine trifluoride gas. **Figure 6(c)** shows that the silicon film on the right half of the reactor wall was removed after 1 min. The right half of the wafer holder is clearly observed. Although the silicon film on the left half of the reactor wall reduces, a thick film still remains.

As shown in **Figure 6(d)**, only a small amount of the polysilicon film remains after 2 min on the left position of the reactor wall. After 3 min, the polysilicon film is perfectly removed, as shown in **Figure 6(e)** which clearly gives an image of the entire wafer holder. The polysilicon film formed on the inner wall of the quartz tube is quickly and perfectly removed at room temperature.

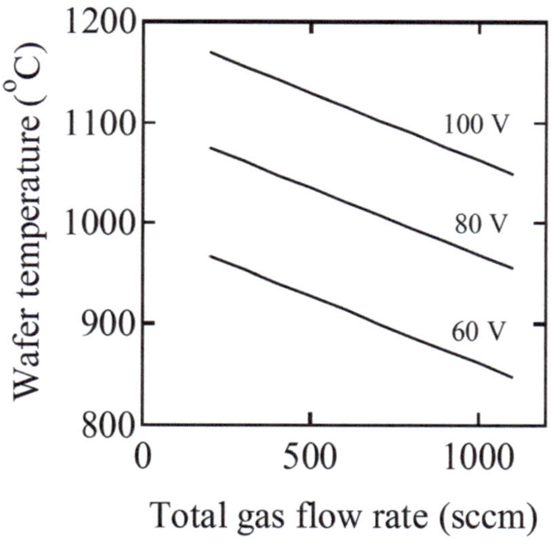

Figure 5. Wafer temperature changing with the lamp voltage and the total gas flow rate.

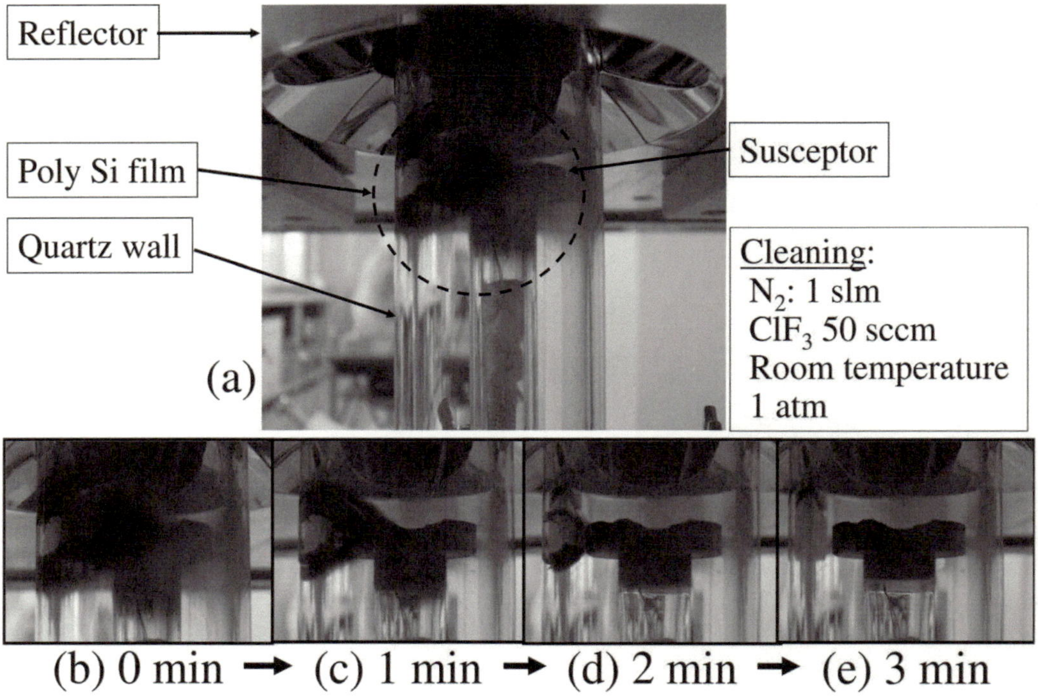

Figure 6. Reactor appearance along the reactor cleaning at room temperature. (a) After the film deposition; (b–e): the decrease of poly Si film.

3. Thermal condition

The wafer temperature is influenced by many parameters [9] of lamp voltage, total gas flow rate and trichlorosilane gas concentration. The details of these thermal influences are evaluated. Particularly, the light absorption and the heat transport by the gases are the major issues. **Table 1** shows the experimental conditions.

3.1. Temperature change caused by trichlorosilane

Figure 7 shows the quartz wafer holder temperature. The hydrogen and trichlorosilane flow rate are 220 and 12 sccm, respectively. The wafer holder temperature slowly becomes 568°C in the ambient consisting of only hydrogen, as shown using dark squares. When the trichlorosilane gas is added from 9 to 13 min, the wafer holder temperature increases, as shown using white squares, and reaches 609°C. Empirically, the wafer surface temperature increase by the trichlorosilane gas is *c.a.* 80 K.

Trichlorosilane has the infrared light absorption [15] at 3.3 and 2.2 μm. The halogen lamp emits the light near 1 μm, the wavelength of which widely distributes [16] to that longer than 2 μm. Thus, the trichlorosilane considerably absorbs the infrared light from the halogen lamps; it increases the temperature of the gas phase.

http://dx.doi.org/10.5772/intechopen.69986

Parameters	Value
Pressure	Atmospheric pressure
Hydrogen gas flow rate	100–1000 sccm
Trichlorosilane gas flow rate	2–60 sccm
Chlorine trifluoride gas flow rate	50 sccm
Substrate	12.5 mm diameter silicon wafer
Electric power	55–100 V
Deposition time	1 min × *n* times or *n* minutes once
Substrate temperature	800–1000°C

Table 1. Silicon CVD conditions.

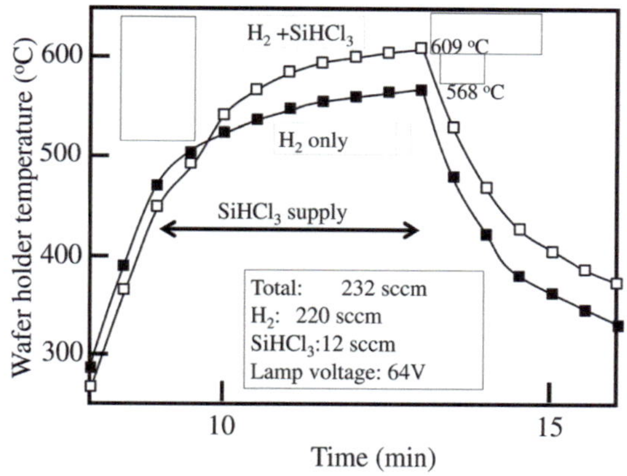

Figure 7. Temperature shift cause by $SiHCl_3$ supply for 4 min (at 64 V).

3.2. Influence of total gas flow rate

Figure 8 shows the wafer holder temperature during the silicon deposition. Dark squares are the temperatures at the hydrogen and trichlorosilane flow rate of 220 and 12 sccm, respectively. The wafer holder temperature increases from 8 min; the temperature increase becomes faster after the addition of trichlorosilane at 10 min. The wafer holder temperature reaches 588°C. The temperatures at the hydrogen and trichlorosilane flow rate of 165 and 9 sccm, respectively, are shown by white squares. The wafer holder temperature reaches 626°C which is higher than that at the higher gas flow rate. The wafer holder temperature increases with the decreasing total gas flow rate.

Although **Figure 7** might show that the decrease in the trichlorosilane gas flow rate induces the temperature decrease by means of less infrared absorption, the wafer holder temperature is actually increased. It is due to the heat transport by the gas flow [9].

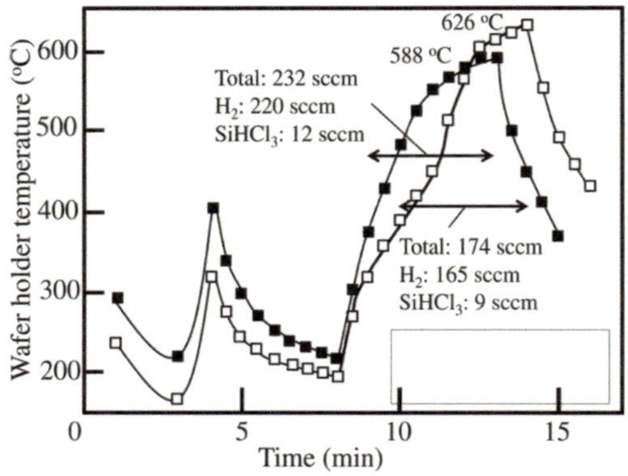

Figure 8. Temperature increase by the decreasing H_2 flow rate (at 65 V).

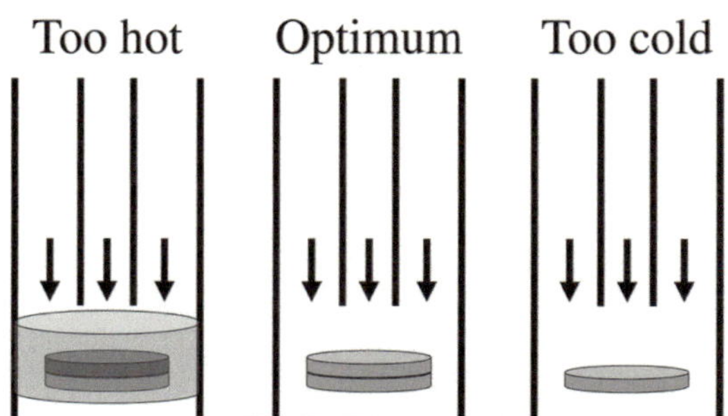

	Too hot	Optimum	Too cold
Grown film	Gray	Specular	None
Wall deposition	Brown	None	None
Cleaning	Yes	No	No

Figure 9. Film growth and wall deposition.

3.3. Temperature and obtained film surface

Figure 9 shows the relationship between the wafer temperature, the film surface and the quartz wall deposition. At a very high temperature, the silicon film deposition occurs at the silicon wafer surface and the quartz tube inner wall surface. The optimum temperatures produce the specular silicon epitaxial film without the tube wall deposition.

4. Reflector influence on rapid heating

In order to achieve the uniform-thick epitaxial film, the thermal conditions are important. The important issues are the infrared ray reflection design [7] and the heat transport through the reflectors set around the wafer. The CVD reactor for the large-diameter wafer has the large infrared lamp module consisting of large and thick reflector plates [9–11]. The reflector surface not only concentrates the infrared rays to the wafer surface but also absorbs the heat coming from the lamps. Then, the reflector becomes high temperature. Through the reflector, various parts and the gas phase, the heat absorbed by the reflector is slowly conducted to the wafer. The temperature of reflector surface and wafer gradually increases during reaching the thermal steady state. Particularly, the massive metallic parts require long time for reaching a thermal steady state. In this case, the heating process requires more electric power and finer heat distribution control. In order to design a quick heat transport through the reflector maintaining low electric power, a thin and slim reflector is expected. The reflector geometry should be optimized for the minimal CVD reactor.

Infrared rays tend to converge to generate a hot spot [7]. For broadening the locally formed heat profile, a silicon carbide susceptor having high heat conductivity is convenient. Additionally, the wafer rotation is popular and effective for averaging the temperature distribution and the film growth rate [17].

In this section for achieving a quick thermal process, the heating behaviour is evaluated using two types of reflectors, that is, the Type-I reflector made of thick mirror plates used in the previous sections and the Type-II reflector made of thin plates. Additionally, the roles of the wafer rotation and the silicon carbide susceptor are explained for preparing a uniformly thick silicon epitaxial film.

4.1. Reactor and reflector design

Figure 10 shows the detail of minimal CVD reactor. Similar to the previous sections, the reactor has the inner and the outer inlet. The distance between the wafer and the inner inlet is 51–56 mm. In this section for improving the heat conduction, the silicon carbide plates are inserted beneath the wafer.

Figure 11 shows the Type-I and Type-II reflectors covered with an electroplated gold film. Three reflectors are horizontally arranged around the quartz tube, as shown in **Figure 10**. **Figure 11(a)** shows the Type-I reflector, which is evaluated in the previous sections. The main body of the Type-I reflector is a 50-mm-thick mirror plate. Its thick body maintains the temperature stable during the long process, because it is not sensitive to fluctuations in the temperature around the reactor. However, it may result in slow heat conduction through it.

As shown in **Figure 12(a)** and **(b)**, through the Type-I reflector and the Type-II reflector, respectively, the heat from the halogen lamps is transported to the silicon wafer. The solid lines indicate the heat conduction from the lamp to the reflector plate. This heat conduction also heats the wafer, with the radiation heat indicated by the dotted lines. Because the heat conduction requires long period to reach long distance, the large and thick Type-I reflector slowly achieves the steady state and makes the thermal process slow and long.

The Type-II reflector is shown in **Figure 11(b)**. This reflector consists of a 5-mm-thick plate which is significantly thinner than that of the Type-I reflector. This thin plate makes the heat transport through itself quicker than that of the Type-I reflector. Particularly, the temperature difference between the inside and outside of the reflector plate is reduced. The thermal process by the Type-II reflector can be quicker than that of the Type-I reflector.

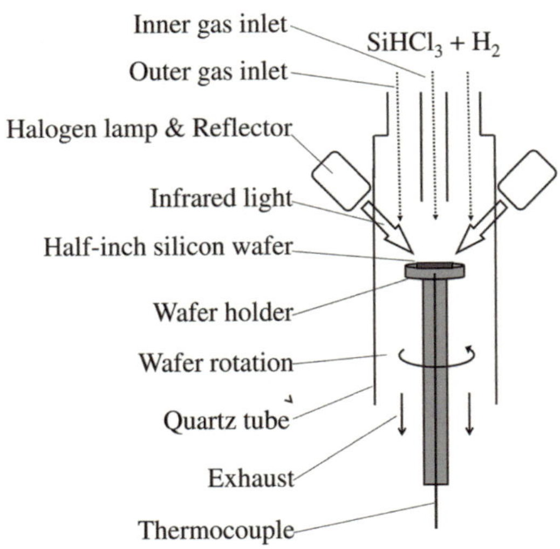

Figure 10. Half-inch silicon CVD reactor for the Minimal Fab.

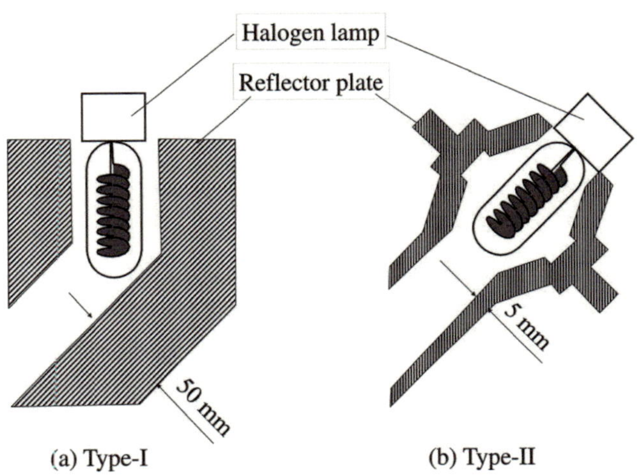

Figure 11. Two reflectors of (a) Type-I and (b) Type-II.

4.2. Process

The gas mixture of hydrogen (H_2) and trichlorosilane ($SiHCl_3$, TCS) is vertically introduced to the silicon wafer in ambient hydrogen at atmospheric pressure, as shown in **Figure 10**. The gas flow rates of the H_2 and TCS are 215 and 9 sccm, respectively. The electric power of 55–65 V is supplied to the halogen lamps. The total electric power is less than 1500 W.

Typically, the epitaxial film formation process has two major steps. Step A removes the native oxide film on the silicon wafer surface at 1100°C for 1 min. Next, the wafer temperature is

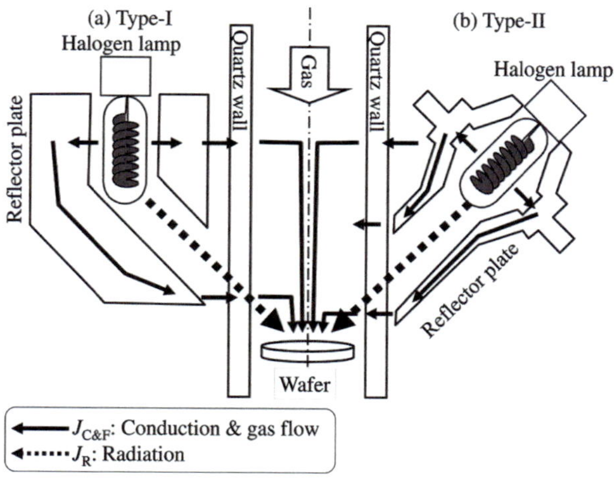

Figure 12. Heat transport from halogen lamp to silicon wafer through (a) Type-I and (b) Type-II reflectors. Dotted lines: heat transport by radiation; solid lines: heat transport by conduction and gas flow.

adjusted to 700–1000°C for Step B, forming the silicon epitaxial film for several minutes by the chemical reaction [13] following Eq. (3).

The process shown in **Figure 13(a)** has the stand-by step, Step C, between Step A and Step B. Step C, waiting for Step B after Step A, realizes the parallel process. The two virtual reactors, the A and B Reactors, are arranged for Steps A and B, respectively. Step C transports the wafer from the A Reactor to the B Reactor at low temperatures. Although the additional period is necessary for increasing and decreasing the temperature, Steps A and B can be simultaneously performed. The parallel process of Steps A and B is expected to become quicker. Thus, in the first part, Step C is intentionally performed using the Type-I and -II reflectors, in order to

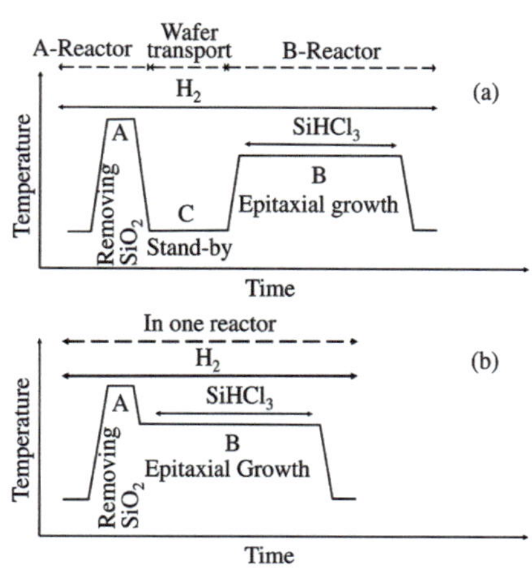

Figure 13. Process of silicon epitaxial growth by (a) Steps A–C and by (b) Steps A and B. Step A removes native oxide film, Step B forms silicon epitaxial film and Step C is for stand-by. Dotted lines show virtual reactor process.

compare their thermal behaviours through exactly the same process. In the second part, the effect of Step C was evaluated.

4.3. Wafer rotation and susceptor

For obtaining the uniform-thick epitaxial film by the classical ways [18], the wafer rotation and the silicon carbide susceptor are employed, as shown in **Figure 14**. Because even the slow wafer rotation has the effect of averaging the film growth rate along the concentric circle [17], the epitaxial film thickness is expected to become flat over the wafer. Silicon carbide has a high thermal conductivity [19] for decreasing the temperature difference over the wafer. The diameter and the thickness of the silicon carbide plate are 16 and 0.58 mm, respectively. Three silicon carbide plates are stacked beneath the silicon wafer.

4.4. Wafer temperature evaluation

The wafer holder temperature, T_{WH}, is measured and evaluated. The silicon wafer temperature, T_W, is obtained from the silicon epitaxial growth rate. At the wafer temperature lower than 1000°C and at the TCS gas concentration higher than 1%, the silicon epitaxial growth rate is governed by the surface chemical reaction. The epitaxial growth rate is expressed using the wafer temperature, T_W, [13]

$$\text{Growth rate } (\mu\text{m/min}) = 1.95 \times 109\, e^{(-26100/T_W)} \qquad (T_W < 1000°\text{C}). \tag{8}$$

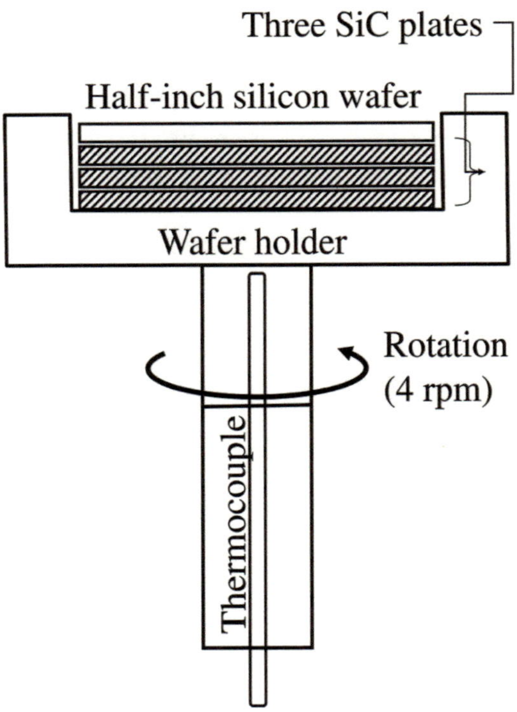

Figure 14. Half-inch wafer and silicon carbide susceptor on rotating wafer holder.

http://dx.doi.org/10.5772/intechopen.69986

When the wafer holder temperature, T_{WH}, is 650°C, the silicon epitaxial film growth rate was about 1.2 μm/min, corresponding to the T_W value near 960°C.

The silicon epitaxial film is formed by the two types of reflectors, as shown in **Figure 15**. The gas flow rates of the TCS and the H_2 were 9 and 215 sccm, respectively. Step C is intentionally introduced for cooling the reflector and for comparing the two types of reflector through the same process. In **Figure 15**, the dotted line and the solid line show the wafer holder temperatures during Step B immediately after Step C, by the Type-I and -II reflectors, respectively.

For the Type-I reflector, the TCS gas is introduced between 0 and 5 min. The wafer holder temperature finally becomes 573°C at 5 min at the halogen lamp voltage of 63 V. The solid line shows the temperature change using the Type-II reflector. The wafer holder temperature reaches 621°C at 4 min at the halogen lamp voltage of 62 V. The Type-II reflector can achieve the wafer temperature higher and faster than the Type-I reflector do.

The wafer holder temperature using the Type-I reflector still increases at 5 min as shown in **Figure 15**. As shown by the dotted line in **Figure 16**, the wafer holder temperature using the Type-I reflector at the halogen lamp voltage of 62 V can reach 553°C at 10 min. This is 70°C lower than that by the Type-II reflector. Additionally, the temperature using the Type-I reflector is entirely lower than that of Type-II through Steps A–C. Thus, the Type-II reflector achieves a quicker heating process.

In **Figures 15** and **16**, the wafer holder temperature increases after introducing the TCS gas at Step B. Because TCS gas absorbs the infrared light [15, 16], the gas phase temperature increases and finally the wafer temperature increases.

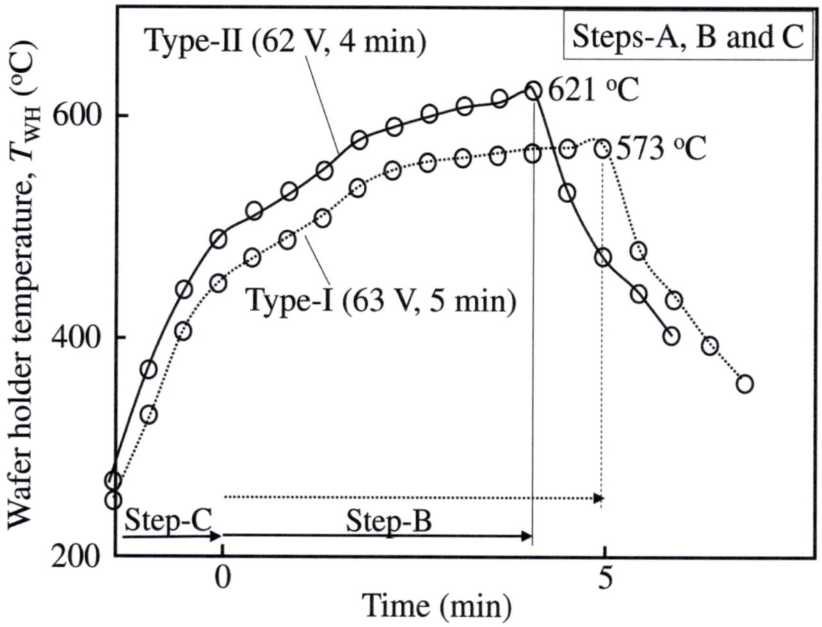

Figure 15. Temperature of wafer holder during Step B immediately after Step C. Dotted line: Type-I reflector at 63 V for 5 min and solid line: Type-II reflector at 62 V for 4 min.

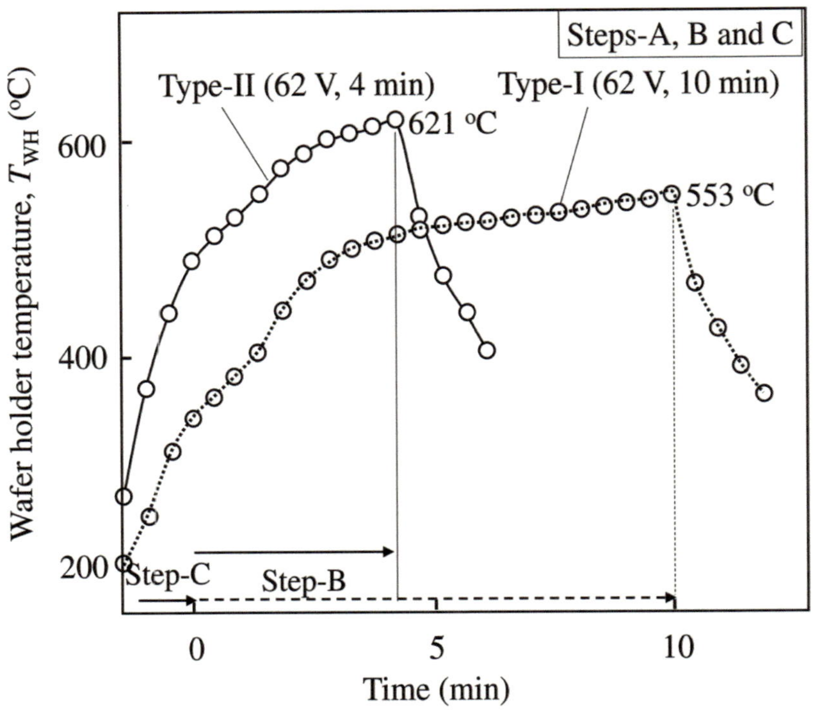

Figure 16. Temperature of wafer holder during Step B after Step C. Dotted line by Type-I reflector at 62 V for 10 min. Solid line by Type-II reflector at 62 V for 4 min.

For the thermal process optimization, the influence of Step C on the wafer temperature is evaluated using the Type-II reflector. **Figure 17** shows the wafer holder temperature with and without Step C at the halogen lamp voltage of 62 V. The flow rates of H_2 gas and TCS gas are 215 and 9 sccm, respectively.

The dotted line shows the wafer holder temperature through Steps A, C and B. In Step A, the wafer holder temperature reaches 650°C. During Step C, the wafer holder temperature is cooled to about 250°C, by decreasing the halogen lamp voltage to 30 from 80 V. The wafer holder temperature is then increased in Step B. At the halogen lamp voltage of 62 V, the wafer holder temperature becomes 621°C at 14 min. However, the wafer holder temperature still increases even at 14 min.

The solid line shows the wafer holder temperature during Steps A and B without Step C. The temperature during Step A is the same as the dotted line. After Step A, the halogen lamp voltage is increased to 62 V. The wafer holder temperature reaches 618°C at 6 min in a steady state. Thus, the process deleting Step C can be quick and stable.

4.5. Wafer rotation and susceptor

Using the Type-II reflector, the silicon epitaxial film is formed on the half-inch silicon wafer surface, by Steps A and B. The TCS and H_2 flow rates are 9 and 215 sccm, respectively, for 4 min.

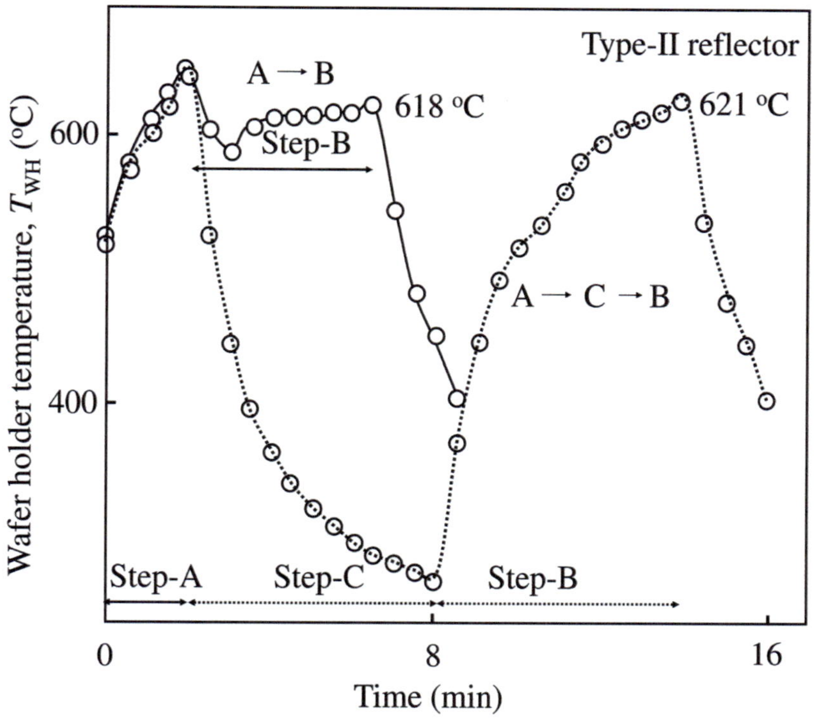

Figure 17. Wafer holder temperature using Type-II reflector. Dotted line: along Steps A, C and B and solid line: along Steps A and B without Step C.

The halogen lamp voltage at Step B is 60 V. The film thickness distribution is evaluated at five points along the longitudinal and transverse lines, using the dotted line and the solid line, respectively, as shown in **Figure 18**. **Figure 18(a)** shows the thickness profile of the epitaxial film which is formed without using the wafer rotation and without using the silicon carbide susceptor. The epitaxial film thickness along the x-axis is from 1.5 to 3.5 μm. By contrast, the epitaxial film thickness along the y-axis is very flat. In this figure, the film thickness shows a decrease from right to left. The epitaxial growth rate is near 0.5 μm/min, which corresponds to the wafer temperature of 950°C following Eq. (4). The epitaxial growth rate in **Figure 18** is governed by the rate of surface chemical reaction [13]. By the relationship, obtained in Section 2 [9], the introduced gas reaches the wafer surface and decreases its temperature. As shown in **Figure 19(a)**, by the asymmetric gas flow direction, the surface temperature is low in the left region. Corresponding to this temperature trend, the epitaxial film in the left region is thinner than that in the right region. Additionally, when the wafer is directly loaded on the quartz wafer holder, an adiabatic-like environment is formed. It produces a locally high-temperature region and the non-uniform thick film, because the infrared rays are easily concentrated to a local spot [20].

The wafer rotation is used for adjusting the asymmetric condition. **Figure 18(b)** shows the epitaxial film thickness profile, when the wafer rotates. The epitaxial film thickness shows a hill and a valley along the x- and y-axes. Although the thickness profile is averaged along the concentric circle of the rotating wafer, rather the complicated thickness profile appears. The

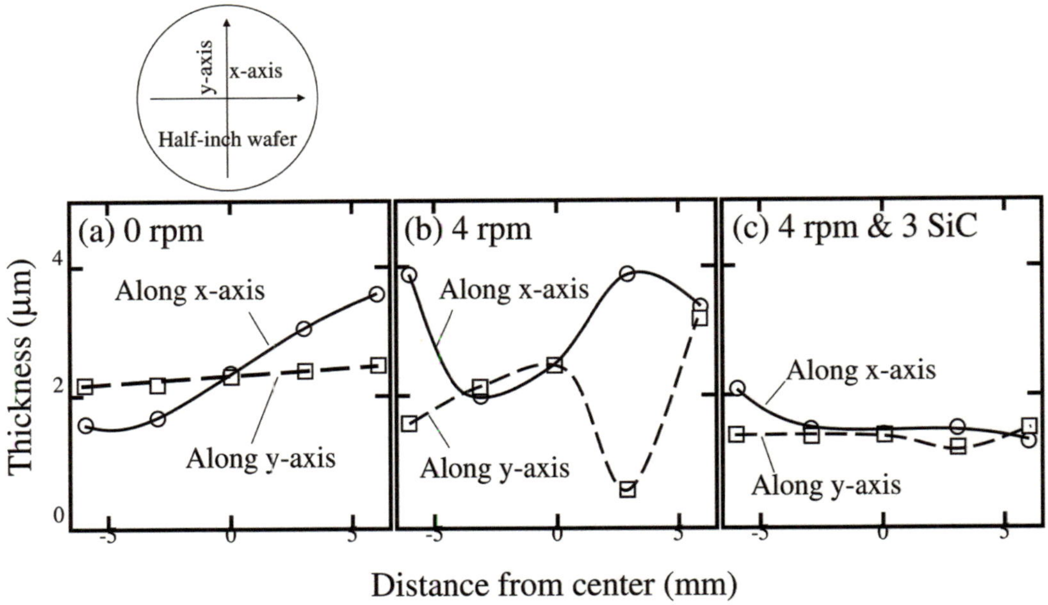

Figure 18. Thickness profile of obtained silicon epitaxial film: (a) without wafer rotation and without silicon carbide susceptor, (b) with wafer rotation and without silicon carbide susceptor and (c) with wafer rotation and with three silicon carbide susceptors.

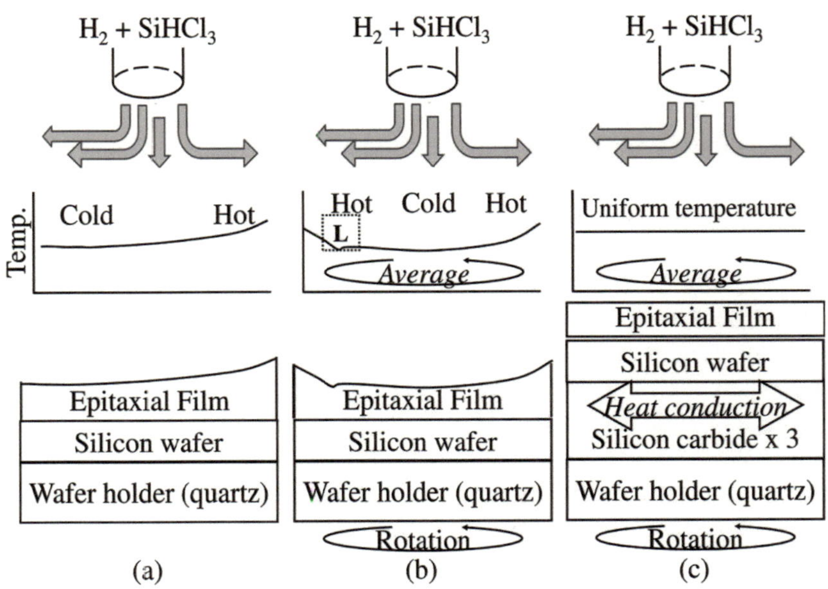

Figure 19. Influence of gas flow from the inlet, wafer rotation and silicon carbide susceptor on the epitaxial film thickness (a) with no wafer rotation and no silicon carbide susceptor, (b) with wafer rotation and no silicon carbide susceptor and (c) with wafer rotation and three silicon carbide susceptors.

local low- or high-temperature spot, denoted by the letter L, still remains. An additional method is necessary for obtaining the flat silicon film.

The locally high and low temperatures over the wafer surface may be produced due to the quartz material having a low thermal conductivity [20]. The heat transport in the horizontal direction is enhanced by the silicon carbide susceptor. **Figure 19(c)** shows that the local non-uniformity of the wafer temperature remaining even using the wafer rotation is reduced by the high heat transport through the silicon carbide susceptor. As shown in **Figure 18(c)**, the epitaxial film thickness becomes very flat along the x- and y-axes.

By using the small, thin and simple geometry of the reactor parts, the quick and flat epitaxial film production is possible. This concept is valid not only for the minimal CVD reactor.

5. Summary

A chemical vapour deposition reactor for the growth of half-inch silicon wafers is designed by employing (i) a vertical gas flow, (ii) rapid thermal operation using concentrated infrared light and (iii) a quick reactor cleaning process. For the rapid heating process, absorption of infrared light and heat transport by the flowing gases are active parameters. Under the optimized conditions, the cleaning-free process is possible. The reactor parts placed near the wafer must be small, slim and thin for quickly heating the wafer. The wafer rotation and the heat-conductive susceptor help to fabricate uniform silicon epitaxial films. Overall, important parameters are listed in **Table 2**. This table includes several parameters which are active and useful for the small-sized reactor.

Parameters	**Value**
Substrate	12.5 mm diameter silicon wafer
Gas flow	Vertical (from top to bottom)
Pressure	Atmospheric pressure
Precursor	Trichlorosilane ($SiHCl_3$)
Hydrogen gas flow rate	Less than 200 sccm
Trichlorosilane gas flow rate	20 sccm
Chlorine trifluoride gas flow rate (Reactor cleaning)	50 sccm in 1000 sccm in N_2
Heating module	Concentrated infrared light (Three halogen lamps) Slim and thin reflector
Electric power	55–65 V, less than 1500 W
Substrate surface cleaning	Near 1000°C in H_2 within 1 min
Deposition time	1–8 min
Substrate temperature	800–1000°C
Substrate rotation	4 rpm

Parameters	Value
Susceptor	Silicon carbide plate
Footprint	30 × 45 cm

Table 2. List of parameters: silicon epitaxial growth for Minimal Fab.

	Minimal CVD reactor	Ordinary CVD reactor
Wafer diameter	12.5 mm, single wafer	150, 200, 300 and 450 mm, multi-(batch) and single wafer
Total gas flow rate	<0.2 slm	>100 slm
Growth rate	Near 1 μm/min	1–8 μm/min
Pressure	1 atm (or reduced pressure)	1 atm or reduced pressure
Precursor	$SiHCl_3$	$SiHCl_3$, SiH_2Cl_2, SiH_4, and so on
Reactor cleaning	ClF_3 gas at higher than RT	HCl gas near 1200°C
Gas flow	Natural convection, Vertical (downward)	Forced flow, Vertical (downward), horizontal, cylinder, pancake, and so on
Heating method	Concentrated light by lamp heating	Lamp, resistant and inductive heating
Electric power	<1500 W	>100,000 W
Footprint	0.3 × 0.45 m^2	Several × Several m^2
Throughput	Several tens wafers/min (Target)	Several wafers/min
Reactor price	> $30,000 (Target)	> $3,000,000

Table 3. Comparison between minimal and ordinary CVD reactor.

The differences of the reactor design between the Minimal Fab and the ordinary CVD are listed in **Table 3**. The film growth on a small wafer using small gas flow rates significantly reduces system cost.

Author details

Ning Li[1], Hitoshi Habuka[1]*, Yuuki Ishida[2,3], Shin-ichi Ikeda[2,3] and Shiro Hara[2,3]

*Address all correspondence to: habuka1@ynu.ac.jp

1 Department of Chemical and Energy Engineering, Yokohama National University, Yokohama, Japan

2 National Institutes of Advanced Science and Technology, AIST Tsukuba Central 2, Tsukuba, Japan

3 Minimal Fab Development Association, AIST Tsukuba Central 2, Tsukuba, Japan

References

[1] http://www.itrs.net/

[2] Thean A. Silicon & beyond CMOS: The path of advanced electronic structure engineering for low-voltage transistors. Solid State Phenomena. 2013;**195**:3

[3] http://unit.aist.go.jp/neri/mini-sys/fabsystem/index.html

[4] Khumpuang S, Hara S. A MOSFET fabrication using a maskless lithography system in clean-localized environment of Minimal Fab. IEEE Transactions on Semiconductor Manufacturing. 2015;**28**(3):393-398

[5] Khumpuang S, Imura F, Hara S. Analyses on cleanroom-free performance and transistor manufacturing cycle time of Minimal Fab. IEEE Transactions on Semiconductor Manufacturing. 2015;**28**(4):551-556

[6] Khumpuang S, Maekawa S, Hara S. Photolithography for Minimal Fab system. IEEJ Transactions on Sensors and Micromachines. 2013;**133**:272-277

[7] Habuka H, Otsuka T, Mayusumi M, Shimada M, Okuyama K. A direct approach to evaluate the thermal condition of a silicon substrate under infrared rays and specular reflectors. Journal of Electrochemical Society. 1999;**146**(2):713-718

[8] Habuka H, Sukenobu T, Koda H, Takeuchi T, Aihara M. Silicon etch rate using chlorine trifluoride. Journal of Electrochemical Society. 2004;**151**:G783

[9] Li N, Habuka H, Ikeda S, Hara S. Silicon chemical vapor deposition process using a half-inch silicon wafer for minimal manufacturing system. Physics Procedia. 2013;**46C**: 230-238

[10] Li N, Habuka H, Ikeda S, Ishida Y, Hara S. Practical thermal condition of silicon CVD reactor for minimal manufacturing. In: Industrial Engineering, Machine Design and Automation (IEMDA 2014) & Computer Science and Application (CCSA 2014). pp. 393-400. 2014 International Conference on Materials Science and Energy Engineering (CMSEE 2014). December 12-14, 2014, Sanya, Hainan, China

[11] Li N, Habuka H, Ikeda S, Ishida Y, Hara S. Practical thermal condition of silicon CVD reactor for minimal manufacturing. In: Proceedings of the Forum on the Science and Technology of Silicon Materials; 19–22 Oct. 2014; Hamamatsu, Japan. pp. 219-221.The Japan Society for the Promotion of Science (2014)

[12] Li N, Habuka H, Ishida Y, Ikeda S-i, Hara S. Reflector influence on rapid heating of minimal manufacturing chemical vapor deposition reactor. ECS Journal of Solid State Science and Technology. 2016;**5**(5):P280-P284. DOI: 10.1149/2.0251605jss

[13] Habuka H, Nagoya T, Mayusumi M, Katayama M, Shimada M, Okuyama K. Model on transport phenomena and epitaxial growth of silicon thin film in $SiHCl_3$-H_2 system under atmospheric pressure. Journal of Crystal Growth. 1996;**169**:61-72

[14] Morishita J, Wu S, Ishihara Y, Kimijima T. Observation of purifier performance to reduce moisture in hydrogen chloride by near infrared laser absorption spectrometry. Journal of Applied Physics. 1997;**36**:L1706

[15] Gibian TG, McKinney DS. Infrared spectra and force constants of chloroform and trichlorosilane. Journal of the American Chemical Society. 1951;**73**:1431-1434

[16] Ushio Electric https://www.ushio.co.jp/jp/feature/halogenHeater/ss05.html (access date: June 23, 2017)

[17] Habuka H, Nagoya T, Katayama M, Shimada M, Okuyama K. Modelling of epitaxial silicon thin-film growth on a rotating substrate in a horizontal single-wafer reactor. Journal of Electrochemical Society. 1995;**142**(12):4272-4278

[18] Crippa D, Rode DL, Masi M. Silicon Epitaxy. San Diego, USA: Academic Press; 2001

[19] Haynes WM. CRC Handbook of Chemistry and Physics. 92nd ed. Boca Raton, USA: CRC Press; 2012

[20] Habuka H, Suzuki T, Sakurai T, Negishi Y, Takeuchi T. Non-empirical design of rapid thermal processing system. Japanese Journal of Applied Physics. 2001;**40**(12):7123-7128

Section 3

III-V Epitaxy

Chapter 7

GaN and InN Hexagonal Microdisks

Chen-Chi Yang, Ikai Lo, Yu-Chi Hsu and
Hong-Yi Yang

Additional information is available at the end of the chapter

http://dx.doi.org/10.5772/intechopen.70120

Abstract

The high-quality GaN microdisks with InGaN/GaN quantum wells (QWs) and InN microdisks were grown on γ-$LiAlO_2$ substrates by plasma-assisted molecular beam epitaxy (PA-MBE). The samples were analysed using scanning electron microscopy, X-ray diffraction, photoluminescence, cathodoluminescence and high-resolution transmission electron microscope. The characteristics of the GaN microdisks and InN microdisks were studied and the effect of growth temperature was evaluated.

Keywords: GaN, InN, microdisk, molecular beam epitaxy

1. Introduction

III-nitride materials have been extensively studied for the applications to high-efficiency lighting sources such as light-emitting diodes (LEDs) or spintronics [1–7]. From the changing of indium content (x), the band-gap of $In_xGa_{1-x}N$ can be tuned from 0.7 to 3.4 eV to cover the whole visible-light spectrum. **Figure 1** shows the diagram of band-gap energies of III-nitrides with bowing parameters [8] versus lattice constants. However, it is difficult to grow high-quality $In_xGa_{1-x}N$/GaN quantum wells (QWs) because of the large lattice mismatch between GaN ($\boldsymbol{a}_{GaN}$ = 0.3189 nm, $\boldsymbol{c}_{GaN}$ = 0.5185 nm) and InN ($\boldsymbol{a}_{InN}$ = 0.35446 nm, $\boldsymbol{c}_{InN}$ = 0.57034 nm) [8, 9]. Furthermore, because of the high volatility of indium atom at high temperature, it is hard to grow a homogenous high-indium-concentration $In_xGa_{1-x}N$/GaN QW thin film by using high-temperature growth techniques (e.g. T > 1000°C) such as vapour phase epitaxy (VPE) or metalorganic chemical vapour deposition (MOCVD). The growth of high-quality $In_xGa_{1-x}N$ epilayer with an indium concentration higher than 20% is regarded as a high challenge [10]. To overcome these difficulties, a plasma-assisted molecular beam epitaxy (PA-MBE) technique was used to grow $In_xGa_{1-x}N$ epilayer at lower temperatures, and some substrates

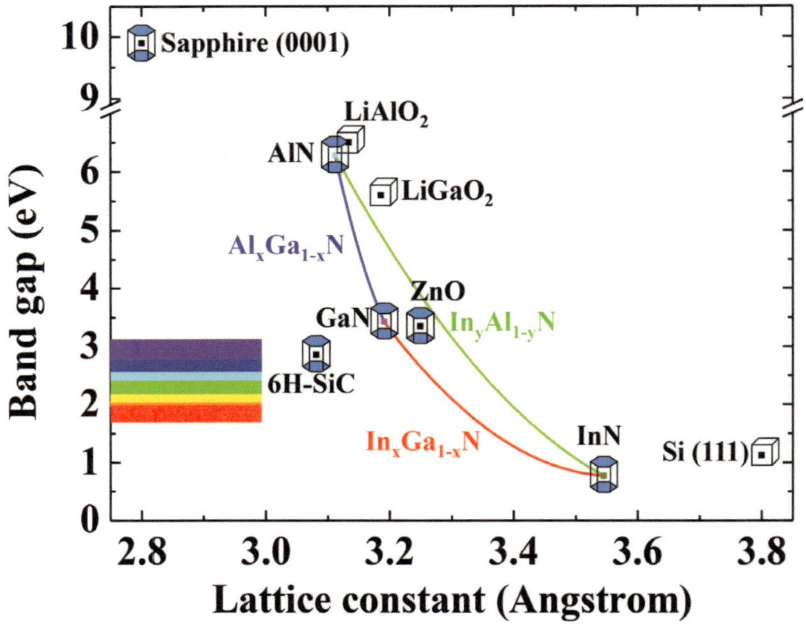

Figure 1. The diagram shows band-gap energies of III-nitrides with bowing parameters versus lattice constants. The substrate materials were also presented.

(such as $LiAlO_2$, $LiGaO_2$, ZnO) were selected to grow GaN epilayers in order to minimize the lattice mismatch between the substrate and GaN layer as compared to the commercial substrates, e.g. sapphire, SiC or Si (1 1 1). From the values of wurtzite GaN on JCPDS files No. 50-0792 and those of γ-$LiAlO_2$ on No. 38-1464, the lattice mismatch between $c/2_{LAO}$ and a_{GaN} is 1.5% indicating that it is suit to grow *c*-plane GaN on the g-$LiAlO_2$ substrate. In this chapter, we show the high-quality epitaxial growth of GaN microdisks with $In_xGa_{1-x}N$/GaN quantum wells and InN microdisks on γ-$LiAlO_2$ lithium aluminium oxide (LAO) substrates at low temperatures by the plasma-assisted molecular beam epitaxy system. Consequently, GaN and InN microdisks provide better opportunities to fabricate the $In_xGa_{1-x}N$/GaN microdisk quantum wells for the application of full-colour micron LED without the sapphire substrate, which is mostly used for the bulk GaN-based quantum wells in commercial LEDs but has a larger lattice mismatch with InN [1, 2].

2. GaN hexagonal microdisks

2.1. Growth of GaN hexagonal microdisks

The sample was grown on a high-quality 1 × 1 cm^2 LAO (1 0 0) substrate by using a low-temperature PA-MBE system (Veeco Applied-GEN 930) with standard effusion cells for Ga-evaporation and an rf-plasma cell with 450 W for N_2-plasma source. The LAO substrate was cut from the crystal ingot, which was fabricated by the traditional Czochralski pulling technique. Before mounting on a holder, the LAO substrate was cleaned with acetone (5 min), isopropanol (5 min) and de-ionized water for a while, and then dried with nitrogen gas immediately. After the chemical cleaning, a thermal treatment was introduced to the LAO substrate in the MBE chamber before epitaxial growth. The LAO substrate was out-gassed at 680°C for 10 min. The temperature was defined by a thermal couple equipped with the backside of the substrate. Thereafter, the substrate temperature was decreased to growth temperatures. The Ga wetting layer was performed on the LAO substrate for 5 min at 630°C, and then the two-step method (i.e. two different N/Ga flux ratios from 28.9 to 139.7, for 35 and 70 min, respectively) was used to fabricate the GaN epi-film at 620°C. The flux ratio was represented by beam equivalent pressure (BEP) of evaporative III-group sources from the standard effusion cell against that of the N_2 source from the rf-plasma cell [11]. In our previous study, we showed the characteristics of *c*-plane GaN $(0\ 0\ 0\bar{1})$ hexagonal microdisks [5]. Besides, we developed a back process to fabricate an electrical contact for the GaN hexagonal microdisk on a transparent p-type GaN template [12]. In this chapter, we have consistently grown a sample of GaN microdisks to demonstrate the self-assembling model [5], as shown in **Figure 2**.

2.2. Characteristics of GaN microdisks

The surface morphology of the GaN microdisk sample was evaluated by the field emission scanning electron microscopy (FE-SEM, SII-3050). **Figure 2** shows SEM images with a tilted

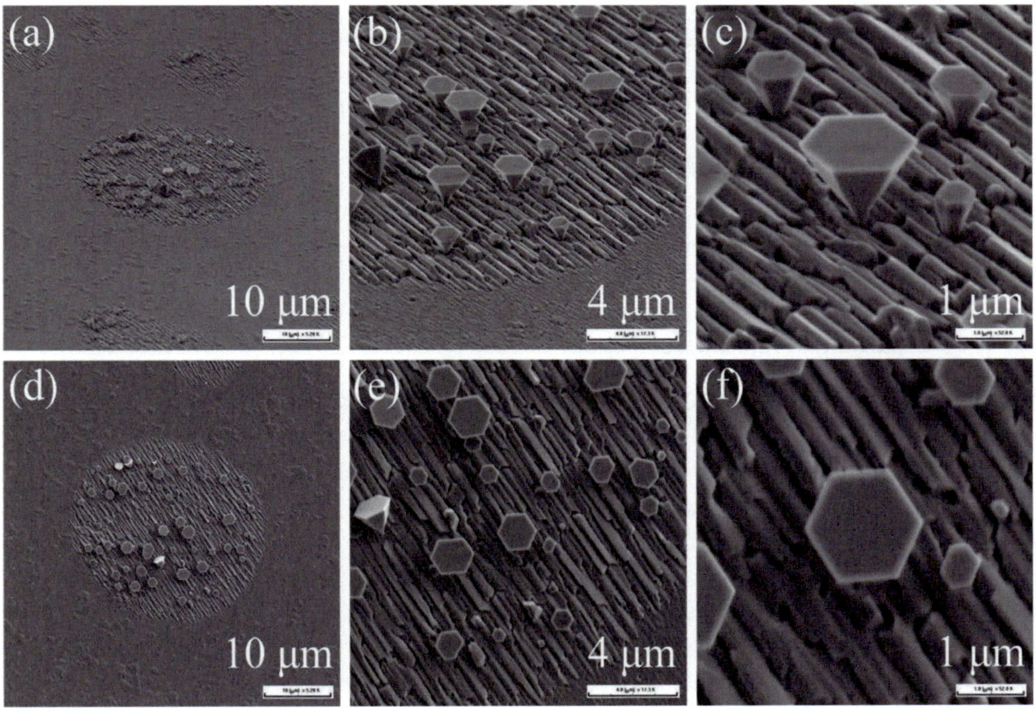

Figure 2. (a)–(c) The tilt-view SEM images of the sample, the scale bars are 10, 4 and 1 µm, respectively. (d)–(f) The top-view SEM images of the sample, the scale bars are 10, 4 and 1 µm, respectively.

angle and a top-view of the sample, respectively. The morphology of the sample exhibited that a two-dimensional (2D) ***M***-plane GaN film and three-dimensional (3D) ***c***-plane GaN hexagonal microdisks were grown on the LAO substrate. The micrographic images of the sample showed that the 2D ***M***-plane GaN epi-film was developed along with the lateral orientation $[1\ 1\ \bar{2}\ 0]_{GaN}$ // $[0\ 0\ 1]_{LAO}$, while the 3D ***c***-plane GaN hexagonal microdisks were grown atop an anionic hexagonal basal plane of LAO. The two-orientation growth of GaN nanopillars on the LAO substrate has been reported in our previous papers [13, 14]. **Figure 2(c)** shows that the neck of contact area between the GaN microdisk and the LAO substrate is small (e.g. less than 200 nm). In addition, the lattice mismatch between $\boldsymbol{c}/2_{LAO}$ and $\boldsymbol{a}_{GaN}$ is only 1.5%. It implies that the GaN microdisk is nearly freestanding as a new substrate for further growth of $In_xGa_{1-x}N$/GaN quantum well (QW) on the top, which ignores the lattice mismatch against the LAO substrate. **Figure 2(f)** shows the enlarged SEM image with a top-view of the GaN hexagonal microdisks shown in the centre of **Figure 2(d)**, and the diameter of the GaN microdisk is about 2.0 μm. Based on the self-assembling model, we will extend to the growth of $In_xGa_{1-x}N$/GaN QW on GaN microdisk and show its characteristics in Section 4.

The optical properties of the GaN microdisk sample were measured by photoluminescence spectroscopy (PL, HORIBA HR800) at room temperature with a light source of He-Cd 325 nm laser. We performed the laser beam focusing on two different spots (S1 and S2) and compared the results with the spot without any microdisk (i.e. mostly ***M***-plane GaN and labelled as background), as shown in **Figure 3**. Two major peaks were obtained for each measurement (S1 or S2). These two major peaks were confirmed by a non-linear Gaussian-function curve fitting with the software Origin (Pro. 8.0). The result of the non-linear Gaussian-function curve fitting showed that the positions of two major peaks for two spots were very consistent. The averaged value for the first peak is (3.385 ± 0.001) eV with the full width at half maximum (FWHM) value equal to (0.128 ± 0.001) eV. It is due to the band-gap transition of wurtzite GaN. The averaged value for the second peak is (2.226 ± 0.003) eV with the FWHM value equal to (0.363 ± 0.033) eV. It is an energy level related to structural defects (e.g. YL in reference [15]) in GaN, so that the FWHM value of GaN is smaller than that of the defect level. The PL intensity corresponding to wurtzite GaN and the defect level indicates that ***c***-plane GaN microdisk is a higher quality structure than ***M***-plane GaN because the intensity of defect level from GaN microdisk is lower than that from ***M***-plane GaN background.

The microstructure of the GaN microdisk sample was analysed by field emission transmission electron microscopy (FE-TEM) (Phillips, model Tecnai F-20) with an electron voltage of 200 kV. The cross-sectional TEM specimen of the sample was prepared by a dual-beam Focus Ion Beam system (FIB, Seiko Inc., SII-3050), on the cleavage plane along the $[1\ \bar{1}\ 0\ 0]$ direction of the ***c***-plane GaN hexagonal microdisk. The FIB was performed with an accelerated voltage of 30 kV to cut the samples roughly and then refined the specimen further by an accelerated voltage of 5 kV. **Figure 4(a)** shows the bright field image with $[1\ 1\ \bar{2}\ 0]_{GaN}$ // $[0\ 0\ 1]_{LAO}$ zone axis. It clearly exhibited that the GaN microdisk was well formed on the LAO substrate. The height for the ***c***-plane GaN hexagonal microdisk from neck to top was about 4.1 μm. The selective area diffraction (SAD) pattern at the top area of the GaN microdisk shown in **Figure 4(b)** clearly showed one single rectangular diffraction pattern at the location of DP01, indicating that the hexagonal microdisk was uniquely formed by the ***c***-plane wurtzite GaN. The ***d***-spacing

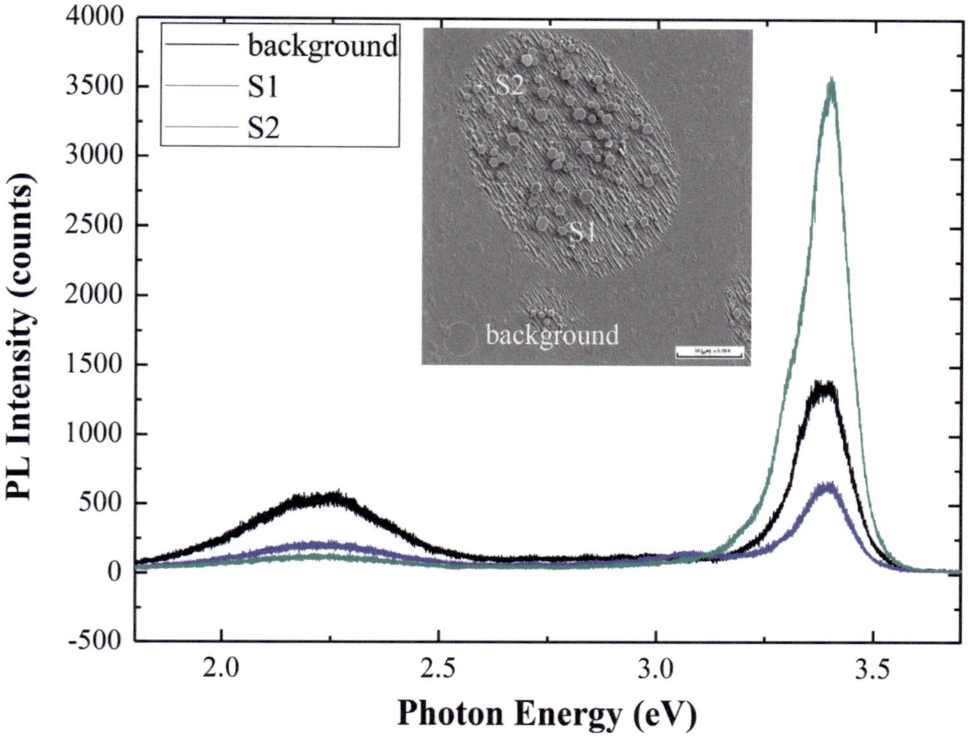

Figure 3. The PL spectra measured at room temperature for different spots (S1 and S2) of the GaN microdisks and ***M***-plane background of the sample.

between the $\{0\ 0\ 0\ \bar{1}\}$ planes of GaN hexagonal microdisk was measured to be $\boldsymbol{d}_c$ = 0.5254 nm and the ***d***-spacing between the $\{1\ \bar{1}\ 0\ 0\}$ planes of GaN hexagonal microdisk was found to be $\boldsymbol{d}_{\mathbf{M}}$ = 0.2785 nm. Compared with the values on JCPDS file No. 50-0792 which are 0.5185 and 0.2762 nm, respectively, the difference between wurtzite GaN microdisk and bulk GaN for $\boldsymbol{d}_c$ and $\boldsymbol{d}_{\mathbf{M}}$ are 1.33 and 0.08%, respectively, revealing that the lattice constant of GaN microdisk is slightly larger than that of bulk GaN. The high-resolution TEM images with the beam direction of $[1\ 1\ \bar{2}\ 0]_{GaN}$ // $[0\ 0\ 1]_{LAO}$ were performed at the areas HR01 and HR02 of the sample, as shown in **Figure 4(a)**. The symmetric hexagonal shape reveals the high-quality crystalline structure of the GaN microdisk, as shown in **Figure 4(c)**. The angle between the edge and the growth direction can be examined directly by the high-resolution TEM image performed at HR02 to be about 28°, as shown in **Figure 4(d)**. The ball-stick model for the standard wurtzite GaN (JCPDS file No. 50-0792) with $a = b = 0.3189$, $c = 0.5185$ and $u = \bar{\alpha}/c = 3/8$ was used to simulate the ***c***-plane GaN hexagonal microdisk in **Figure 4(e)**, where blue balls represented Ga atoms and red balls represented N atoms. The *c*-plane GaN $(0\ 0\ 0\ \bar{1})$ hexagonal microdisk was built up with the capture of N atoms by the $\bar{\beta}$-dangling bonds of the most-outside Ga atoms and then the capture of Ga atoms by $\bar{\alpha}$-bonds of N atoms to form the microdisk [5]. The lateral over-growth along the $(1\,\bar{1}\,0\ 0)$ direction was extended to one $\boldsymbol{d}_{\mathbf{M}}$-spacing for each unit step-layer (i.e. $\boldsymbol{d}_c$-spacing), resulting in the angle of 28° off the *c*-axis. Based on the ball-stick model, the laterally extensive width along the $[1\ \bar{1}\ 0\ 0]_{GaN}$ direction per unit step-layer was equal to $\frac{\sqrt{3}}{2}a$. The edge was

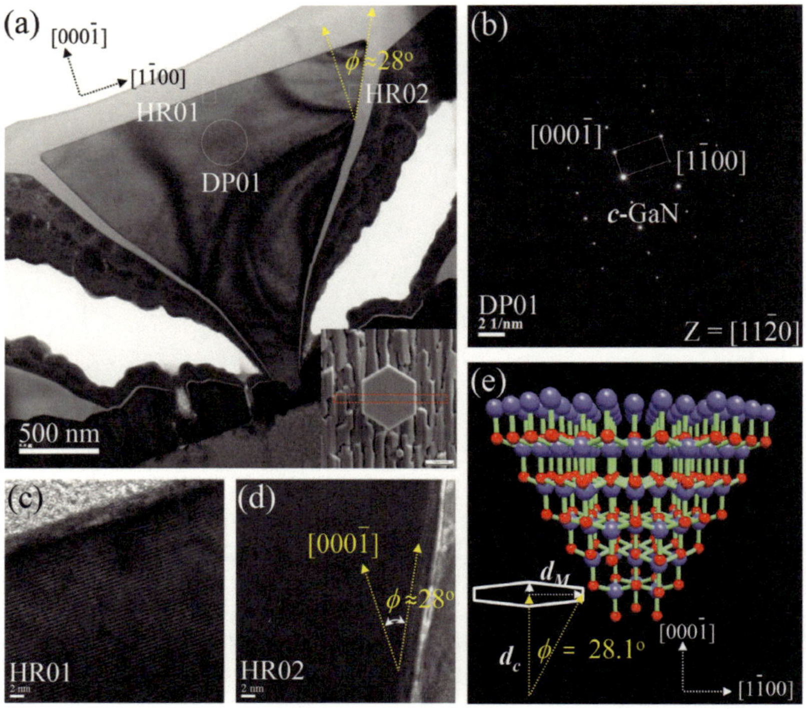

Figure 4. TEM analyses of the GaN hexagonal microdisk: (a) the bright field image with $[1\,1\,\bar{2}\,0]_{GaN}$ // $[0\,0\,1]_{LAO}$ zone axis. The selective area diffraction patterns taken at the point shown in (a) are presented in (b), the scale bars are 2 (1/nm). The high-resolution TEM images taken at the points shown in (a) are presented in (c) and (d), the scale bars are 2 nm and (e) the ball-stick model for GaN hexagonal microdisk.

then tilted off the *c*-axis $[0\ 0\ 0\ \bar{1}]$ direction by the angle, $\phi = \tan^{-1}(\frac{\sqrt{3}}{2}a/c) = 28.1°$, where $\frac{\sqrt{3}}{2}a$ is equal to one d_{M}, as shown in **Figure 4(e)**. We also calculated the angle from the measured SAD data at the GaN hexagonal microdisk in **Figure 4(b)**, and obtained that the ***d***-spacing between the $\{0\ 0\ 0\ \bar{1}\}$ planes was $d_c = 0.5254$ nm and the ***d***-spacing between the $\{1\ \bar{1}\ 0\ 0\}$ planes was $d_M = 0.2785$ nm, resulting in $\phi = \tan^{-1}(d_M/d_c) = 28.0°$, which was in good agreement with the model predicted. The angle between the edge and the growth direction can be examined directly by the high-resolution TEM image performed at HR02 to be about 28°, as shown in **Figure 4(d)**.

3. InN hexagonal microdisks

3.1. Growth of InN hexagonal microdisks

The two-orientation growth of GaN nanopillars on the LAO substrate has been reported in our previous papers [13, 14] and reconfirmed in Section 2. In this section, we applied the two-orientation growth mechanism to grow the 2D ***M***-plane InN epi-film and 3D ***c***-plane InN hexagonal thin disks on the LAO substrate with the InGaN buffer layer at low growth temperature (470°C). The InN microdisk sample was grown on a high-quality 1×1 cm^2 LAO (1 0 0) substrate with an InGaN buffer layer between them by a low-temperature PA-MBE system (Veeco Applied-GEN 930). The LAO substrate was cut from the crystal ingot, which was fabricated by the traditional Czochralski pulling technique. The growth details were described completely in our previous paper which was published in *AIP Advances* [16].

3.2. Characteristics of InN microdisks

The crystal structure of the InN microdisk sample was characterized by the high-resolution X-ray diffraction (XRD; Bede D1) measurement and is shown in **Figure 5(a)**. From the result of X-ray diffraction pattern (i.e. the peak at 2θ = 31.69°), we estimated the indium content of $In_xGa_{1-x}N$ on the basis of Vegard's law to be about 20% [17]. The peaks at 2θ = 29.07, 31.31, 32.29 and 34.69° represented the X-ray diffraction patterns from the ***M***-plane InN ($1\bar{1}00$), ***c***-plane InN ($000\bar{2}$), ***M***-plane GaN ($1\bar{1}00$) and LAO (1 0 0), respectively. These peak positions at the X-ray diffraction patterns were obtained and matched with those data of the standard wurtzite structure bulk InN (JCPDS file No. 50-1239) by the asymmetric double sigmoidal linear curve fitting with the software Quick Graph (Version 2.0). The ***d***-spacing between the {$000\bar{2}$} planes of InN was evaluated to be $\boldsymbol{d}_{0002}$ = 0.28216 nm from the Bragg's law ($2d\sin\theta = n\lambda$) with Cu $K_{\alpha 1}$wavelength λ = 0.1540562 nm. The lattice constant of wurtzite InN microdisk is smaller than that of bulk InN by comparing with the value on JCPDS file, $\boldsymbol{d}_{0002}$ = 0.28528 nm, and the difference between InN microdisk and bulk InN is 1.09%.

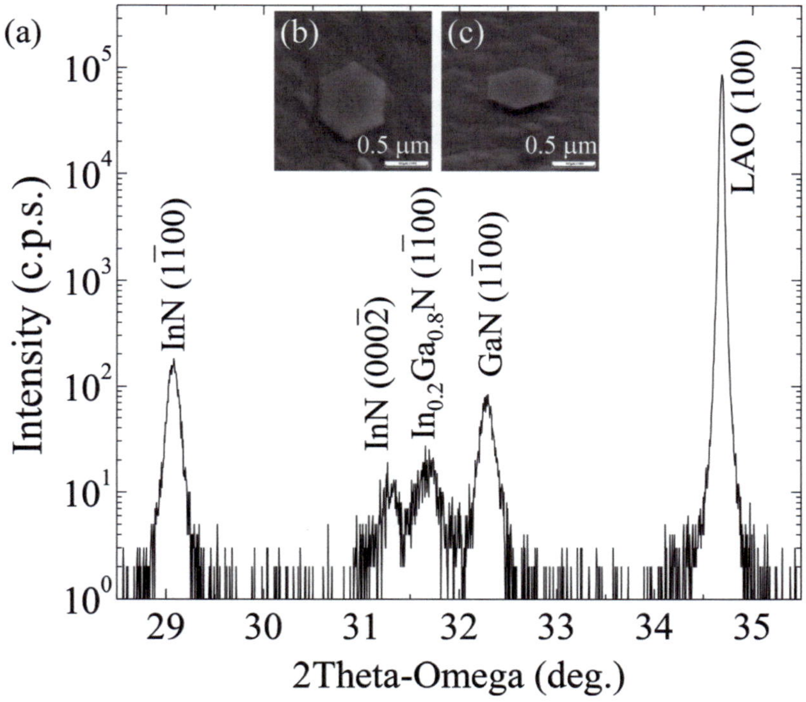

Figure 5. (a) The X-ray 2Theta-Omega scan of the sample. (b) The top-view SEM image of InN hexagonal thin disk, the scale bar is 0.5 μm. (c) The tilt-view SEM image of InN hexagonal thin disk, the scale bar is 0.5 μm.

The surface morphology of the InN microdisk sample was evaluated by the field emission scanning electron microscopy (FE-SEM, SII-3050). **Figure 5(b)** shows the top-view SEM image of the sample, and the diameter of the InN microdisk was about 0.96 μm. The morphology of the sample exhibited that 2D ***M***-plane InN epi-film and 3D ***c***-plane InN hexagonal microdisks were grown on the LAO substrate. **Figure 5(c)** shows the tilt-view SEM image of the InN microdisk. The micrographic images of the sample showed that the 3D ***c***-plane InN hexagonal microdisks and nanopillars were grown atop an anionic hexagonal basal plane of LAO, while the 2D ***M***-plane InN epi-film was developed along with the lateral orientation $[1\,1\,\bar{2}\,0]_{InN}$ // $[0\,0\,1]_{LAO}$. Compared with the shape of GaN microdisk, InN microdisk was thinner than GaN microdisk.

The microstructure of the sample was analysed by field emission transmission electron microscopy (FE-TEM) (Phillips, model Tecnai F-20) with an electron voltage of 200 kV. The cross-sectional TEM specimen of the sample was prepared by a dual-beam FIB system (Seiko Inc., SII-3050), on the cleavage plane along the $[1\,\bar{1}\,0\,0]$ direction of the *c*-plane InN hexagonal thin disk. The FIB was performed with an accelerated voltage of 30 kV to cut the samples roughly and then refined the specimen further by an accelerated voltage of 5 kV. **Figure 6(a)** shows the bright field image with the $[1\,1\,\bar{2}\,0]_{InN}$ // $[0\,0\,1]_{LAO}$ zone axis. It clearly exhibited that InN was well-formed on the InGaN buffer layer, and the InGaN buffer layer was well-established on the GaN epi-layer. The thicknesses of ***M***-plane InN, ***M***-plane InGaN and ***M***-plane GaN were measured to be about 265, 51 and 137 nm, respectively. The height for the *c*-plane InN hexagonal thin disk from neck to top was about 188 nm. The *c*-plane wurtzite structure was followed up to the neck area and formed a uniform *c*-plane InGaN pyramid-shaped structure. Outside the pyramid-shaped structure, the wave-shaped structures were produced by the staking faults between the misfit *c*-plane wurtzite structures of InGaN and InN. The wave-shaped structures became uniform and then the *c*-plane wurtzite structure was followed further to form the InN hexagonal microdisk. The selective area diffraction (SAD) pattern at the top area of hexagonal thin disk shown in **Figure 6(b)** clearly showed one single rectangular diffraction pattern at the location of DP01, indicating that the hexagonal thin disk was uniquely formed by the ***c***-plane wurtzite InN. The ***d***-spacing between the $\{0\,0\,0\,\bar{1}\}$ planes of InN hexagonal thin disk was measured to be $\boldsymbol{d}_c = 0.5687$ nm and the ***d***-spacing between the $\{1\,\bar{1}\,0\,0\}$ planes of InN hexagonal thin disk was $\boldsymbol{d}_M = 0.3025$ nm. Compared with the values on JCPDS file No. 50-1239, which are 0.5703 and 0.30647 nm, respectively, the difference between wurtzite InN microdisk and bulk InN for $\boldsymbol{d}_c$ and $\boldsymbol{d}_M$ are 0.28 and 1.24%, respectively, revealing that the lattice constant of wurtzite InN microdisk is smaller than that of bulk InN. The angle between edge and growth direction can be examined directly by the high-resolution TEM image performed at HR01 to be about 73°, as shown in **Figure 6(c)**. To establish the growth mechanism of the thin InN hexagonal microdisk, we demonstrated a ball-stick model for the self-assembled thin InN microdisk. The ball-stick model for the standard wurtzite InN (JCPDS file No. 50-1239) with $a = b = 0.3537$, $c = 0.5703$ and $u = \bar{a}/c = 3/8$ was used to simulate the *c*-plane InN hexagonal microdisk, as shown in **Figure 6(d)** and **(e)**, where blue balls represented In atoms and red balls represented N atoms. In the case of InN thin disk, when the growth temperature lowered to 470°C, the ***c***-plane InN $(0\,0\,0\,\bar{1})$ hexagonal thin disk was built up with the capture of N atoms by the $\bar{\beta}$-dangling bonds of the most-outside In atoms and then the lateral over-growth occurred; and the capture of In atoms by $\bar{\beta}$-dangling bonds of N atoms to form

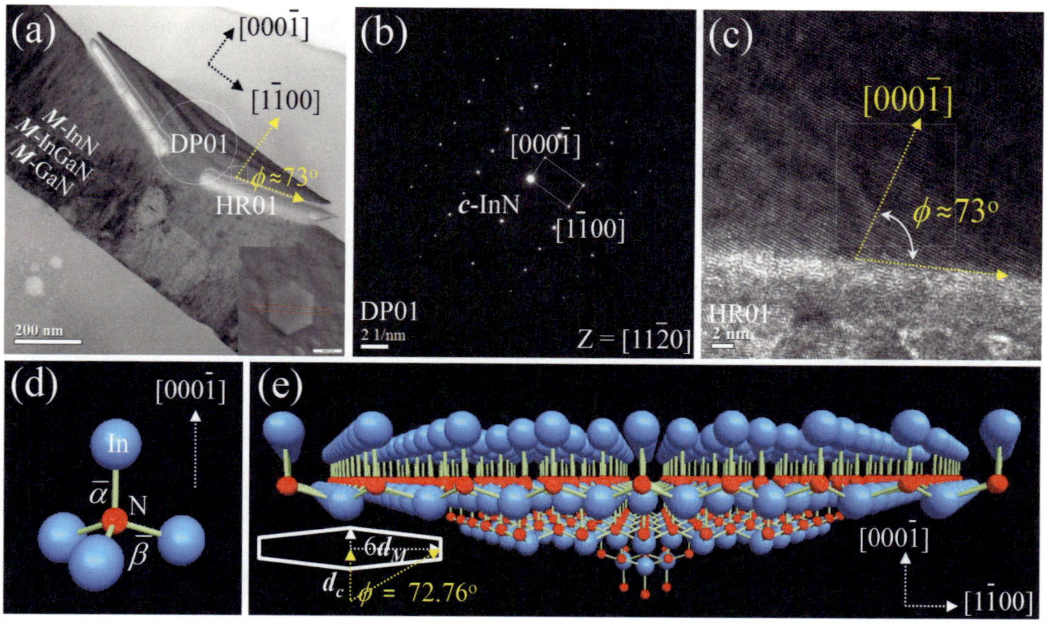

Figure 6. TEM analyses of the InN hexagonal thin disk: (a) the bright field image with $[1\,1\,\bar{2}\,0\,1]_{InN}$ // $[0\,0\,1]_{LAO}$ zone axis. The selective area diffraction patterns taken at the points shown in (a) are presented in (b), the scale bars are 2 (1/nm). The high-resolution TEM images taken at the point shown in (a) are presented in (c), the scale bars are 2 nm. The ball-stick model for InN epilayer: (d) the chemical bonds of the $(0\,0\,0\,\bar{1})$ surface and (e) the hexagonal thin disk.

the thin microdisk. The lateral over-growth along the $(1\,\bar{1}\,0\,0)$ direction was extended to six $\boldsymbol{d}_M$-spacings for each unit step-layer (i.e. $\boldsymbol{d}_c$-spacing), resulting in the angle of 73° off the *c*-axis. Based on the ball-stick model, the laterally extensive width along the $[1\,\bar{1}\,0\,0]_{InN}$ direction per unit step-layer was equal to $3\sqrt{3}\,a$. The edge was then tilted off the *c*-axis $[0\,0\,0\,\bar{1}]$ direction by the angle, $\phi = \tan^{-1}(3\sqrt{3}\,a/c) = 72.76°$, where $3\sqrt{3}\,a$ is equal to $6\boldsymbol{d}_M$, as shown in **Figure 6(e)**. We also calculated the angle from the measured SAD data at the InN hexagonal thin disk in **Figure 6(b)**, and obtained that the ***d***-spacing between the $\{0\,0\,0\,\bar{1}\}$ planes was $\boldsymbol{d}_c = 0.5687$ nm and the ***d***-spacing between the $\{1\,\bar{1}\,0\,0\}$ planes was $\boldsymbol{d}_M = 0.3025$ nm, resulting in $\phi = \tan^{-1}(6\boldsymbol{d}_M/\boldsymbol{d}_c) = 72.60°$, which was in good agreement with the model predicted (**Table 1**).

4. InGaN/GaN quantum well

4.1. Growth

The growth mechanism of the awl-shaped GaN microdisk is divergently self-assembled, indicating that the hexagonal neck area for initial nucleation between GaN microdisk and LAO substrate is very small (diameter ~100 nm), and the strain due to the lattice-mismatch between GaN and LAO substrate will not be delivered to the awl-shaped GaN microdisk at the top. This is the way that the GaN microdisk can be grown in balance with a good awl-shape of

	Growth temperature	Growth mechanism (lateral over-growth)	Oblique angle	Application
GaN microdisk	620°C	One d_M/d_c	28°	Base for InGaN/GaN QW
InN microdisk	470°C	Six d_M/d_c	73°	Base for InGaN/GaN QW

Table 1. Comparison of properties of GaN and InN microdisks.

hexagonal disk. The experimental results revealed that the awl-shaped GaN microdisk exhibited a high-quality single crystal. Therefore, the awl-shaped GaN microdisk can be regarded as a nearly freestanding substrate (strain-free) to grow the $In_xGa_{1-x}N$/GaN multiple quantum wells (MQWs) on its top. The $In_xGa_{1-x}N$/GaN double quantum well (DQWs) microdisk sample was grown on a high-quality 1 × 1 cm² LAO (1 0 0) substrate by low-temperature PA-MBE system (Veeco Applied-GEN 930). The LAO substrate was cut from the crystal ingot, fabricated by the traditional Czochralski pulling technique. The growth details were described completely in our previous paper which was published in *Applied Physics Letters* [18].

4.2. Characteristics of InGaN/GaN microdisks

The surface morphology of the InGaN/GaN microdisk sample was evaluated by the field emission scanning electron microscopy (FE-SEM, SII-3050). **Figure 7** shows SEM images with a tilted-angle view and a top view of the sample, respectively. The surface morphology of the sample was formed by the two-orientation growth mechanism. Comparing with the surface morphology of GaN microdisks, the shape of the as-grown InGaN/GaN DQW microdisks still maintains the hexagonal shape. **Figure 7(f)** shows the enlarged SEM image with a top view of the GaN hexagonal microdisk, which is shown in the centre of **Figure 7(d)**, and the diameter of the centre GaN microdisk is about 1.96 μm.

The optical properties of the sample were measured by photoluminescence (PL, HORIBA HR800) at room temperature with a light source of He-Cd 325 nm laser. We performed the laser beam focusing on three different spots (S1–S3) and compared the results with the spot without any microdisk (i.e. mostly ***M***-plane GaN and labelled as background), as shown in **Figure 8**. Two major peaks were obtained for each measurement (S1–S3). These two major peaks were confirmed by a non-linear Gaussian-function curve fitting with the software Origin (Pro. 8.0). The result of the non-linear Gaussian-function curve fitting showed that the positions of two major peaks for three spots were very consistent. The averaged value for the first peak is (2.199 ± 0.001) eV with the FWHM value equal to (0.410 ± 0.005) eV. It is due to the band-gap transition of InGaN wells. According to Vegard's law [19] with the bowing effect of bulk $In_xGa_{1-x}N$: $E_g(x) = [3.42 - x*2.65 - x*(1 - x)*2.4]$ eV [8], we estimated the content of indium in the $In_xGa_{1-x}N$ DQWs, which is found to be about 28%. We note that the bowing factor needs to be modified slightly for a quantum well, and the energy-shift due to the quantum confinement will also result in a small deviation to the indium concentration of the $In_xGa_{1-x}N$/GaN DQWs. The averaged value for the second peak is (3.380 ± 0.001) eV with

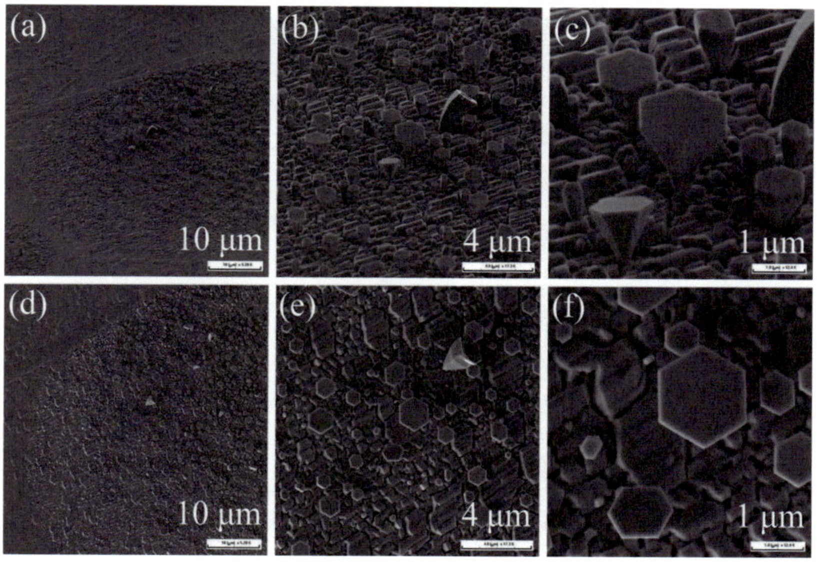

Figure 7. (a)–(c) The tilt-view SEM images of the sample, the scale bars are 10, 4 and 1 μm, respectively. (d)–(f) The top-view SEM images of the sample, the scale bars are 10, 4 and 1 μm, respectively.

the FWHM value equal to (0.043 ± 0.001) eV. It is due to the band-gap transition of wurtzite GaN. The FWHM value of GaN is smaller than that of $In_xGa_{1-x}N$ QDWs because the defect density in GaN is much less than that in InGaN. The PL intensity from $In_xGa_{1-x}N$ quantum wells is higher than that from GaN due to the effect of quantum confinement. It also dominates the PL spectrum of background area tremendously.

An $In_xGa_{1-x}N$/GaN microdisk was used to evaluate the details of light-emitting area by cathodoluminescence (CL, JEOL JSM-6330) and secondary electron images (SEI) measurements. The result of CL measurement is shown in **Figure 9**. We obtained the CL spectrum by detecting the photon energy from 700 (e.g. 1.77 eV) to 340 nm (e.g. 3.64 eV) with an accelerative voltage of 10 kV and an extraction voltage of photon-electric magnitude tube of 1100 V. Under such a condition, we observed two major peaks at (2.221 ± 0.0001) eV with the FWHM value equal to (0.182 ± 0.001) and (3.398 ± 0.0001) eV with the FWHM value equal to (0.046 ± 0.001) eV. We also observed the smallest peak at (2.805 ± 0.002) eV with the FWHM value equal to (0.110 ± 0.001) eV. The insets of **Figure 9** show the CL images corresponding to the three peak energies. The CL images show that the peak for the wavelength of 364 nm (e.g. 3.407 eV) is mainly emitted from wurtzite GaN microdisks. The peak for the wavelength of 560 nm (e.g. 2.214 eV) is the highest peak and mainly emitted from the $In_xGa_{1-x}N$ quantum wells in the $In_xGa_{1-x}N$/GaN microdisks. We also observed the edge of microdisk is brighter than the central area. It might arise from the enhanced emission by total internal reflection in the DQWs structure [20]. The optical properties of these two major peaks are consistent with the results of PL spectra, but the smallest peak for the wavelength of 450 nm (e.g. 2.756 eV) is not observed in the PL spectra. The peak at 2.805 eV was attributed to the energy level related to the structural defects (e.g. Y_{10} in reference [15]) in microdisks.

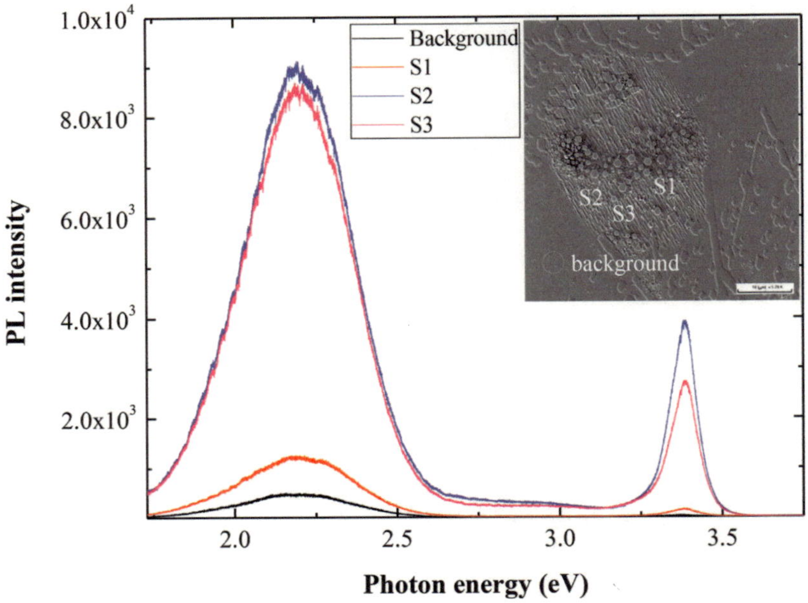

Figure 8. The PL spectra measured at room temperature for different spots (S1–S3) of the $In_xGa_{1-x}N$/GaN DQWs and the *M*-plane background of the sample.

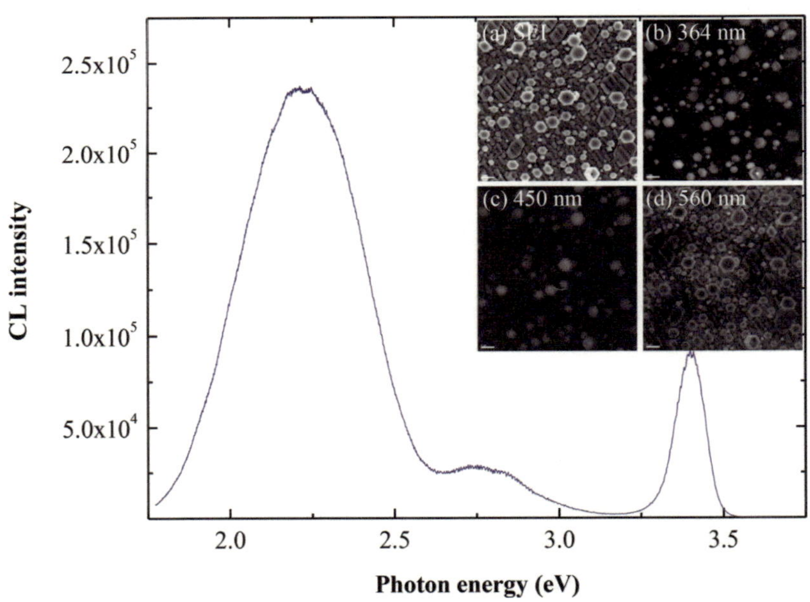

Figure 9. The CL spectrum measured at room temperature for the $In_xGa_{1-x}N$/GaN QW of the sample. The inset (a) shows the SEM image. The inserts (b), (c) and (d) show the CL images for wavelength of 364 nm, 433 nm, and 550 nm to the peak energies of CL spectrum.

The microstructure of the $In_xGa_{1-x}N$/GaN microdisk sample was analysed by field emission transmission electron microscopy (FE-TEM) (Phillips, model Tecnai F-20) with an electron voltage of 200 kV. The cross-sectional TEM specimen of the sample was prepared by a dual-beam FIB system (Seiko Inc., SII-3050), on the cleavage plane along the $[1\bar{1}00]$ direction of the *c*-plane GaN hexagonal microdisk. The FIB was performed with an accelerated voltage of 30 kV to cut the samples roughly and then refined the specimen further by an accelerated voltage of 5 kV. **Figure 10(a)** and **(b)** shows the bright field image with the $[11\bar{2}0]_{GaN}$ // $[001]_{LAO}$ zone axis, and we found that the height of the microdisk is about 2.12 μm. **Figure 10(c)** shows the scanning transmission electron microscopy (STEM) images of the $In_xGa_{1-x}N$/GaN MQWs grown on the GaN microdisk. From the high contract image between $In_xGa_{1-x}N$ wells and GaN barriers, as shown in **Figure 10(f)**, the thicknesses of $In_xGa_{1-x}N$ well and GaN barrier evaluated from the STEM image were found to be about 2.75 and 17 nm, respectively. The quantum-well-thickness of 2.75 nm offers a good quantum confinement for the charged carriers and photon emission in the $In_xGa_{1-x}N$/GaN DQWs. It is consistent with the results of PL and CL spectra. The selective area diffraction (SAD) pattern at the top area of GaN microdisk shown in **Figure 10(d)** clearly showed one single rectangular diffraction pattern at the location of DP01, indicating that the hexagonal microdisk was uniquely formed by the ***c***-plane wurtzite GaN. The ***d***-spacing between the $\{000\bar{1}\}$ planes of GaN hexagonal microdisk was measured to be $d_c = 0.5172$ nm and the ***d***-spacing between the $\{1\bar{1}00\}$ planes of GaN hexagonal

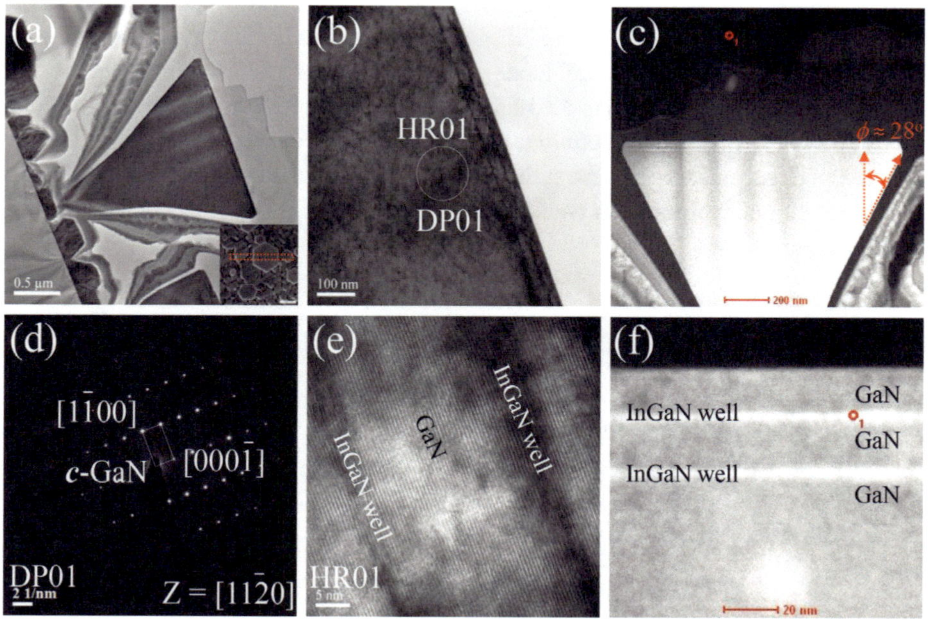

Figure 10. TEM analyses of the InGaN/GaN QW microdisk: (a) and (b) show the bright field images with $[1\ 1\ \bar{2}\ 0]_{GaN}$ // $[0\ 0\ 1]_{LAO}$ zone axis, the scale bars are 0.5 and 100 nm, respectively. (c) The STEM image taken at the top area of InGaN/GaN QW, the scale bar is 200 nm. The selective area diffraction patterns taken at the points shown in (b) are presented in (d), the scale bar is 2 (1/nm). The high-resolution TEM images taken at the point shown in (b) are presented in (e), the scale bars are 2 nm. (f) The enlarged STEM image of InGaN/GaN QW (the scale bar is 20 nm).

microdisk was d_M = 0.2764 nm. The angle between edge and growth direction can be obtained by $\phi = \tan^{-1}(d_M/d_c) = 28.1°$, which was in good agreement with the model predicted. The angle between the edge and the growth direction examined directly by the STEM image can be found about 28°, as shown in **Figure 10(c)**. The high-quality crystalline micro structures for the $In_xGa_{1-x}N$/GaN DQWs were reconfirmed by the high-resolution TEM images, as shown in **Figure 10(e)**. It showed that the $In_xGa_{1-x}N$ wells were well-stacked on the high-quality GaN microdisk to form well-assembled crystalline $In_xGa_{1-x}N$/GaN DQWs with some minor structural defects (e.g. dislocations or stacking-faults) occurred in the $In_xGa_{1-x}N$ wells. The HR results indicated that the density of structural defects in $In_xGa_{1-x}N$ wells is greater than that in GaN layers. It supported the results of FWHM analyses of PL spectra, indicating that FWHM (0.173 eV) for the peak of 2.192 eV from $In_xGa_{1-x}N$ wells was greater than FWHM (0.043 eV) for the peak of 3.383 eV from the wurtzite GaN layer. The structural defeats in $In_xGa_{1-x}N$ wells yield the peak of 2.805 eV in CL measurement.

5. Conclusion

We have grown GaN and InN hexagonal microdisks on the LAO substrates at low temperatures (GaN at 630°C and InN at 470°C) by PA-MBE. From the SEM images and TEM analyses, we found that 3D *c*-plane hexagonal microdisks and 2D ***M***-plane epi-film were grown on the LAO substrate. From TEM analyses, the oblique angle of GaN and InN hexagonal microdisks can be examined directly to be about 28 and 73°, respectively. The lateral over-growth mechanism causes the bigger oblique angle of InN hexagonal thin disks at low growth temperature. From PL and CL analyses, we found that the high-intensity light of 367-nm wavelength (3.380 eV) and 566-nm wavelength (2.192 eV) emitted from the GaN microdisks and $In_xGa_{1-x}N$/GaN quantum wells, respectively. Therefore, the stain-free microdisk provides an opportunity to fabricate $In_xGa_{1-x}N$/GaN microdisk quantum well for the application of full-colour micron LED without the sapphire substrate.

Acknowledgements

The project was supported by the Ministry of Science and Technology of Taiwan and the Core Facilities Laboratory for Nanoscience and Nanotechnology in Kaohsiung and Pintung Area.

Author details

Chen-Chi Yang, Ikai Lo*, Yu-Chi Hsu and Hong-Yi Yang

*Address all correspondence to: ikailo@mail.phys.nsysu.edu.tw

Department of Physics, Center for Nanoscience and Nanotechnology, National Sun Yat-Sen University, Kaohsiung, Taiwan

References

[1] Nakamura S, Senoh M, Iwasa N, Nagahama SI, Yamada T, Mukai T. Superbright green InGaN single-quantum-well-structure light-emitting diodes. Japanese Journal of Applied Physics. 1995;**34**(Part 2, Number 10B):L1332-L1335. DOI: 10.1143/JJAP.34.L1332

[2] Nakamura S, Pearton S, Fasol G, editors. The Blue Laser Diode. 2nd ed. The Complete Story. Springer ed. Berlin: Springer Science & Business Media; 2000. p. 367. DOI: 10.1007/978-3-662-04156-7

[3] Lo I, Tsai JK, Yao WJ, Ho PC, Tu LW, Chang TC, Elhamri S, Mitchel WC, Hsieh KY, Huang JH, Huang HL, Tsai WC. Spin splitting in modulation-doped $Al_xG_{a1-x}N$/GaN heterostructures. Physics Review B. 2002;**65**(16):161306. DOI: 10.1103/PhysRevB.65.161306

[4] Lo I, Wang WT, Gau MH, Tsai JK, Tsay SF, Chiang JC. Gate-controlled spin splitting in GaN/AlN quantum wells. Applied Physics Letters. 2006;**88**:082108. DOI: 10.1063/1.2178505

[5] Lo I, Hsieh CH, Hsu YC, Pang WY, Chou C. Self-assembled GaN hexagonal micropyramid and microdisk. Applied Physics Letters. 2009;**94**:062105. DOI: 10.1063/1.3079078

[6] Fred Schubert E, Kim JK. Solid-state light sources getting smart. Science. 2005;**308**(5726):1274-1278. DOI: 10.1126/science.1108712

[7] Ponce FA, Bour DP. Nitride-based semiconductors for blue and green light-emitting devices. Nature. 1997;**386**(6623):351-359. DOI: 10.1038/386351a0

[8] Vurgaftman I, Meyer JR. Band parameters for nitrogen-containing semiconductors. Journal of Applied Physics. 2003;**94**:3675. DOI: 10.1063/1.1600519

[9] Madelung O, editor. Semiconductors: Group IV Elements and III–V Compounds. New York: Spring; 1991. p. 163. DOI: 10.1007/978-3-642-45681-7

[10] El-Masry NA, Piner EL, Liu SX, Bedair SM. Phase separation in InGaN grown by metalorganic chemical vapor deposition. Applied Physics Letters. 1998;**72**:40. DOI: 10.1063/1.120639

[11] Tsai JK, Lo I, Chuang KL, Tu LW, Huang JH, Hsieh CH, Hsieh KY. Effect of N to Ga flux ratio on the GaN surface morphologies grown at high temperature by plasma-assisted molecular-beam epitaxy. Journal of Applied Physics. 2004;**95**:460. DOI: 10.1063/1.1634388 95: 460

[12] Lo I, Wang YC, Hsu YC, Shih CH, Pang WY, You ST, Hu CH, Chou MMC, Hsu GZL. Electrical contact for wurtzite GaN microdisks. Applied Physics Letters. 2014;**105**(8):082101. DOI: 10.1063/1.4894080

[13] Hsieh CH, Lo I, Gau MH, Chen YL, Chou MC, Pang WY, Chang YI, Hsu YC, Sham MW, Chiang JC, Tsai JK. Self-assembled c-plane GaN nanopillars on γ-$LiAlO_2$ substrate grown by plasma-assisted molecular-beam epitaxy. Japanese Journal of Applied Physics. 2008;**47**(2R):891. DOI: 10.1143/JJAP.47.891

[14] Lo I, Hsieh CH, Chen YL, Pang WY, Hsu YC,Chiang JC, Chou MC, Tsai JK, Schaadt DM. Line defects of M-plane GaN grown on g-$LiAlO_2$ by plasma-assisted molecular beam epitaxy. Applied Physics Letters. 2008;**92**:202106. DOI: 10.1063/1.2924288

[15] Reshchikov MA, Morkoc HJ. Luminescence properties of defects in GaN. Journal of Applied Physics. 2005;**97**:061301. DOI: 10.1063/1.1868059

[16] Yang CC, Lo I, Hu CH, Huang HC, Chou MMC. Growth of InN hexagonal microdisks. AIP Advances. 2015;**6**:085015. DOI: 10.1063/1.4961699

[17] Denton AR, Ashcroft NW. Vegar's law. Physical Review A. 1991;**43**(6):3161. DOI: 10.1103/PhysRevA.43.3161

[18] Hsu YC, Lo I, Shih CH, Pang WY, Hu CH, Wang YC, Tsai CD, Chou MMC, Gary Z. Green light emission by InGaN/GaN multiple-quantum-well microdisks. Applied Physics Letters. 2014;**104**:102105. DOI: 10.1063/1.4868417

[19] Vegard L. Die Konstitution der Mischkristalle und die Raumfüllung der Atome. Zeitschrift Fur Physik. 1921;**5**:17-26. DOI: 10.1007/BF01349680

[20] Choi S, Ton-That C, Phillips MR, Aharonovich I. Observation of whispering gallery modes from hexagonal ZnO microdisks using cathodoluminescence spectroscopy. Applied Physics Letters. 2013;**103**:171102. DOI: 10.1063/1.482648s1

Chapter 8

Heterostructures of III-Nitride Semiconductors for Optical and Electronic Applications

Basanta Roul, Greeshma Chandan,
Shruti Mukundan and Saluru Baba Krupanidhi

Additional information is available at the end of the chapter

http://dx.doi.org/10.5772/intechopen.70219

Abstract

III-Nitride-based heterostructures are well suited for the fabrication of various optoelectronic devices such as light-emitting diodes (LEDs), laser diodes (LDs), high-power/-high-frequency field-effect transistors (FETs), and tandem solar cells because of their inherent properties. However, the heterostructures grown along polar direction are affected by the presence of internal electric field induced by the existence of intrinsic spontaneous and piezoelectric polarizations. The internal electric field is deleterious for optoelectronic devices as it causes a spatial separation of electron and hole wave functions in the quantum wells, which thereby decreases the emission efficiency. The growth of III-nitride heterostructures in nonpolar or semipolar directions is an alternative option to minimize the piezoelectric polarization. The heterostructures grown on these orientations are receiving a lot of focus due to their potential improvement on the efficiency of optoelectronic devices. In the present chapter, the growth of polar and nonpolar III-nitride heterostructures using molecular beam epitaxy (MBE) system and their characterizations are discussed. The transport properties of the III-nitride heterostructure-based Schottky junctions are also included. In addition, their applications toward UV and IR detectors are discussed.

Keywords: heterostructures, InN/GaN, InGaN/Si, InGaN/GaN, HRXRD

1. Introduction

There has been remarkable progress in the development of group III-nitride-based heterostructures because of their potential application in fabricating various optical and electronic devices such as light-emitting diodes (LEDs), laser diodes (LDs), tandem solar cells, field-effect transistors (FETs), and Schottky junctions. Heterostructures are ubiquitous of semiconductor devices, and most of semiconductor devices have two or more semiconductor materials. A heterostructure is formed between two layers of dissimilar semiconductors having unequal energy bandgap. In order to realize high-performance devices, growth of device quality heterostructures is required. Yoshida et al. reported improved cathodoluminescence efficiency of the GaN layer grown on sapphire using AlN as a buffer layer [1]. Later Akasaki et al. [2] and Nakamura et al. [3] had employed a two-step growth method, where a nucleation layer was grown at low temperature followed by the GaN layer at high temperatures. Nakamura et al. fabricated the first blue LED consisting of a p-GaN/n-InGaN/n-GaN double heterostructure in 1993 [4] for which he won the Nobel Prize in 2014 and the first violet laser consisting of InGaN/GaN/AlGaN heterostructure in 1996 [5]. Similarly, Khan et al. [6] achieved the first breakthrough in the field of high mobility transistors based on AlGaN/GaN heterostructure in 1994.

So far most of indium gallium nitride (InGaN)-based LEDs are built along Ga-polar (0001) orientation, which is susceptible to the strong internal electric field induced by the spontaneous and the piezoelectric polarization in wurtzite III-nitrides. The effect of polarization will be explained in detail in the coming section. There have been concerted efforts in exploring III-nitride materials and devices along nonpolar and semipolar crystallographic orientations [7, 8]. The two major challenges in the field of InGaN-based LEDs are the "efficiency droop" under a high injection current density and the "green gap" in the plot of efficiency versus emission wavelength [9]. Very promising reports of LEDs and laser diodes on nonpolar and semipolar GaN bulk substrates, in the longer wavelength of green and yellow, tend to validate the concept of crystallographic engineering [10]. However, nonpolar and semipolar GaN bulk substrates are presently very small in size and expensive in price. Therefore, most of the current research in this field is focused on the growth of high-quality epilayers on nonnative substrates which are available in large wafer sizes and also cost-effective than the native substrates, thus paving the way for the commercialization of devices based on nonpolar and semipolar GaN [11]. However, the lattice mismatch along different directions poses difficulty in the hetero-epitaxial growth of nonpolar (a- and m-plane) and semipolar GaN on the foreign substrates, often resulting in nonuniform nucleation. This leads to the growth of GaN with a defective microstructure, arising due to the formation of basal-plane stacking faults (BSFs) and partial dislocations (PDs) [12, 13].

The absence of inversion symmetry in wurtzite GaN along the [0001] direction indicates that [0001] and [000−1] directions are not equivalent. Along this crystal direction if one face ends with Ga atoms, the other face will end with N atoms in the place of Ga atom and vice versa. This asymmetric arrangement of Ga and N atoms along the [0001] direction gives rise to charge polarity and thus can be referred as microscopic dipoles. The presence of polarity and lack of inversion symmetry lead to the generation of macroscopic polarization along [0001] direction and are usually referred to as spontaneous polarization. The word "spontaneous" means that it is arising only due to the crystallographic arrangement of the end faces but not due to the strain [14]. In III-nitrides, asymmetry of inversion is present only along the c-axis. Hence, P_{SP} is parallel to this direction, and c-plane nitrides are therefore called polar nitride materials. The c-axis direction consists of two opposite stacking sequences of atomic layering, resulting in either cation-face (metal) or anion-face (nitrogen) epitaxy. Ga and In polarity is in the [0001] direction, and N-polarity is in the [000−1] direction as shown in **Figure 1** [15]. The

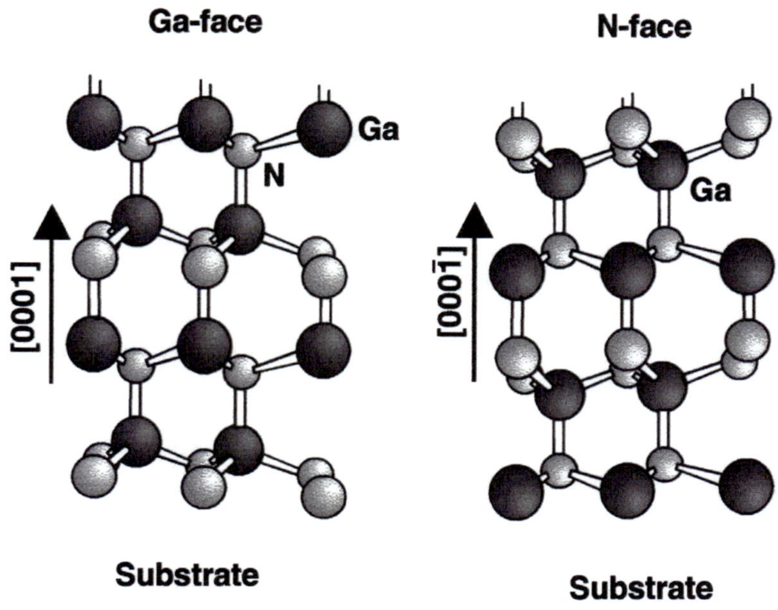

Figure 1. Ball and stick model illustrating crystal structure of wurtzite Ga-polarity and N-polarity GaN. Reprinted with permission from [15].

presence of external stress due to lattice mismatch in the films grown on foreign substrates, or in heterostructures, results in additional, piezoelectric contribution to the polarization. The total polarization, therefore, can be presented as the sum of the two components:

$$P_{total} = P_{spon} + P_{piezo} \tag{1}$$

The type of stress determines the direction of piezoelectric polarization. When a layer of smaller lattice constant than that of GaN, such as AlGaN, is grown on GaN, the grown AlGaN layer experiences tensile stress. Whereas when a layer with larger lattice constant than GaN such as InGaN is grown on GaN, the resulting strain in the InGaN layer is compressive. The piezoelectric field generated as a result of the tensile strain in the case of AlGaN/GaN and compressive strain in the case of InGaN/GaN is parallel and antiparallel, respectively. The overall polarization effect in InGaN is therefore smaller compared to AlGaN [16]. The polarization field plays a very pivotal role in GaN-based LED devices. Currently, all highly efficient blue or UV LED devices are based on multi-quantum well (MQW) structures. In a MQW structure, very few atomic layers of a narrow bandgap material, referred to as wells, are sandwiched between thicker wide bandgap materials, referred to as barriers. In quantum well (QW) structures, the charge carriers are confined to wells with high-energy barriers on either side, thus preventing the charge carriers from escaping without recombining with their counterparts and thus increasing the probability of radiative recombination. The presence of spontaneous and piezoelectric polarization in QW leads to asymmetry in the electric-field profile and results in bending of the conduction and valence bands, thus spatially separating the charge carriers. Due to this spatial separation of the charge carriers, the overlapping of charge carrier wave functions is substantially reduced resulting in lower recombination probability. The bending of bands also leads to bandgap shrinkage. As a result the emitted radiation is red-shifted [17]. This process is referred to as quantum-confined Stark effect and is undesirable.

Apart from the above-mentioned detrimental effects, the effective width of QWs is reduced as a consequence of tilt in the band edges which leads to a higher charge carrier density and may eventually lead to nonradiative Auger recombination. The effective barrier height is also lowered due to band bending, which means the carrier confinement is weakened with increasing bias voltage leading to carrier leakage. It is believed that these two non-radiative recombination mechanisms are responsible for the reduced efficiency in GaN-based LEDs when operated at higher current [18]. Besides wurtzite nitrides, cubic nitrides also have similar bandgaps and are free from the spontaneous polarization. However, due to the instability of cubic nitrides and the poor quality, they are less preferred for device applications. Growth of wurtzite materials that have either no polarization field or reduced one in the growth direction can solve these problems. **Figure 2(b–i)** shows the wurtzite III-nitride planes which are perpendicular or inclined to the [0001] direction [19]. There are two surfaces perpendicular to the

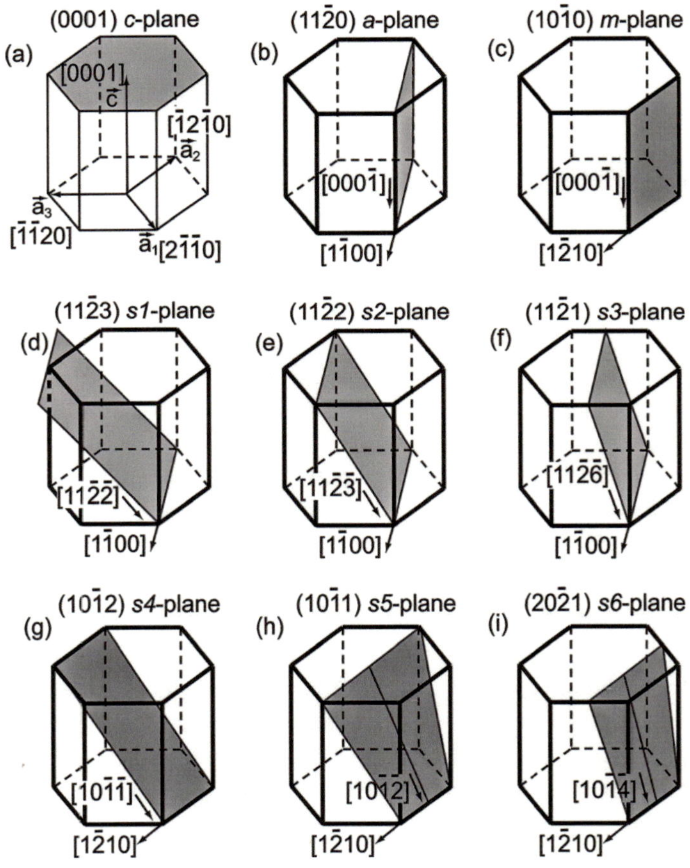

Figure 2. Schematic views of (a) polar c-plane; (b) and (c) nonpolar a- and m-planes; (d)–(i) various semipolar s-planes. Reprinted with permission from [19].

c-axis, which have equal number of Group III and V atoms and are called nonpolar surfaces. Alternatively, inclined surfaces such as (1 0 −1 −3), (1 0 −1 −1), and (1 1 −2 2) are known to have lower polarization fields and are often called semipolar planes [20].

Compared to the conventionally oriented c-plane GaN, nonpolar and semipolar planes were considered "unstable" for a long time. Rough and faceted surfaces have been a prolonged problem for device fabrication on these directions [21, 22]. In 2000, Waltereit et al. first demonstrated planer m-plane GaN growth via molecular beam epitaxy [7]. This demonstration was followed by Craven et al.'s metal organic chemical vapor deposition growth of a-plane GaN films in 2002 [23]. While considerable progress started after this breakthrough in the growth of thin films in nonpolar orientation, thick-film or bulk growth of this orientation continued to be elusive, hence limiting the performance of nonpolar GaN-based devices due to lack of suitable substrates. Achieving low defect density, nonpolar oriented film is a trending problem and exponential progress is seen in this field over the past decade. In the present chapter, the growth of polar and nonpolar III-nitride heterostructures using molecular beam epitaxy (MBE) and their characterizations will be presented. The transport properties of the III-nitride heterostructure-based Schottky junctions will be presented. In addition, applications toward UV and IR detectors will be discussed.

2. Growth and characterizations of III-nitride heterostructures

2.1. Polar InN/GaN heterostructures

The InN/GaN heterostructure system has several advantages which includes the high rate of optical phonon emission in InN (2.5×10^{13} s^{-1}), high peak value of the steady-state electron drift velocity in InN (5×10^{7} cm s^{-1}), and large conduction band offset. However, the fabrication of high-quality InN/GaN heterostructures is a challenging issue due to the difficulty in continuous growth of InN and GaN films because of large difference in the optimum growth temperature between them (InN ~500°C and GaN ~750°C). In addition, due to the low dissociation temperature of InN, it is very difficult to get high-quality InN/GaN heterostructure at the growth temperature of InN. In addition, the large lattice mismatch between InN and GaN (11%) results in a poor interface. Several groups [24–26] have studied the interfaces of the heterostructures like InN/GaN, GaN/ZnO, and GaN/AlN. InN/GaN MQWs with 1 and fractional monolayers of InN were proposed and experimentally demonstrated by Yoshikawa et al. [27]. Similarly, InN/GaN single-quantum well and double heterostructures were fabricated by the PAMBE [28]. In this section, InN/GaN heterostructures were grown using MBE system. InN thin films of thicknesses around 300 nm were epitaxially grown on 4 μm-GaN/Al_2O_3 (0 0 0 1) templates at different substrate temperatures. Thermal cleaning of the GaN templates was carried out at 700°C for 5 min in the presence of nitrogen plasma. Following that, a step growth method was employed to grow high-quality InN epilayers. The first step involved the growth of low-temperature InN nucleation layer at 400°C for 15 min, which resulted in the formation of thin buffer layer ~20 nm. Subsequently, the substrate temperature was increased to 450°C (sample A), 470°C (sample B), 500°C (sample C), and 530°C (sample D) to grow active InN epilayers. The nitrogen flow rate and the forward RF power of the plasma source were set to 0.5 sccm and 350 W. The indium BEP was 1.53×10^{-7} mbar.

Figure 3(a) shows a 2θ-ω scan of InN films grown at different growth temperatures [29]. The peaks at 2θ = 31.3 and 65.5° are attributed to the (0002) and (0004) planes of the InN epilayers, whereas the peaks at 2θ = 34.56 and 72.81° are attributed to the (0002) and (0004) planes of the GaN templates. The peak at 2θ = 41.69° corresponds to the (0006) plane of sapphire substrate. The InN films grown at low (450°C) and high (530°C) temperatures show a peak at around 2θ = 33° indicating the presence of In metal. The presence of In metal could be due to the low migration of In at low temperature and more dissociation of InN at high temperature. The structural quality of the films was evaluated from the full width at half maximum (FWHM) of (0002) InN X-ray rocking curve (XRC). The rocking curves of the (0002) InN reflection are shown in **Figure 3(b)** [29]. Growth of InN epilayers at high temperature improved crystal quality, i.e., the FWHM of the (0002) InN XRC decreased from 532.8 to 450 arcsec corresponding to growth temperatures 450 and 500°C. The low migration velocities of In and N atoms at low growth temperature are the most probable reason for the relatively inferior crystal quality [30]. On the other hand, growth of InN at higher temperature (530°C) resulted in pronounced dissociation of InN, thus leading to high FWHM of 716.4 arcsec. The screw dislocation density of InN films, as calculated from the FWHM of the rocking curves, was found to be 2.27×10^8, 2.03×10^8, 1.62×10^8, and 4.12×10^8 cm^{-2} for samples A, B, C, and D, respectively. In addition, InN films show n-type conductivity with carrier concentrations in the order of 10^{18} to 10^{19} cm^{-3}. As growth temperature increases from 450 to 500°C, the carrier concentration decreases from 1.57×10^{19} to 3.10×10^{18} cm^{-3}. The high carrier concentration at low growth temperature may be due to relatively poor crystallinity. On the other hand, the carrier concentration increased to 1.10×10^{19} cm^{-3} at high growth temperature of 530°C, which may be associated with the large dissociation of InN.

The room temperature optical absorption spectra squared as a function of growth temperature is shown in **Figure 3(c)** [29]. The absorption spectrum exhibits characteristic interference fringes due to the underlying thicker GaN epilayer. The absorption edge of the InN films was estimated by extrapolating the linear part of the squared absorption down to the photon energy axis. The energy corresponding to this absorption edge is the amount of energy needed by an electron to make a vertical transition from the upper valence band to the Fermi surface in the conduction band. Therefore, this energy can be considered as the Fermi-level energy in the conduction band. A strong shift was observed in the absorption edge with a change in carrier concentration. This is usually referred as Burstein-Moss shift which is a commonly observed phenomenon in narrow-gap semiconductors owing to non-parabolic conduction band [31].

2.2. Nonpolar InN/GaN heterostructures

Growth of nonpolar a-plane InN/GaN heterostructures has been an important subject recently due to its potential improvement on the efficiency of III-nitride-based optoelectronic devices [32, 33]. However, growth of high-quality nonpolar InN/GaN heterostructures is challenging due to the low thermal decomposition temperature of InN film and high equilibrium vapor pressure of nitrogen. Despite the growth of high-quality nonpolar GaN films [23, 34, 35],

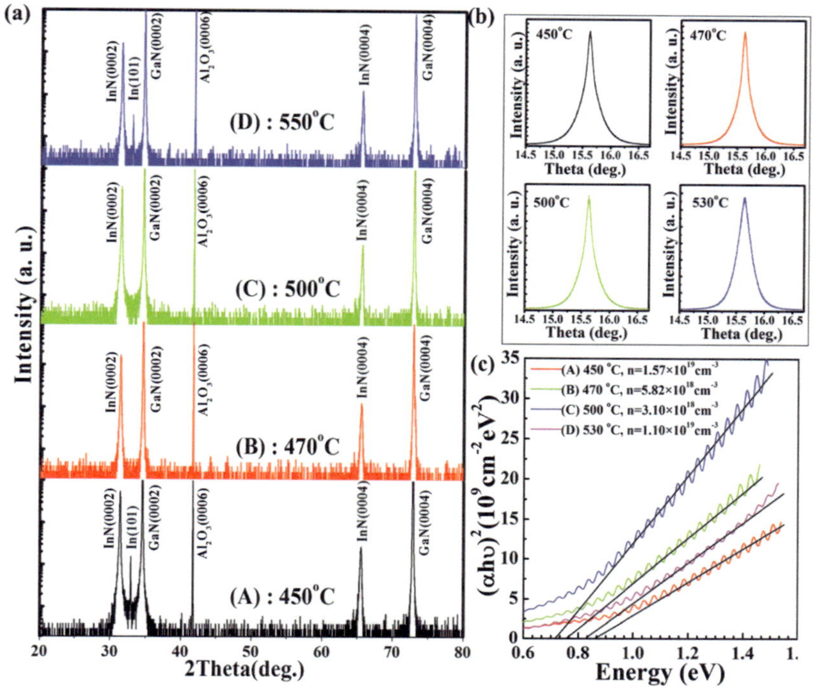

Figure 3. (a) 2θ-ω HRXRD scanning curve of InN films grown at different growth temperatures, (b) the XRC of the (0002) reflections of InN films grown at different growth temperatures, and (c) optical absorption spectra of InN films. Reprinted with permission from [29].

there are very few reports on the growth of nonpolar InN films [36, 37]. The earlier reports on the InN on r-sapphire substrates indicate the growth of cubic (001) [38, 39] and polar (0001) InN [40]. The nonpolar a-plane InN was demonstrated by using GaN buffer layer on r-plane sapphire [41, 42]. In this section, nonpolar (1 1 −2 0) a-InN/GaN heterostructures were grown on r-plane and m-plane sapphire substrates, respectively, using MBE system. Prior to the growth, thermal cleaning of the sapphire substrate was carried out at 850°C inside MBE chamber for 30 min under ultrahigh vacuum. RF power and flow rate were kept constant at 350 W and 0.5 sccm, respectively. (1 1 −2 0) a-GaN buffer layer was grown at 760°C temperature. The growth of InN epilayers was carried out using two-step growth processes: growth of low-temperature InN buffer layer (~20 nm) on GaN under layer at 400°C and growth of InN epilayers at different temperatures. The growth temperature of (1 1 −2 0) a-InN varied from 470 to 530°C.

Figure 4(a) show 2θ-ω scans of nonpolar (1 1 −2 0) a-plane InN/GaN heterostructures grown on r-plane sapphire substrate at different growth temperatures [42]. The peaks at 2θ = 57.7 and 51.61° are assigned to the a-plane GaN and a-plane InN reflections, respectively, along with that from the r-plane sapphire substrate. The sample grown at 530°C shows the peak at 32.9° which corresponds to In (101) reflections. **Figure 4(b)** shows the RHEED patterns in the azimuths [0001] for a-plane GaN and a-plane InN [43]. The Bragg spots appear with weak streaky lines in the RHEED patterns observed for a-plane GaN that confirms the reasonable smooth surface [44]. Spotty nature of a-plane InN RHEED pattern indicates the 3D growth of

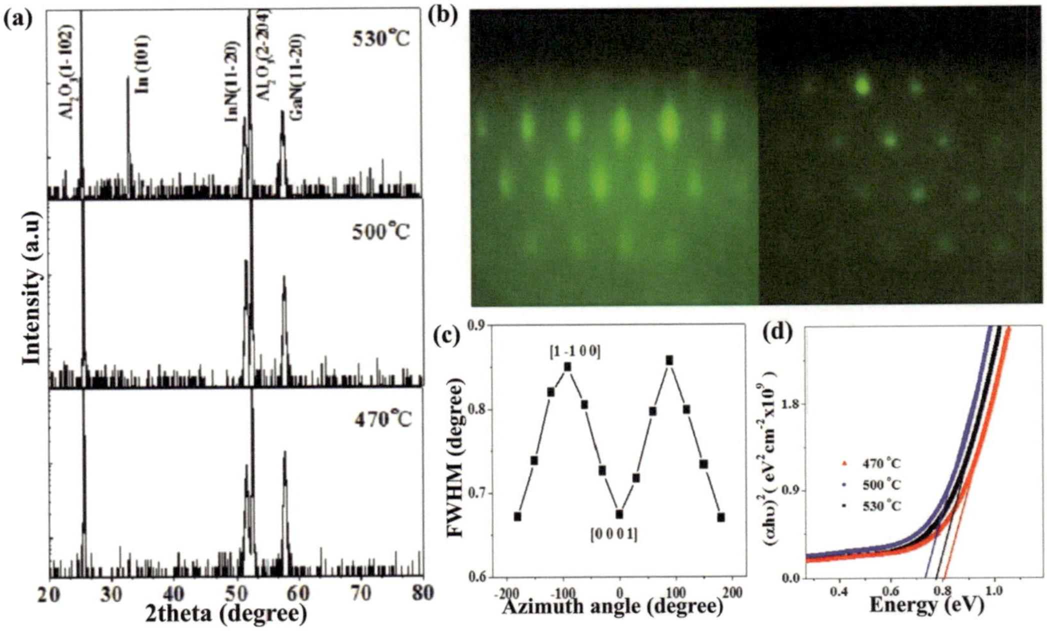

Figure 4. (a) 2θ-ω HRXRD scans of nonpolar a-plane InN film grown on GaN/r-Al_2O_3 substrate at different temperatures, (b) RHEED patterns of a-plane (11−20) GaN and a-plane (11−20) InN taken along (0 0 0 1) azimuth, (c) rocking curve FWHM of nonpolar a-plane InN grown at 500°C with different azimuth angles, and (d) optical absorption spectra of nonpolar a-plane InN. Reprinted with permission from [42, 43].

InN [43]. The X-ray rocking was carried out to see the structural quality of the samples. The FWHM of the rocking curve along with different azimuth angles has been plotted in **Figure 4(c)** for InN film grown at 500°C [42]. The variation of FWHM was found to be M-type with respect to the azimuth angles. The FWHM of the (1 1 −2 0) reflection was found to be strongly dependent on the azimuth angle with respect to the scattering plane. Azimuth angle was defined as zero when the incident beam is parallel to the [0001] direction. Darakchieva et al. reported similar kind of M-type behavior in a-plane InN epilayers grown on r-sapphire substrates with GaN buffer layer [45]. The FWHM of (1 1 −2 0) reflection along [0 0 0 1] and [1 −1 0 0] directions was determined, and the values are 0.67 and 0.85°, respectively.

Absorption spectroscopy was used to determine the bandgap of nonpolar a-plane InN films and is shown in **Figure 4(d)** [42]. The direct optical bandgap for InN can be investigated by fitting the absorption data. The bandgap of InN epilayers grown at 470, 500, and 530°C as obtained by fitting the absorption data is found to be 0.81, 0.74, and 0.78 eV, respectively. This shows that the bandgap of the samples grown at low temperatures is blue shifted with respect to the bulk, which could be due to the high background carrier concentration [46]. The carrier concentration of InN films was estimated by using Hall measurements, and the carrier concentrations was found to be in the order of 10^{18} to 10^{19} cm^{-3} with growth temperature. The film grown at 500°C showed the lowest carrier concentration, whereas the film grown at 470°C showed a higher carrier concentration due to the poor crystallinity. The carrier concentration in the film grown at 530°C was also found to be higher and could be as a result of high dissociation rate of InN at that temperature [47].

2.3. Polar InGan/GaN heterostructures

Indium gallium nitride (InGaN), a ternary compound of III-nitride semiconductors, has received considerable attention due to its potential applications in optoelectronic devices [48–51]. The choice of InGaN as an active layer in high-performance optoelectronic devices is due to the advantage one gets in tuning the energy bandgap from visible to near-ultraviolet region by changing the In composition. The most challenging issues in InGaN-based nitride semiconductors include the spatial fluctuation of indium composition and the generation of dislocations at the interface of InGaN-based heterostructures due to the limited solubility of indium atom into GaN because of the difference in the In-N and Ga-N bond length [52–54]. Because of dislocations the non-radiative recombination increases, leading to the rapid decrease in the performance of the InGaN-based devices [55, 56].

InGaN films of thicknesses around 200 nm were grown on 4 μm GaN/Al_2O_3 (0 0 0 1) templates using PAMBE. After the templates were chemically degreased, they were outgassed at 700°C for 5 min in the presence of nitrogen plasma. Followed by thermal cleaning, the InGaN films were grown in a single step. The indium composition was varied by changing the In/Ga flux ratio and substrate temperature, whereas the nitrogen plasma conditions were constant, flow rate ~0.5 SCCM and forward RF power of 350 W. **Figure 5(a)** shows a 2θ-ω scan of as-grown InGaN films [57]. Well-resolved peaks corresponding to InGaN (0002) reflection was observed along with the GaN and Al_2O_3 peaks in all the samples. The peaks at 2θ = 34.56 and 41.69° attribute to the (0002) planes of the GaN films and (0006) plane of the sapphire substrate, respectively. The indium concentration in the as-grown InGaN films was determined by linearly interpolating the peak position of (0002) plane from their end binaries, by assuming that Vegard's law is valid [58, 59]. The indium concentration obtained for InGaN films with different growth conditions is given in **Table 1**. For a constant In/Ga flux ratio (0.61), the decrease in growth temperature 600 to 550°C leads to the suppression of spinoidal decomposition. The transformation of multiple peaks to a single peak corresponding to InGaN (0002) planes is the evidence. The samples grown at high temperature, ~560°C, with high In/Ga flux ratio (0.99) showed similar single peaks. However, the presence of In (101) peak states that metallic In was also present. Further, it was also observed that at growth temperature ~560°C, the increase in In/Ga ratio from 0.61 to 0.99 not only led to increase in the In incorporation but also decreased the spinoidal decomposition. Therefore, we could obtain high-quality, metal-free, single composition InGaN epilayers with 23% indium (sample E) on GaN/c-Al_2O_3 when grown at 550°C with In/Ga ~0.61.

The X-ray rocking curve (XRC) was carried out to see the structural quality of the InGaN film. **Figure 5(b)** shows the XRC of the (0002) InGaN reflection of sample E [57], and the corresponding FWHM was found to be 390 arcsec, indicating the high quality of the as-grown InGaN film. This value is comparable to the values in literature [60, 61]. Though this value indicates the quality of bulk film, it doesn't tell us much about the interface dislocations. The influence of such interface defects on the barrier height, and the ideality factor will be discussed in further sections. **Figure 5(c)** shows the room temperature PL spectra of sample E [57]. As we can see from the PL spectra, the emission peak at 2.48 eV corresponds to the free excitonic transition between valence and conduction bands of InGaN film. Also, one can notice the presence of fringes around this peak, which are a result of Fabry-Perot interference, hence

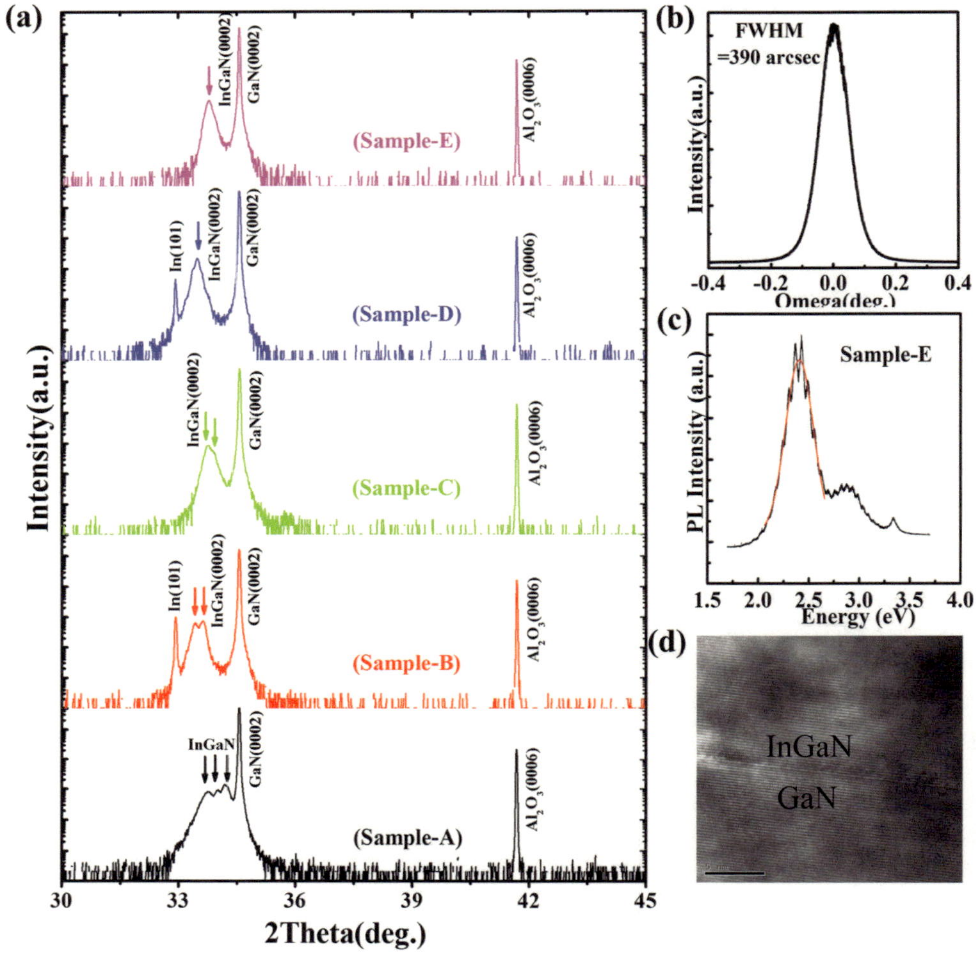

Figure 5. (a) 2θ-ω HRXRD scanning curve of InGaN films grown on GaN, (b) X-ray rocking curve of the (0002) plane of InGaN film (sample-E), (c) the room temperature photoluminescence spectra of InGaN film grown on GaN (sample E), and (d) HRTEM of InGaN/GaN interface. Reprinted with permission from [57].

Sample ID	In/Ga flux ratios	Growth temperature (°C)	In composition
Sample A	0.61	600	11, 16, 24%
Sample B	0.99	540	28, 35%
Sample C	0.61	560	20, 24%
Sample D	0.99	560	33% (In)
Sample E	0.61	550	23%

Table 1. Growth parameters used for the growth of InGaN films.

suggesting a smooth and abrupt interface. The film also exhibits a peak at 3.41 eV, which corresponds to underlying GaN layer. A second peak at 2.94 eV might correspond to the thin InGaN strain-relieving layer with ~6% of In [62, 63]. The indium composition of the InGaN layer was calculated using Vegard's law and was found to be 22% in sample E, which is in correlation with the value as estimated by HRXRD. **Figure 5(d)** is the high-resolution transmission electron microscopy (HRTEM), which showing an extremely sharp interface and the growth plane (0001) is perpendicular to the growth direction. High-quality interface as seen from the HRTEM image justifies the less defect densities and high crystallinity of the InGaN films.

2.4. Nonpolar InGaN/GaN heterostructures

InGaN alloys of various compositions are being optimized as materials for the fabrication of light-emitting diodes which are active in the entire visible spectrum extending up to ultraviolet wavelengths [32, 64]. Indium-rich InGaN alloys are now being considered potential candidates for longer wavelength emitters, thermionic emitters, multi-junction solar cells, etc. [65–67]. A concern with polar heterostructures is the intrinsic and strong polarization fields resident in the lattice. To overcome such polarization effect, substrates oriented in nonpolar directions, i.e., (1 0 −1 0) m-plane or (1 1 −2 0) a-plane, are used. Devices grown on these orientations are receiving a lot of focus due to this enhanced behavior. However, a slight compromise in terms of quality has to be made because of the large mismatch in lattice constants, thermal expansion coefficients, and elastic constants of InGaN and GaN. In addition, the large lattice constant mismatch between GaN and InN (~11%) results in a phase separation in InGaN alloys which has been theoretically predicted and experimentally observed [68]. This makes the growth of InGaN very challenging, especially for higher concentration of indium (>20%) [58]. Growth parameters such as growth temperature, growth rate, and flux ratio are seen to drastically affect the indium incorporation in InGaN films. During growth of the InGaN alloys, the evaporation of indium species from the surface will be suppressed at lower temperatures and leads to higher growth rates as the indium species become trapped by the growing layer [69, 70]. As the indium content in the InGaN alloy increases, the material quality degrades due to phase separation, inhomogeneity of solid solution, and indium metal droplets. Few reports are available in the literature for the study of the nonpolar InGaN-based multiple quantum well (MQW) structures and other devices. Song et al. studied the effect of periodicity of a-plane InGaN/GaN multiple quantum wells on the output power of the LEDs grown by MOCVD [71].

In this section, $In_xGa_{1-x}N$ films of thicknesses around 125 nm were grown on 125 nm a-GaN (11−20)/r-Al_2O_3 (1−102) by PAMBE system. A two-step growth process was employed to grow a-GaN film, which constituted the growth of a 20 nm thin low-temperature GaN buffer layer at 500°C and a ~125nm GaN epilayer at 760°C. The nitrogen plasma RF power and N_2 flow rate were kept constant at 350 W and 1 sccm, respectively, throughout the growth duration. The gallium beam equivalent pressure (BEP) was kept at 5.6×10^{-7} mbar, corresponding to the growth in the slightly nitrogen-rich region. $In_xGa_{1-x}N$ films were grown on top of the a-GaN layer at the growth conditions as tabulated in **Table 2**. Three $In_xGa_{1-x}N$ films (samples A, B, and C) with different indium compositions were grown by varying the In/Ga flux ratios and growth temperatures. During the $In_xGa_{1-x}N$ growth, the nitrogen flow rate and RF power of the plasma source were kept at 0.5 sccm and 350 W, respectively.

Figure 6(a) shows the 2θ-ω HRXRD scan of the grown films, which confirmed the growth of nonpolar a-InGaN epilayer oriented in the (1 1 −2 0) direction and a-GaN epilayer oriented in

Sample ID	Gallium BEP (mbar)	Indium BEP (mbar)	Growth temp. (°C)
(A)	1.23×10^{-7}	1.17×10^{-7}	550
(B)	1.23×10^{-7}	8.87×10^{-8}	550
(C)	1.23×10^{-7}	8.87×10^{-8}	540

Table 2. Growth conditions for a-$In_xGa_{1-x}N$/a-GaN/r-sapphire substrate.

the (1 1 −2 0) direction on (1 −1 0 2) r-plane sapphire. The peak at $2\theta = 56.64°$ for sample A, $2\theta = 56.59°$ for sample B, and $2\theta = 56.36°$ for sample C was assigned to (1 1 −2 0) $In_xGa_{1-x}N$ which corresponds to the different composition of indium in the grown films. The calculated compositions of the samples were $In_{0.17}Ga_{0.83}N$ (sample A), $In_{0.19}Ga_{0.81}N$ (sample B), and $In_{0.22}Ga_{0.78}N$ (sample C). We observe that by decreasing the indium flux during the growth from 1.17×10^{-7} to 8.87×10^{-8}, we could incorporate more indium into the $In_xGa_{1-x}N$ alloy. Higher indium incorporation was also observed when the substrate temperature was decreased from 550C to 540°C keeping the indium flux constant. Growth parameters play a very critical role in controlling the composition of the $In_xGa_{1-x}N$ films. For the given set of growth parameters, we have not observed any phase separation or indium segregation in the grown $In_xGa_{1-x}N$ films.

X-ray rocking curve (ω) analysis was carried out to determine the crystal quality of the as-grown structures. Rocking curves along different azimuth angles have been recorded, and RC FWHM versus azimuth angle has been plotted for all the samples as shown in **Figure 6(b)**. The RC FWHM of the reflection along the normal varied with the azimuth angle and showed an M-type behavior [72]. The measured FWHM value of (1 1 −2 0) reflection of $In_{0.22}Ga_{0.78}N$ along [0001] direction defined as azimuth angle 0°and along [1−100] direction defined as azimuth angle 90° was found to be 0.532 and 0.703°, respectively. The RC FWHM values of (1 1 −2 0) GaN reflections were found to be 0.47 and 0.52° along azimuth angle 0° corresponding to reflection along [0001] direction and 90° corresponding to reflection along [1−100] direction, respectively. The reason behind the broadening of InGaN rocking curves could be attributed to the presence of defects such as partial dislocations and stacking faults, thus suggesting that the crystalline quality of nonpolar InGaN films is reduced with the increase in indium incorporation [73]. Cross-sectional plan view of TEM image obtained in bright field is shown in **Figure 6(c)**, which shows a clear interface of the a-InGaN/a-GaN/r-sapphire structure. The thickness of the each layer grown is confirmed from the image to be around 125 nm each. Basal stacking faults (BSF) are visible as thin lines, and they arise from the low-temperature nucleation layer. Due to the anisotropic nature of the growing surface, the nonpolar GaN typically has a high density of BSFs. The room temperature PL spectrum of $In_xGa_{1-x}N$ films is shown in **Figure 6(d)**. The position of the near-band-edge emission (NBE) of the $In_xGa_{1-x}N$ as observed for the three samples were 2.67, 2.59, and 2.56 eV for samples A, B, and C, respectively. Using Vegard's law, the values of indium fraction were found to be 0.20, 0.22, and 0.23 for samples A, B, and C, respectively.

Reciprocal space maps (RSMs) were recorded to look into the microstructure and strain present in the films. The reciprocal lattice points (RLPs) will be elongated along the Q_x axis if the broadening is caused by the limited mosaic block dimensions [74]. The RLPs will be tilted

http://dx.doi.org/10.5772/intechopen.70219

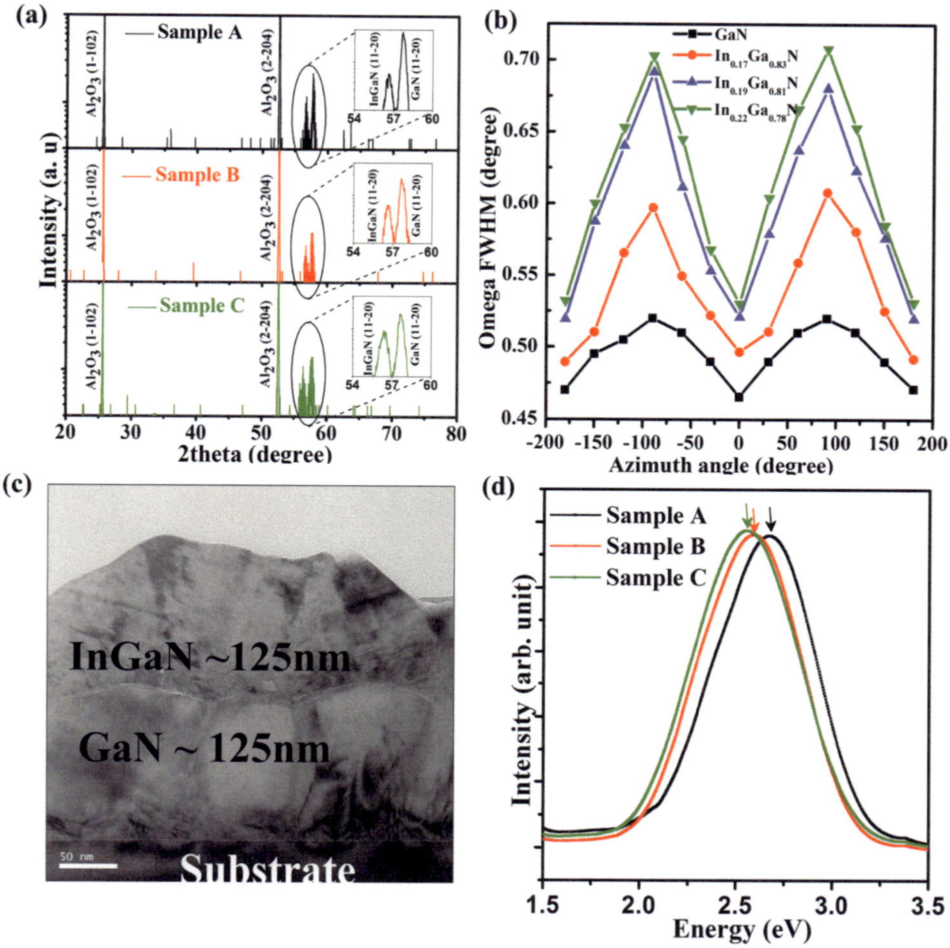

Figure 6. (a) HRXRD 2θ-ω scan of nonpolar (11−20) a-InGaN epilayer grown on (11−20) a-GaN/(1−102) r-plane sapphire, (b) FWHM of the rocking curve along with variation in azimuth angles for InGaN and GaN layer, (c) bright-field plan-view cross-sectional image of a-InGaN/a-GaN/r-sapphire substrate, and (d) room temperature PL spectrum of $In_xGa_{1-x}N$ films.

in the reciprocal space if additional mosaic tilt exists in the sample. From **Figure 7(a)–(c)**, the RLPs along the symmetric (1 1 −2 0) reflection of all samples are all broadened in the Q_x direction with negligible inclination, indicating that the dominant broadening mechanism for it is the limited mosaic block size. Asymmetric reflection are expected to be elongated parallel to the lateral scattering vector if the material experiences broadening due to short-sized mosaic blocks and could be also inclined if an additional mosaicity tilting exists in the material [75]. To understand the strain in the film with respect to the substrate, asymmetric RSM along (1 0 −1 0) a-GaN scans were obtained and are shown in **Figure 7(d)–(f)**. Shift of the center of each peak along the Q_z axis in an asymmetric scan gives a direct evidence of the strain present between the layers. Splitting of the GaN peak for all compositions implies a formation of a thin layer of high gallium composition $In_xGa_{1-x}N$ layer which is highly strained with respect to the GaN

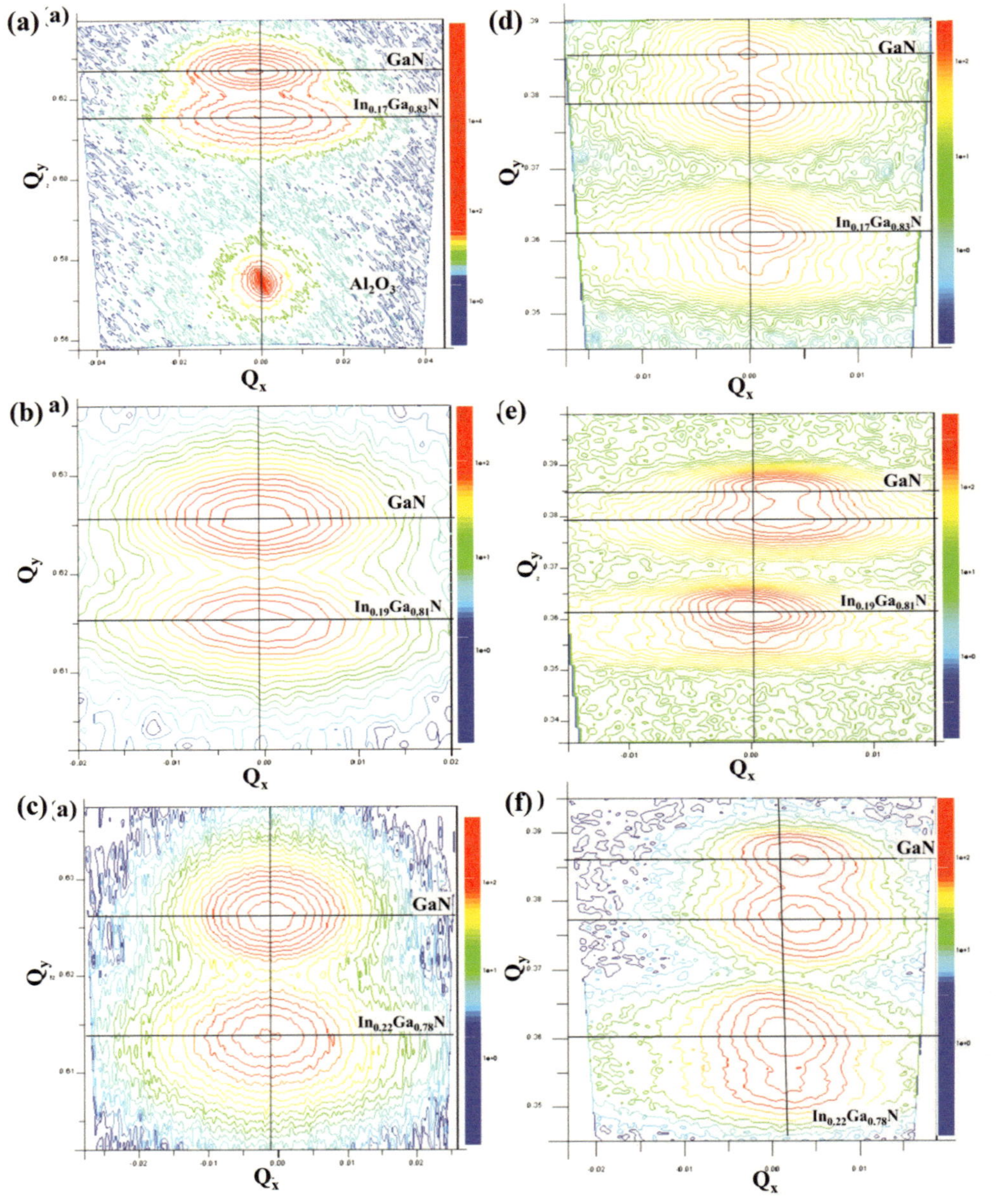

Figure 7. (a)–(c) RSM of symmetric (1 1 −2 0) of $In_xGa_{1-x}N$/GaN/r-sapphire and (d)–(f) RSM of asymmetric (1 0 −1 0) $In_xGa_{1-x}N$/GaN layers.

layer. Hence, from the asymmetric RSM scans, it is understood that after a thin layer of $In_xGa_{1-x}N$ acting as a buffer layer, the $In_xGa_{1-x}N$ film with In concentration of 17, 19 and 22% was formed in samples A, B, and C, respectively. As the In concentration is increased, the InGaN film is showing signs of relaxation which indirectly means increase in the dislocation density, reducing the crystallinity of the layer which is consistent with the observation of the variation of FWHM in rocking curve measurements.

2.5. InGaN/Si heterostructures

n-n heterojunctions have found use in different applications such as photodetectors, light-emitting diodes, solar cells, injection lasers, etc. [76–80]. So far the most extensively studied material systems are ZnO/Si, $In_xGa_{1-x}As$, $In_xGa_{1-x}Sb$, Ge-Si, etc. [81, 82]. Very few groups have reported studies on isotype heterojunctions of $In_xGa_{1-x}N$ system [83, 84]. The large lattice mismatch of InGaN with silicon substrates has hindered the realization of silicon-based commercial devices and is limited to sapphire which happens to be an insulator. Therefore, obtaining smooth and abrupt heterojunctions with minimum density of interface defects by overcoming the lattice mismatch issue has been a concern of great interest for many researchers across the globe. Several attempts have been made to demonstrate low power consuming or self-powered photodetectors [85, 86]. There are a few reports on the UV photodetection using InGaN as an active layer [87]. In this work, we report the development of a UV photodetector which is operated at zero bias. The device comprises of a simple n-InGaN/n-Si heterojunction. The role of interface defects originating due to the large lattice mismatch, such as traps, in modulating the built-in electric field driven photoresponse has been discussed. The n-Si (111) substrates of 1×1 cm^2 in size were cleaned chemically by trichloroethylene, acetone, and methanol and were dipped in 5% HF for 60 s to remove the native oxide prior to loading in the growth chamber. Thermal cleaning of Si (111) was carried out at 900°C for the removal of native oxide layer. The substrate temperature was further reduced to 550°C, and growth was carried out for 2 h without any intermediate steps. The indium (In) beam equivalent pressure (BEP) and gallium (Ga) BEP, nitrogen flow, and plasma power were kept at 8.52×10^{-8} mbar, 1.2×10^{-7} mbar, 1 sccm, and 350 W, respectively. Various experimental techniques were employed to study the as-grown samples as mentioned in upcoming sections. Circular aluminum contacts with diameters of 600 μm were then deposited by thermal evaporation on the InGaN films and Si (111) substrate with the help of a physical mask to study the current-voltage and photoresponse studies.

Figure 8(a) shows the 2θ-ω HRXRD scan of InGaN epilayers on Si (111) substrates [88]. The peaks at 2θ = 28.45 and 58.86° attribute to Si (111) and Si (222) reflections, respectively. The peaks at 2θ = 34.05 and 71.89° correspond to (0002) and (0004) reflections of InGaN. Also, no secondary phases or residual indium or InN phases were found, confirming that the films are single crystalline. The In composition was determined to be 15% from the linear interpolation of the 2θ peak positions of (0002) GaN (34.59°) and (0002) InN (31.22°). In **Figure 8(b)** the rocking curve FWHM value (degree) of asymmetric reflection (2.16) is higher than that of the symmetric reflection (2.05), which attributes to the presence of large edge threading dislocations (TDs) [73].

The room temperature PL spectra of the $In_xGa_{1-x}N$ films are shown in **Figure 8(c)**. Near-band-edge emission peaks are observed at 502.12 nm corresponding to the bandgap of 2.46 eV. Due to stoke shift of the PL spectra and discrepancies in the exact value of InN, bandgap [89] further adds ambiguity to the exact In content determination from PL spectra. A small hump is observed around 425 nm which might be due to the initial layers with a large number of dislocations arising due to the large lattice mismatch of Si and InGaN films [90]. One can also correlate the presence of a large number of defects from the FWHM values of both symmetric

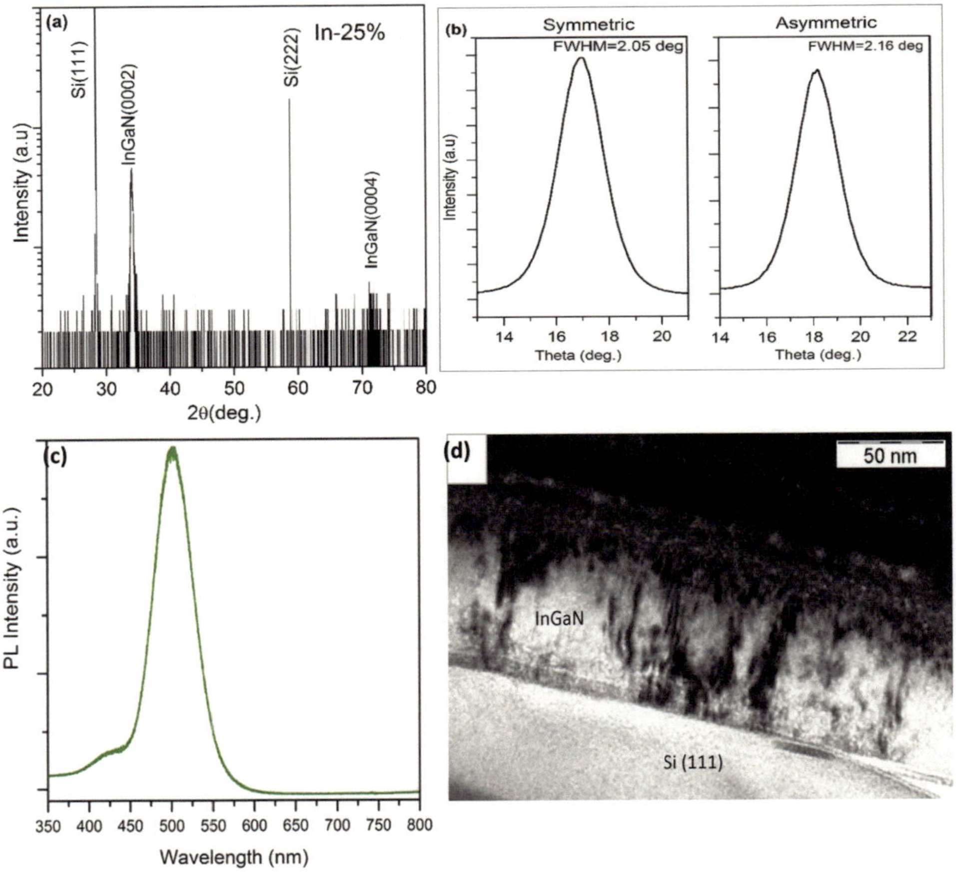

Figure 8. (a) HRXRD 2θ-ω scan of InGaN on bare Si (111), (b) X-ray rocking curve of (0002) and (10−11) reflections, (c) photoluminescence spectra of InGaN/Si, and (d) cross-sectional TEM of InGaN/Si heterostructure. Reprinted with permission from [88].

and asymmetric XRCs. The contrast at the interface in the TEM image in **Figure 8(d)** clearly indicates the presence of initial low-contrast Ga-rich layers which possibly attribute to the ~425 nm peak in PL spectra. The thickness was estimated to be ~100 nm from the TEM image. The other dark regions are formed as a result from the Ga ion beam damage during sample thinning.

3. Transport properties of III-nitride heterostructure-based Schottky junctions

3.1. Polar and nonpolar InN/GaN heterostructures.

The semiconductor heterostructure exhibits the Schottky barrier at the interface due to the formation of conduction band offset because of their different bandgap values. The concepts on the band offset are directly transferable to the Schottky barrier height problems. Hence, it is vital to study the transport properties of the semiconductor heterostructure-based Schottky junctions. The InN/GaN heterostructure has a large conduction band offset, thus ensuring effective blockage of the conduction current over the barriers [91, 92], which help for the development of electronic devices operating in THz frequency range [91]. Hence, studying the transport properties across the InN/GaN interface is very important from device view point. Chen et al. [93] studied temperature-dependent current-voltage characteristics of InN/GaN-based Schottky junctions in the range of 300–400K and found that the barrier height (~1.25 eV) and the ideality factor (~1.25) are nearly temperature independent. Similarly, Wang et al. [94] employed capacitance-voltage measurement technique to determine the Schottky barrier height to be 0.94 eV at room temperature. This section presents the study of the temperature-dependent electrical transport properties of InN/GaN heterostructures and observed the temperature-dependent barrier height and the ideality factor. **Figure 9(a)** shows the room temperature *J-V*(current density-voltage) characteristics for the junction. The junction between InN and GaN exhibits a rectifying behavior which suggests an existence of Schottky barrier height at the junction [95, 96].

To investigate further, we have studied the temperature-dependent *J-V (J-V-T)* characteristics ranging from 200 to 500 K and is given in **Figure 9(b)** [95]. In the present case, the GaN film is highly doped with silicon ($N_D \sim 1.4 \times 10^{18}$ cm^{-3}) resulting in a lower barrier height at the InN/GaN interface, which is due to the tunneling of charge carriers through the interface. This kind of behavior reveals that the current transport is primarily dominated by thermionic field emission (TFE) mechanism, where the carriers tunnel from GaN to InN. According to the transport theory, TFE dominates only when $E_{00}/kT \approx 1$, where E_{00} is the characteristic tunneling energy which determines the nature of conduction mechanism. When $E_{00}/kT \approx 1$, the Boltzmann distribution tail of thermionic emission drops off by a factor of $exp(-1)$, which is much faster than the decrease rate of the tunneling probability. On the other hand, thermionic emission (TE) is predominant when $E_{00}/kT << 1$ because the tunneling probability drops off faster than TE [97]. In the present case, the value of the barrier height and the ideality factor (η)

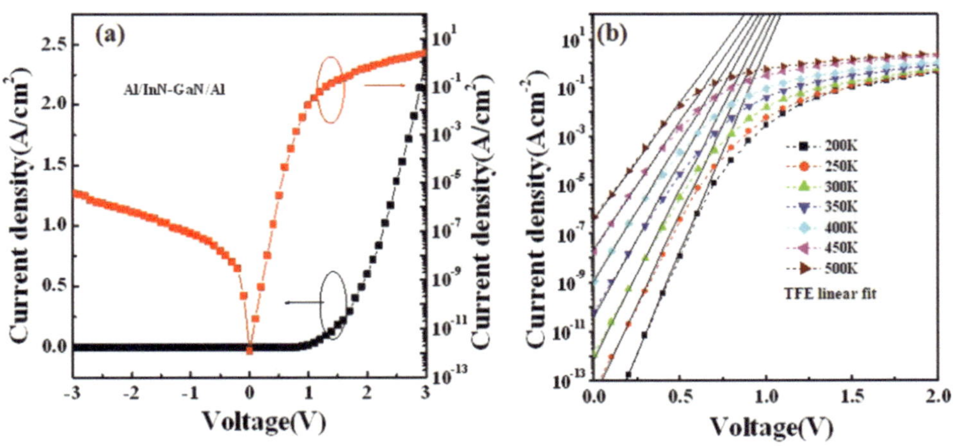

Figure 9. (a) The room temperature *J–V* characteristics of InN/GaN Schottky junction and (b) the forward *J–V* characteristics with TFE fitting as a function of measurement temperature. Reprinted with permission from [95, 96].

were calculated by fitting a line in the linear region of the forward J-V curves using the TFE equation and are shown in **Figure 9(b)**. From the abovementioned analysis, the barrier height and the ideality factor are found to be temperature dependent. Thus, our results indicate the presence of inhomogeneity at the interface due to the presence of various types of defects, which is why we observe the temperature-dependent behavior of barrier height [98, 99]. From the fitting the value of E_{00}/kT is observed to be nearly one, suggesting that the TFE conduction mechanism would be considered to be a more realistic model for the analysis of the electronic transport in polar InN/GaN heterostructure.

An optoelectronic device based on nonpolar III-nitride heterostructure has been an important subject due to its potential improvement on the efficiency. However, the transport behavior of nonpolar a-plane InN/GaN heterostructure interfaces is limited. In this section, we will discuss our results on the transport properties of nonpolar a-plane InN/GaN heterostructure. The inset of **Figure 10(a)** shows the schematic diagram of InN/GaN heterostructure Schottky junction. **Figure 10(a)** shows the room temperature *I-V* characteristic of nonpolar a-plane InN/GaN heterostructure Schottky junction [43]. The rectifying behavior of the *I-V* curve indicates the existence of Schottky barrier at the nonpolar InN/GaN interface. The forward bias *J-V*

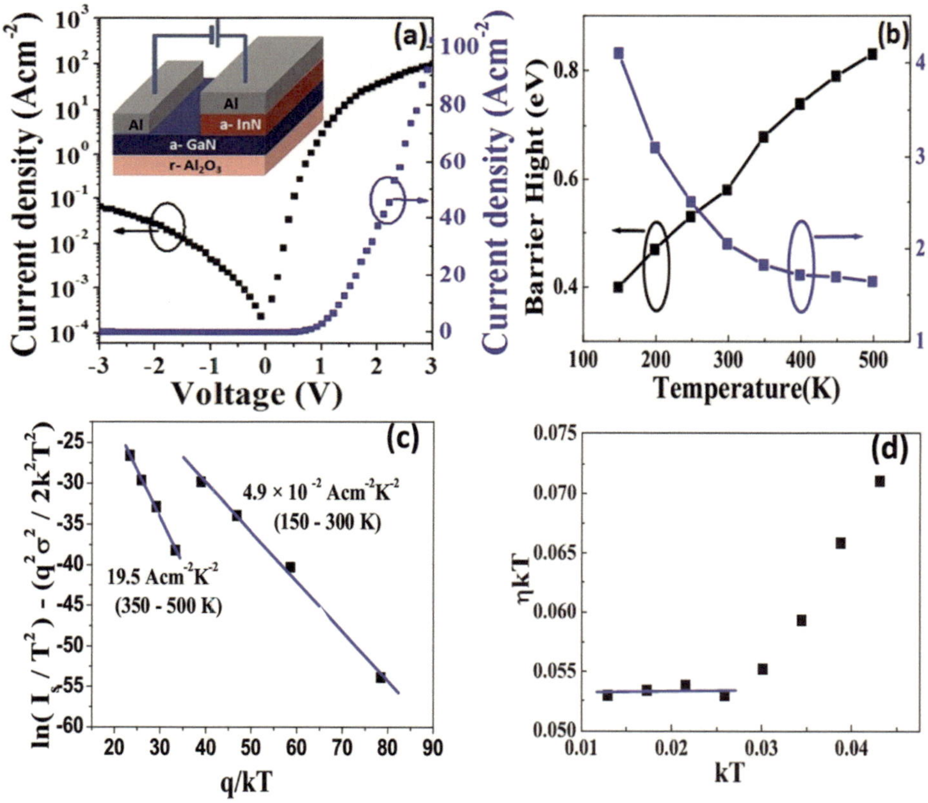

Figure 10. (a) Room temperature *I–V* characteristics of the nonpolar *a*-plane InN/GaN heterostructures, (b) temperature-dependent barrier height and the ideality factor, (c) modified Richardson plot of $ln(Is/T^2)-q^2\sigma^2/2k^2T^2$ versus q/kT, and (d) plot of ηkT as a function of kT. Reprinted with permission from [43, 96].

characteristics of the InN/GaN heterostructure were measured at different temperatures. The values of the barrier height and ideality factor were estimated by fitting the linear region of the forward *J-V* curves using thermionic emission model. It was found that the ideality factor and the barrier height values range from 1.65 and 0.83 eV (500 K) to 4.1 and 0.4 eV (150 K), respectively, as shown in **Figure 10(b)**. The variation of barrier height and the ideality factor with the measurement temperature indicates the presence of inhomogeneous nature of non-polar InN/GaN interface.

The inhomogeneous nature of barrier height at the InN/GaN interface could arise due to the presence of various types of defects at the interface [98, 99]. The inhomogeneous nature at the interface was explained by using Richardson plot of saturation current. In the Richardson plot, we could identify two separate temperature ranges, i.e., 150–300 and 350–500 K from the slopes, and the values of Richardson's constants (A^*) were found to be much lower than the theoretical value of 24 $Acm^{-2}K^{-2}$ for n-GaN. This type of barrier height inhomogeneity at the InN/GaN interface can be explained by considering the Gaussian distribution of barrier heights at the interface [100, 101] and can be written as

$$P(\varphi_b) = \frac{1}{\sigma_s\sqrt{2\pi}}\exp\left[-\frac{(\varphi_b - \overline{\varphi_b})^2}{2\sigma_s^2}\right] \tag{2}$$

where $1/\sigma_s\sqrt{2\pi}$ is the normalization constant and $\overline{\varphi_b}$ and σ_s are the mean and standard deviation of barrier height, respectively. Considering Gaussian distribution of barrier height, the effective barrier height, φ_b, given by the expression

$$\varphi_b = \overline{\varphi_{b0}} - \frac{q\sigma_s^2}{2kT} \tag{3}$$

Here, $\overline{\varphi_{b0}}$ is the zero bias mean barrier height. Considering the barrier height inhomogeneities, the conventional Richardson plot is modified as follows:

$$\ln\left(\frac{I_s}{T^2}\right) - \left(\frac{q^2\sigma_s^2}{2k^2T^2}\right) = \ln(AA^*) - \frac{q\overline{\varphi_{b0}}}{kT} \tag{4}$$

Figure 10(c) shows the modified Richardson plot. In the first region (300–500 K), the values of $\overline{\varphi_b}$ and A^* were found to be 1.15 eV and 19.5 $A/cm^2\,K^2$, respectively. The calculated Richardson constant value in the temperature range of 350–500 K is very close to the theoretical value of 24 $A/cm^2\,K^2$ for n-type GaN. This indicates that at higher temperatures (350–500 K) the current transport is dominated by thermionic emission mechanism. The values of $\overline{\varphi_b}$ and A^* for polar c-plane InN/GaN heterojunction were found to be 1.6 eV and 25.8 $A/cm^2\,K^2$, respectively [102]. The reduced barrier height in nonpolar a-plane InN/GaN heterojunction when compared to the polar c-plane InN/GaN can be attributed to the absence of polarization field at the interface. The value of A^* is largely deviated in the temperature range of 150–300 K indicating the reduced dominance of thermionic emission. **Figure 10(d)** shows $E_0 = \eta kT$ versus kT plot for the nonpolar a-InN/GaN Schottky diode. The value, E_0, seems to be independent of temperature at

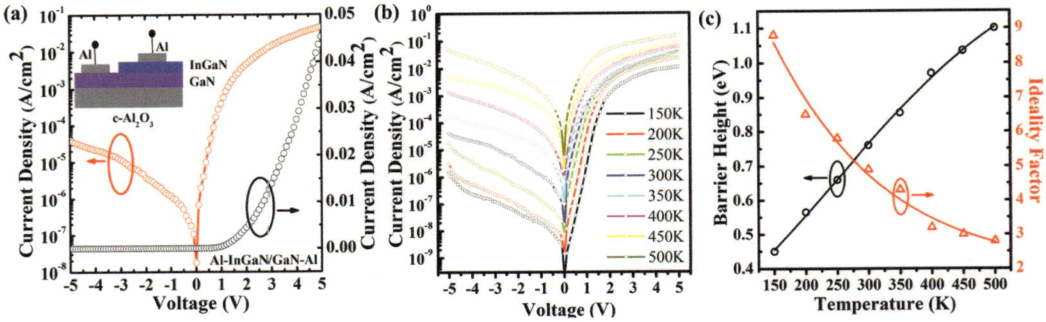

Figure 11. (a) Room temperature *J-V* characteristics of the InGaN/GaN heterostructure, (b) temperature-dependent *J-V* characteristics of the InGaN/GaN heterostructure, and (c) variation of the barrier height and the ideality factor with measurement temperature. Reprinted with permission from [104].

low temperatures, thus indicating the dominance of field emission in the range of 150–500 K [103]. It can be understood that the carriers lack the required energy to surmount the barrier at the low temperature and thus tunnel through the defect states at the interface.

3.2. Polar InGaN/GaN heterostructures

In this section, we have grown InGaN/GaN heterostructures using plasma-assisted molecular beam epitaxy and studied the temperature-dependent electrical transport characteristics. **Figure 11(a)** shows the room temperature *J-V* characteristics of the InGaN/GaN heterostructure in both linear and semilog scale [104]. The schematic diagram of the device structure has been shown in the inset of **Figure 11(a)**. The rectifying nature of *J-V* characteristic of InGaN/GaN heterostructure suggested the presence of a Schottky barrier at the interface. The temperature-dependent *J-V* characteristic of the heterostructure is shown in **Figure 11(b)**. It is observed that, as the measurement temperature increases, the forward bias current increases, which indicate that the current transport across the Schottky junction is governed by the thermionic emission mechanism. Then, the values of barrier height and ideality factor were calculated by using thermionic emission model. **Figure 11(c)** shows the variation of barrier height and ideality factor with measurement temperature [104]. It is observed that there is a temperature-dependent variation of both barrier height and ideality factor. The temperature dependence of φ_b indicates that the barrier height is inhomogeneous in nature, which may be due to various types of defects present at the InGaN/GaN interface. Moreover, the observed ideality factor greater than unity indicates a nonideal nature of the heterojunction, which is attributed to the presence of interface defect states.

4. InGaN/Si heterostructure-based UV and IR photodetectors

Room temperature *I-V* measurements were performed on InGaN/Si (111) heterojunctions with Si biased positively and were shown in **Figure 3(a)**. From **Figure 3(a)**, it is observed that the device is showing rectifying characteristics, both in the dark (room temperature as well as low temperature) and under ultraviolet radiation, which are consistent with the n-n isotype heterojunctions of other materials as reported by others [81, 82, 105]. *I-V* characteristics were obtained on Al/InGaN/Si(111)/Al, Al/InGaN/Al, and Al/Si (111)/Al. The behavior of Al/InGaN/Al and Al/Si (111)/Al was ohmic and that of the Al/InGaN/Si (111)/Al junction was rectifying, thus confirming that the rectifying characteristic is primarily arising from the n-InGaN/n-Si isotype heterojunction. Although it seems to be like a leaky rectifying behavior, subsequent low-temperature current-voltage measurements were carried out to further confirm the rectifying behavior. Hall measurements were carried out, and negligible changes were observed at low temperatures in conductivity, mobility, carrier concentration, etc. The room temperature conductivity, mobility, and bulk carrier concentration of the InGaN layer were found to be ~201 $(\Omega\ \text{cm})^{-1}$, ~17 cm^2/Vs, and ~8 × 10^{19} cm^{-3}. The high background electron concentration (~10^{19}–10^{21}) is a well-observed characteristic of the undoped InGaN and InN due to the nitrogen vacancies in the bulk and along the edge dislocations at the interface and In vacancy/N antisite complexes [106] which explains the source of doping in our case. A detailed study on the current-transport mechanisms can be found elsewhere [88].

Although the device is not perfectly rectifying at room temperature, the interesting characteristics observed were in the region of zero bias. An abrupt increase in photocurrent in the presence of UV radiation was observed at zero bias than either reverse or positive biases and is shown in the inset of **Figure 12(a)**. Ager et al. [83] have shown the operation of similar devices in photovoltaic mode. However, the InGaN films grown on their samples were much thicker compared to the present work and also used several buffer layers. The photocurrent response and stability were studied from the on-off cycles of a UV lamp at zero bias and different voltages and are shown in **Figure 12(b)**. The order of increase in the photocurrent magnitude is higher in case of zero bias (>1.5) than that of the positive and reverse bias. **Figure 12(c)** and **(d)** shows the growth and decay responses which can be described as [107]

$$I(t) = I_{dark} + A\left[1 - \exp\left\{\frac{-(t - t_0)}{\tau_g}\right\}\right] \tag{5}$$

$$I(t) = I_{dark} + A\left[\exp\left\{\frac{-(t - t_0)}{\tau_g}\right\}\right] \tag{6}$$

for growth and decay, respectively, where I_{dark} is dark current; A is scaling constant; t_0 is the time when UV lamp was switched on or off for growth or decay, respectively; and τ_g and τ_d are growth and decay times. The response times obtained from the above equations are 20 and 33 ms for growth and decay. The responsivity (R_λ) of this photodetector is calculated from, $R_\lambda = I_\lambda / P_\lambda S$, where I_λ is photocurrent, P_λ is the incident power of UV lamp of wavelength, and S is the area of illuminated junction which is 0.09 cm^2. The external quantum efficiency (EQE) is given as $EQE = hcR_\lambda / e\lambda$, [108]. The values obtained for spectral responsivity and external quantum efficiency are 0.0942 A/W and 32.4%, respectively, which are better than the values reported in the literature [85] for such devices. The responsivities and external quantum efficiencies in case of different biases were found to be better than zero bias (**Table 3**). From **Figure 12(b)**, it can be seen that the photocurrents obtained at different biases do not overlap, i.e., the photocurrent

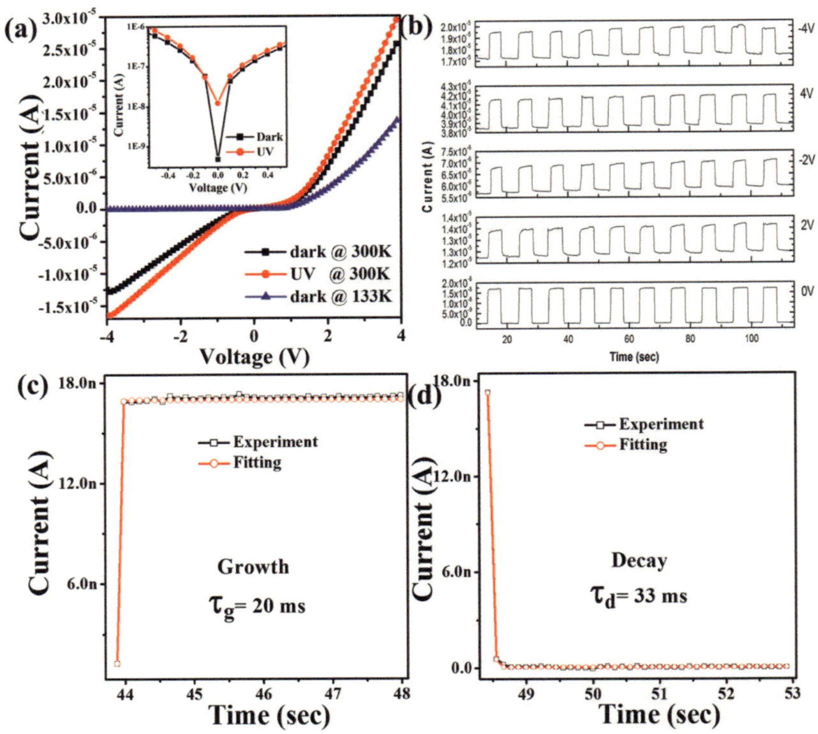

Figure 12. (a) Room temperature current-voltage characteristics of *n*-InGaN/*n*-Si heterojunction in dark at room temperature and at 133 K and under UV radiation at room temperature, (b) photoresponse at different working voltages, (c) time response of photocurrent growth, and (d) decay from the fitting of experimental values. Reprinted with permission from [88].

Voltage (V)	Responsivity (A/W)	EQE (%)
0	0.0942	32.4
2	0.6217	213.8
4	1.7097	588.1
−2	0.5746	197.6
−4	1.2575	432.5

Table 3. Bias-dependent responsivities and external quantum efficiencies.

range in each case is distinct from that of others, suggesting that such devices can be used for switching purposes and logical operations as well.

The mechanism for this type of behavior in n-n isotype heterojunctions can be explained with the help of a model proposed by Yawata et al. [77]. When the electron and hole pairs are generated, the electrons are swept away from the junctions due to the built-in electric field, whereas the holes are trapped in the notch. The holes being positively charged neutralize the electrons at interface and eventually lower the barrier height. For an intense illumination, the concentration of holes trapped at the notch is increased, thus drastically lowering the barrier

http://dx.doi.org/10.5772/intechopen.70219

height and resulting in abrupt increase of electron flow. However, in case of a bias (positive or negative as explained above), the electrons at the interface tunnel through the depletion region leaving behind holes which are eventually refilled by electrons from the other side. The forward and negative bias characteristics from the current-voltage plots are in good agreement with the proposed mechanism.

5. Conclusions

In summary, we presented the studies on the growth, characterizations, and transport properties of III-nitride-based heterostructures. Here, discussion has been carried out on the growth of polar InN/GaN, nonpolar InN/GaN, polar InGaN/GaN, nonpolar InGaN/GaN, and InGaN/Si heterostructures by using MBE system followed by their characterizations. Moreover, we have presented the electrical transport properties across the heterostructures interface. In addition, UV and IR photodetection studies on InGaN/Si heterostructures have been discussed.

Author details

Basanta Roul[1,2], Greeshma Chandan[1], Shruti Mukundan[1] and Saluru Baba Krupanidhi[1*]

*Address all correspondence to: sbkenator@gmail.com

1 Materials Research Centre, Indian Institute of Science, Bangalore, India

2 Central Research Laboratory, Bharat Electronics, Bangalore, India

References

[1] Yoshida S, Misawa S, Gonda S. Improvements on the electrical and luminescent properties of reactive molecular beam epitaxially grown GaN films by using AlN-coated sapphire substrates. Applied Physics Letters. 1983;**42**:427

[2] Akasaki I, Amano H, Koide Y, Hiramatsu H, Sawak N. Effects of ain buffer layer on crystallographic structure and on electrical and optical properties of GaN and $Ga_{1-x}Al_xN$ ($0 < x \leqq 0.4$) films grown on sapphire substrate by MOVPE. Journal of Crystal Growth. 1989;**98**:209

[3] Nakamura S. GaN growth using GaN buffer layer. Japanese Journal of Applied Physics. 1991;**30**:L1705

[4] Nakamura S, Senoh M, Mukai T. P-GaN/N-InGaN/N-GaN double-heterostructure blue-light-emitting diodes. Japanese Journal of Applied Physics. 1993;**32**:L8-L11

[5] Nakamura S, Senoh M, Nagahama S, Iwasa N, Yamada T, Matsushita T, Kiyoku H, Sugimoto Y. InGaN-based multi-quantum-well-structure laser diodes. Japanese Journal of Applied Physics. 1996;**35**:L74

[6] Khan A, Kuzina MJN, Olson DT, Schaff WJ, Burm JW, Shur MS. Microwave performance of a 0.25 μm gate AlGaN/GaN heterostructure field effect transistor. Applied Physics Letters. 1994;**65**:1121-1122

[7] Waltereit P, Brandt O, Trampert A, Grahn HT, Menniger J, Ramsteiner M, Reiche M, Ploog KH. Nitride semiconductors free of electrostatic fields for efficient white light-emitting diodes. Nature. 2000;**406**:865

[8] Park SH, Ahn D. Depolarization effects in (11-22)-oriented InGaN/GaN quantum well structures Applied Physics Letters. 2007;**90**:013505

[9] Krames MR, Shchekin OB, Mueller-Mach R, Mueller GO, Zhou L, Harbers G, Craford MG. Status and future of high-power light-emitting diodes for solid-state lighting. Journal of Display Technology. 2007;**3**:160

[10] Sato H, Chung RB, Hirasawa H, Fellows, Masui H, Wu F, Saito M, Fujito K, Speck JS, DenBaars SP, Nakamura S. Optical properties of yellow light-emitting diodes grown on semipolar (11-22) bulk GaN substrates. Applied Physics Letters. 2008;**92**:221110

[11] Sun Q, Han J. Springer Series in Materials Science. 2012:156

[12] Craven MD, Wu F, Chakraborty A, Imer B, Mishra UK, DenBaars SP, Speck JS. Microstructural evolution of *a*-plane GaN grown on *a*-plane SiC by metalorganic chemical vapor deposition. Applied Physics Letters. 2004;**84**:1281

[13] Sun Q, Yerino CD, Zhang Y, Cho YS, Kwon SY, Kong BH, Cho HK, Lee IH, Han J. Effect of NH_3 flow rate on *m*-plane GaN growth on m-plane SiC by metalorganic chemical vapor deposition. Journal of Crystal Growth. 2009;**311**:3824

[14] Trybus E. Molecular beam epitaxy growth of indium nitride and indium gallium nitride materials for photovoltaic applications [PhD thesis]. 2009

[15] Ambacher O, Smart J, Shealy JR, Weimann NG, Chu K, Murphy M, Schaff WJ, Eastman LF, Dimitrov R, Wittmer L, Stutzmann M, Rieger W, Hilsenbeck J. Two-dimensional electron gases induced by spontaneous and piezoelectric polarization charges in N- and Ga-face AlGaN/GaN heterostructures. Journal of Applied Physics. 1999;**85**:3222

[16] Khromov S. Doping effects on the structural and optical properties of GaN [PhD thesis]. 2013

[17] Monemar B, Pozina G. Group III-nitride based hetero and quantum structures. Progress in Quantum Electronics. 2000;**24**:239

[18] Shen YC, Mueller GO, Watanabe S, Gardner NF, Munkholm A, Krames MR. Auger recombination in InGaN measured by photoluminescence. Applied Physics Letters. 2007;**91**:141101

[19] Romanov AE, Young EC, Wu F, Tyagi A, Gallinat CS, Nakamura S, DenBaars SP, Speck JS. Basal plane misfit dislocations and stress relaxation in III-nitride semipolar heteroepitaxy. Journal of Applied Physics. 2011;**109**:103522

[20] Nye JF. Physical Properties of Crystals. New York: Oxford University Press; 1985

[21] Sano M, Aoki M, Epitaxial growth of undoped and Mg-doped GaN. Japanese Journal of Applied Physics. 1976;**15**:1943

[22] Madar R, Michel D, et al. Growth anisotropy in the GaN/Al2O3 system. Journal of Crystal Growth. 1977;**40**:239

[23] Craven MD, Lim SH, et al. Structural characterization of nonpolar (11-20) *a*-plane GaN thin films grown on (1-102) *r*-plane sapphire. Applied Physics Letters. 2002;**81**:469

[24] Kehagias T, Iliopoulos E, Delimitis A, Nouet G, Dimakis E, Georgakilas A, Komninou P. Interfacial structure of MBE grown InN on GaN. Physica Status Solidi (a). 2005; **202**:777

[25] Von Pezold J, Bristowe PD. Atomic structure and electronic properties of the GaN/ZnO (0001) interface. Journal of Material Science. 2005;**40**:3051

[26] Bernardini F, Fiorentini V. Macroscopic polarization and band offsets at nitride heterojunctions. Physical Review B. 1998;**57**:R9427

[27] Yoshikawa A, Che SB, YamaguchiW, Saito H,Wang XQ, Ishitani Y, Hwang ES. Proposal and achievement of novel structure InN/GaN multiple quantum wells consisting of 1 ML and fractional monolayer InN wells inserted in GaN matrix. Applied Physics Letters. 2007;**90**:073101

[28] Che SB, Terashima W, Ohkubo T, Yoshitani M, Hashimoto N, Akasaka K, Ishitani Y, Yoshikawa A. InN/GaN SQW and DH structures grown by radio frequency plasma-assisted MBE. Physica Status Solidi (c). 2005;**2**:2258

[29] Roul B, Kumar M, Rajpalke MK, Bhat TN, Kalghatgi AT, Krupanidhi SB. Effect of carrier concentration of InN on the transport behavior of InN/GaN heterostructure based Schottky junctions. Solid State Communication. 2012;**152**:1771

[30] Lin JC, Su YK, Chang SJ, Lan WH, Chen WR, Cheng YC, Lin WJ, Tzeng YC, Shin HY, Chang CM. InN grown on GaN/sapphire templates at different temperatures by MOCVD. Optical Materials. 2007;**30**:517

[31] Wu J. When group-III nitrides go infrared: New properties and perspectives. Journal of Applied Physics. 2009;**106**:011101

[32] Ponce FA, Bour DP. Nitride-based semiconductors for blue and green light-emitting devices. Nature. 1997;**386**:351

[33] Simeonov D, Feltin E, Buhlmann HJ, Zhu T, Castiglia A, Mosca M, Carlin JF, Butté R, Grandjean N. Blue lasing at room temperature in high quality factor GaN/AlInN microdisks with InGaN quantum wells. Applied Physics Letters. 2007;**90**:061106

[34] Rajpalke MK, Bhat TN, Roul B, Kumar M, Misra P, Kukreja LM, Sinha N, Krupanidhi SB. Growth temperature induced effects in non-polar *a*-plane GaN on *r*-plane sapphire substrate by RF-MBE. Journal of Crystal Growth. 2011;**314**:5

[35] Rajpalke MK, Roul B, Kumar M, Bhat TN, Sinha N, Krupanidhi SB. Structural and optical properties of nonpolar (11–20) *a*-plane GaN grown on (1–102) *r*-plane sapphire substrate by plasma-assisted molecular beam epitaxy. Scripta Materials. 2011;**65**:33

[36] Lu H, Schaff WJ, Eastman LF, Wu J, Walukiewicz W, Cimalla V, Ambacher O. Growth of *a*-plane InN on *r*-plane sapphire with a GaN buffer by molecular-beam epitaxy. Applied Physics Letters. 2003;**83**:1136

[37] Grandal J, Sánchez-García MA, Calleja E, Gallardo E, Calleja JM, Luna E, Trampert A, Jahn A. InN nanocolumns grown by plasma-assisted molecular beam epitaxy on AA-plane GaN templates. Applied Physics Letters. 2009;**94**:221908

[38] Cimalla V, Pezoldt J, Ecke G, Kosiba R, Ambacher O, Spiess L, Teichert G, Lu H, Schaff WJ. Growth of cubic InN on rr-plane sapphire. Applied Physics Letters. 2003;**83**:3468

[39] Tsuyuguchi A, Teraki K, Koizumi T, Wada J, Araki T, Nanishi Y, Naoi H. Institute of Physics Conference Series. 2005;**184**:239

[40] Weber ZL, Lu H, Schaff WJ, Kryliouk O, Park HJ, Mangum J, Anderson T. Comparison of structural perfection of InN layers and InN nanorods grown on the c- and r-planes of Al_2O_3. Physica Status Solidi (c). 2007;**4**:2469

[41] Ajagunna AO, Iliopoulos E, Tsiakatouras G, Tsagaraki K, Androulidaki M, Georgakilas A. Epitaxial growth, electrical and optical properties of a-plane InN on r-plane sapphire. Journal of Applied Physics. 2010;**107**:024506

[42] Rajpalke MK, Roul B, Bhat TN, Kumar M, Sinha N, Jali VM, Krupanidhi SB. Effects of growth temperature on nonpolar a-plane InN grown by molecular beam epitaxy. Physica Status Solidi (c). 2014;**11**:932

[43] Rajpalke MK, Bhat TN, Roul B, Kumar M, Krupanidhi SB. Current transport in nonpolar a-plane InN/GaN heterostructures Schottky junction. Journal of Applied Physics. 2012;**112**:023706

[44] Schuber R, Chou MMC, Vincze P, Schimmel Th, Schaadt DM. Growth of *A*-plane GaN on (0 1 0) $LiGaO_2$ by plasma-assisted MBE. Journal of Crystal Growth. 2010;**312**:1665

[45] Darakchieva V, Xie MY, Franco N, Giuliani F, Nunes B, Alves E, Hsiao CL, Chen LC, Yamaguchi T, Takagi Y, Kawashima K, Nanishi Y. Structural anisotropy of nonpolar and semipolar InN epitaxial layers. Journal of Applied Physics. 2010;**108**:073529

[46] Wu J,WalukiewiczW, ShanW, Yu KM, Ager JW III, Haller EE, Lu H, Schaff WJ. Effects of the narrow band gap on the properties of InN. Physical Review B. 2002;**66**:201403

[47] Zhang PF, Liu XL, Zhang RQ, Fan HB, Song HP, Wei HY, Jiao CM, Yang SY, Zhu QS, Wang ZG. Valence band offset of MgO/InN heterojunction measured by X-ray photoelectron spectroscopy. Applied Physics Letters. 2008;**92**:042906

[48] Nakamura S, Senoh M, Iwasa N, Nagahama S. High-power InGaN single-quantum-well-structure blue and violet light-emitting diodes. Applied Physics Letters. 1995;**67**:1868

[49] Laboutin O, Cao Y, Johnson W, Wang R, Li G, Jena D, Xing H. InGaN channel high electron mobility transistor structures grown by metal organic chemical vapor deposition. Applied Physics Letters. 2012;**100**:121909

[50] Hasan T, Kaysir R, Islam S, Bhuiyan AG, Islam R, Hashimoto A, Yamamoto A. 2DEG properties in InGaN/InN/InGaN-based double channel HEMTs. Physica Status Solidi (c). 2010;**7**:1997

[51] Sang L, Liao M, Koide Y, Sumiya M. High-temperature ultraviolet detection based on InGaN Schottky photodiodes. Applied Physics Letters. 2011;**99**:031115

[52] Cheong MG, Suh EK, Lee HJ, Dawson M. Growth and properties of InGaN/GaN quantum wells and blue light emitting diodes by metal–organic chemical vapour deposition. Semiconductor Science and Technology. 2002;**17**:446

[53] Kim IH, Park HS, Park YJ, Kim T. Formation of V-shaped pits in InGaN/GaN multiquantum wells and bulk InGaN films. Applied Physics Letters. 1998;**73**:1634

[54] Görgensa L, Ambacher O, Stutzmann M, Miskys C, Scholz F, Off J. Characterization of InGaN thin films using high-resolution X-ray diffraction. Applied Physics Letters. 2000;**76**:577

[55] Cao XA, Topol K, Sandvik FS, Teetsov FS, Sandvik PM, LeBoeuf SF, Ebong A, Kretchmer J, Stokes EB, Arthur S, Keloyeros AE, Walker D. Influence of defects on electrical and optical characteristics of GaN/InGaN-based light-emitting diodes. Proceedings of SPIE. 2002;**4776**:105

[56] Kim J, Cho YH, Ko DS, Li XS, Won JY, Lee E, Park SH, Kim JY, Kim S. Influence of V-pits on the efficiency droop in InGaN/GaN quantum wells. Optics Express.2014;**22**:A857

[57] Sinha N, Roul B, Mukundan S, Chandan G, Mohan L, Jali VM, Krupanidhi SB. Growth and electrical transport properties of InGaN/GaN heterostructures grown by PAMBE. Materials Research Bulletin. 2014;**61**:539-543

[58] El-Masry NA, Piner EL, Liu SX, Bedair SM. Phase separation in InGaN grown by metalorganic chemical vapor deposition. Applied Physics Letters. 1998;**72**:40

[59] Singh R, Doppalapudi D, Moustakas TD, Romano LT. Phase separation in InGaN thick films and formation of InGaN/GaN double heterostructures in the entire alloy composition. Applied Physics Letters.1997;**70**:1089

[60] Jmerik VN, Mizerov AM, Shubina TV, Yagovkina M, Listoshin VB, Sitnikova AA, Ivanov SV, Kim MH, Koike M, Kim BJ.Plasma-assisted MBE of InGaN epilayers with atomically smooth and nanocolumnar morphology, grown on MOVPE GaN/Al_2O_3 templates Journal of Crystal Growth. 2007;**3–302**:469

[61] Wang H, Jiang DS, Jahn U, Zhu JJ, Zhao DG, Liu ZS, Zhang SM, Qiu YX, Yang H. Investigation on the strain relaxation of InGaN layer and its effects on the InGaN structural and optical properties. Physica B. 2010;**405**:4668

[62] Chang YL, Wang JL, Li F, Mi Z. High efficiency green, yellow, and amber emission from InGaN/GaN dot-in-a-wire heterostructures on Si(111). Applied Physics Letters. 2010;**96**:013106

[63] Kehagias T, Dimitrakopulos GP, Becker P, Kioseoglou J, Furtmayr F, Koukoula T, Hausler I, Chernikov A, Chatterjee S, Karakostas T, Solowan HM, Schwarz UT, Eickhoff M, Komninou P. Nanostructure and strain in InGaN/GaN superlattices grown in GaN nanowires. Nanotechnology. 2013;**24**:435702

[64] Powell RC, Lee NE, Greene JE. Growth of GaN(0001)1×1 on Al_2O_3(0001) by gas-source molecular beam epitaxy. Applied Physics Letters. 1992;**60**:2505

[65] Pantha BN, Dahal R, Li J, Lin JY, Jiang HX, Pomrenke G. Thermoelectric properties of $In_xGa_{1-x}N$ alloys. Applied Physics Letters. 2008;**92**:042112

[66] McLaughlin DVP, Pearce JM. Progress in indium gallium nitride materials for solar photovoltaic energy conversion. Metallurgical and Materials Transactions A. 2013;**44**:1947

[67] Junqiao W. When group-III nitrides go infrared: New properties and perspectives. Journal of Applied Physics. 2009;**106**:011101

[68] Karpov SYu. Suppression of phase separation in InGaN due to elastic strain. MRS Internet Journal of Nitride Semiconductor Research. 1988;**3**:e16

[69] Keller S, Keller BP, Kapolnek D, Abare AC, Masui H, Coldren LA, Mishra UK, DenBaars SP. Growth and characterization of bulk InGaN films and quantum wells. Applied Physics Letters. 1996;**68**:3147

[70] Keller S, Keller BP, Kapolnek D, DenBaars SP, Shmagin IK, Kolbas RM, Krishnankutty S. Growth of bulk InGaN films and quantum wells by atmospheric pressure metalorganic chemical vapour deposition. Journal of Crystal Growth. 1997;**170**:349

[71] Song KM, Kim JM, Shin CS, Hwangand Dae SM, Yoon H. Growth and characterization of *a*-plane InGaN/GaN multiple quantum well LEDs grown on r-plane sapphire. Semiconductors and Science Technology. 2012;**27**:015011

[72] Paskova T, Kroeger R, Figge S, Hommel D, Darakchieva V, Monemar B, Preble E, Hanser A, Williams NM, Tutor M. High-quality bulk a-plane GaN sliced from boules in comparison to heteroepitaxially grown thick films on r-plane sapphire. Applied Physics Letters. 2006;**89**:051914

[73] Heying B,Wu XH, Keller S, Li Y, Kapolnek D, Keller BP, DenBaars SP, Speck JS. Role of threading dislocation structure on the x-ray diffraction peak widths in epitaxial GaN films. Applied Physics Letters. 1996;**68**:643

[74] Chierchia R, Bottcher T, Figge S, Diesselberg M, Heinke H, Hommel D. Mosaicity of GaN epitaxial layers: Simulation and experiment. Physica Status Solidi (b). 2001;**228**:403

[75] Wang H, Chen C, Gong Z, Zhang J, Gaevski M, Su M, Yang J, Asif Kahn M. Anisotropic structural characteristics of (1120) GaN templates and coalesced epitaxial lateral overgrown films deposited on (1012) sapphire. Applied Physics Letters. 2004;**84**:499

[76] Kim HY, Kim JH, Kim YJ, Chae KH, Whang CN, Song JH, Im S. Photoresponse of Si detector based on n-ZnO/p-Si and n-ZnO/n-Si structures. Optical Materials. 2001;**17**:141

[77] Yawata S, Anderson RL. Optical modulation of current in Ge-Si N-N heterojunctions. Physica Status Solidi. 1965;**12**:297

[78] Tan ST, Sun XW, Zhao JL, Iwan S, Cen ZH, Chen TP, Ye JD, Lo GQ, Kwong DL, Teo KL. Ultraviolet and visible electroluminescence from n-ZnO/SiO_x/(n,p)-Si heterostructured light-emitting diodes. Applied Physics Letters. 2008;**93**:013506

[79] Hudait MK, Krupanidhi SB. Growth, optical, and electron transport studies across isotype n-GaAs/n-Ge heterojunctions. Journal of Vacuum Science & Technology B: Microelectronics and Nanometer Structures. 1999;**17**:1003

[80] Milnes AG. Semiconductor Heterojunction Topics—Introduction and Overview. Solid-State Electronics. 1986;**29**:99

[81] Tansley TL, Owen SJT. Conductivity of Si-ZnO p-n and n-n heterojunctions. Journal of Applied Physics. 1984;**55**:454

[82] Srivastava AK, Zyskind JL, Lum RM, Dutt BV, Klingert JK. Electrical characteristics of InAsSb/GaSb heterojunctions. Applied Physics Letters.1986;**49**:41

[83] Ager JW, Reichertz LA, Cui Y, Romanyuk YE, Kreier D, Leone SR, Man K, Yu M, Schaff WJ, Walukiewicz W. Electrical properties of InGaN-Si heterojunctions. Physica Status Solidi (c). 2009;**6**(S2):S413

[84] Kumar P, Soto Rodriguez PED, Gómez VJ, Alvi NH, Calleja E, Nötzel R. First demonstration of direct growth of planar high-in-composition InGaN layers on Si. Applied Physics Express. 2013;**6**:035501

[85] Hatch SM, Briscoe J, Dunn S. A self-powered ZnO-nanorod/CuSCN UV photodetector exhibiting rapid response. Advanced Materials. 2013;**25**:867

[86] Bie YQ, Liao ZM, Zhang HZ, Li GR, Ye Y, Zhou YB, Xu J, Qin ZX, Dai L, Yu DP. Self-powered, ultrafast, visible-blind UV detection and optical logical operation based on ZnO/GaN nanoscale p-n junctions. Advanced Materials. 2011;**23**:649

[87] Ohsawa J, Kozawa T, Fujishima O, Itoh H. Narrow-band 400 nm MSM photodetectors using a thin InGaN layer on a GaN/sapphire structure. Physica Status Solidi (c). 2006;**3**:2278

[88] Chandan G, Mukundan S, Mohan L, Roul B, Krupanidhi SB. Trap modulated photoresponse of InGaN/Si isotype heterojunction at zero-bias. Journal of Applied Physics. 2015;**118**:024503

[89] Orsal G, El Gmili Y, Fressengeas N, Streque J, Djerboub R, Moudakir T, Sundaram T, Ougazzaden A, Salvestrini JP. Bandgap energy bowing parameter of strained and relaxed InGaN layers. Optical Materials Express. 2014;**4**:1030

[90] Wu J, Walukiewicz W, Yu KM, Ager JW, Haller EE, Lu H, Schaff WJ. Small band gap bowing in $In_{1-x}Ga_xN$ alloys. Applied Physics Letters. 2002;**80**:4741

[91] Reklaitis A. Terahertz-frequency InN/GaN heterostructure-barrier varactor diodes. Journal of Physics: Condensed Matter. 2008;**20**:384202

[92] Tsen KT, Poweleit C, Ferry DK, Lu H, Schaff WJ. Observation of large electron drift velocities in InN by ultrafast Raman spectroscopy. Applied Physics Letters. 2005;**86**:222103

[93] Chen NC, Chang PH, Wang YN, Peng HC, Lien WC, Shih CF, Chang CA, Wu GM. Schottky behavior at InN–GaN interface. Applied Physics Letters. 2005;**87**:212111

[94] Wang K, Lian C, Su. N, Jena D, Timler J. Conduction band offset at the InN⊚GaN heterojunction. Applied Physics Letters. 2007;**91**:232117

[95] Roul B, Rajpalke MK, Bhat TN, Kumar M, Sinha N, Kalghatgi AT, Krupanidhi SB. Temperature dependent electrical transport behavior of InN/GaN heterostructure based Schottky diodes. Journal of Applied Physics. 2011;**109**:044502

[96] Roul B, Kumar M, Rajpalke MK, Bhat TN, Krupanidhi SB. Binary group III-nitride based heterostructures: band offsets and transport properties. Journal of Physics D: Applied Physics. 2015;**48**:423001

[97] Tung RT. Recent advances in Schottky barrier concepts. Material Science Engineering R. 2001;**35**:1

[98] Hattab A, Perrossier JL, Meyer F, Barthula M, Osten HJ, Griesche J. Schottky barrier inhomogeneities at contacts to carbon-containing silicon/germanium alloys. Material Science Engineering B. 2002;**89**:284

[99] Cimilli FE, Saglam M, Turut A. Determination of the lateral barrier height of inhomogeneous Au/n-type InP/In Schottky barrier diodes. Semiconductors and Science Technology. 2007;**22**:851

[100] Singh A. Characterization of interface states at Ni/*n*CdF2 Schottky barrier type diodes and the effect of CdF_2 surface preparation. Solid-State Electronics. 1985;**28**:223

[101] McCafferty PG, Sellai A, Dawson P, Elabd H. Barrier characteristics of PtSi/p-Si schottky diodes as determined from I-V-T measurements. Solid-State Electronics. 1996;**39**:583

[102] Roul B, Bhat TN, Kumar M, Rajpalke MK, Sinha N, Kalghatgi AT, Krupanidhi SB. Barrier height inhomogeneities in InN/GaN heterostructure based Schottky junctions. Solid State Communication. 2011;**151**:1420

[103] Arslan E, Altındal S, Ozcelik S, Ozbay E. Tunneling current via dislocations in Schottky diodes on AlInN/AlN/GaN heterostructures. Semiconductors and Science Technology. 2009;**24**:075003

[104] Roul B, Mukundan S, Chandan G, Mohan L, Krupanidhi SB. Barrier height inhomogeneity in electrical transport characteristics of InGaN/GaN heterostructure interfaces. AIP Advances. 2015;**5**:037130

[105] Milnes AG, Oldham WG. Semiconductor heterojunction topics—Introduction and overview. Solid State Electronics. 1963;**6**:121

[106] Pantha BN, Wang H, Khan N, Lin JY, Jiang HX. Origin of background electron concentration in $In_xGa_{1-x}N$ alloys. Physical Review B. 2011;**84**:7

[107] Ghosh S, Sarker BK, Chunder A, Zhai L, Khondaker SI. Position dependent photodetector from large area reduced graphene oxide thin films. Applied Physics Letters. 2010; **96**:163109

[108] Chitara B, Krupanidhi SB, Rao CNR. Solution processed reduced graphene oxide ultraviolet detector. Applied Physics Letters. 2011;**99**:113114

Epitaxy and Device Properties of InGaAs Photodetectors with Relatively High Lattice Mismatch

Xing-you Chen, Yi Gu and Yong-gang Zhang

Additional information is available at the end of the chapter

http://dx.doi.org/10.5772/intechopen.70259

Abstract

In this chapter, our works on the developments of wavelength-extended InGaAs photodetectors with cutoff wavelength >1.7 μm are reviewed. Various InGaAs/InAlAs p-i-n heterojunction structures have been grown on InP and GaAs substrates by gas source molecular beam epitaxy, some details on the InGaAs photodetector structures and the techniques of metamorphic buffer layer such as linearly, step, and one-step continuously InAlAs graded buffer, and dislocation restraint methods of compositional overshoot and digital alloy are introduced. The material characteristics and device properties were evaluated by atomic force microscopy, high-resolution X-ray diffraction and reciprocal space mapping, cross-sectional transmission electron microscopy, and current-voltage measurements, etc. The results provide clues to the development of metamorphic device structures on lattice-mismatched material systems.

Keywords: InGaAs, photodetector, metamorphic, lattice mismatch, X-ray diffraction, atomic force microscopy, photoluminescence

1. Introduction

InGaAs photodetectors (PDs) and focal plane arrays (FPAs) are attracting particular interests as they can be tailored to cover one of the atmospheric windows of 1–3 μm in shortwave infrared band. The $In_{0.53}Ga_{0.47}As$ PDs grown lattice-matched to InP are commercially mature with cutoff wavelength at 1.7 μm. Wavelength-extended $In_xGa_{1-x}As/InP$ ($0.53 < x < 1$) PDs with cutoff wavelength more than 1.7 μm have been extensively investigated over the past decades due to their important applications in spatial remote sensing, earth observation, environmental monitoring, etc. [1, 2]. However, for application in longer wavelength region, the indium content should be increased, which increases the lattice mismatch between high indium content InGaAs and InP substrate. For instance, the $In_{0.83}Ga_{0.17}As$ PD with cutoff wavelength of 2.5 μm possesses a lattice mismatch of about +2.1% with respect to InP. In that case, the dark current increases several orders of magnitude, which severely hinders the development of wavelength-extended InGaAs PDs in the infrared range. To suppress the dark current of InGaAs PD, it is a prerequisite to optimize the InGaAs material with high indium (In) content. Therefore, high In content $In_xGa_{1-x}As$ has been grown on InP, GaAs, and Si substrate by using techniques such as molecular beam epitaxy (MBE) and metal organic chemical vapor deposition (MOCVD). During the process of material growth, dislocation restriction techniques such as beryllium (Be) doping [3], dilute nitride [4], strained or strain-compensated superlattice (SL), and strain-driven quantum dots (QDs) were adopted to reduce the threading dislocation density (TDD) in the metamorphic buffer layer (MBL). Several sets of results with different x values of the $In_xGa_{1-x}As$ layers are listed in **Table 1**. Our own results are also included. It can be seen that the TDD could be quite different for InGaAs with various In contents due to different lattice mismatches with different substrates.

In this chapter, the InGaAs/InAlAs p-i-n heterojunction structure was always used for the growth of wavelength-extended InGaAs PD with high indium content because of higher quantum efficiency than homostructure. Both the MBL schemes and experimental parameters have

X value	Growth method	Substrate	TDD in $In_xGa_{1-x}As$ (cm^{-2})	Reference
0.68	MOCVD	InP (0 0 1)	~10^6 (XTEM)	Ji et al. [5]
0.82	MOCVD	InP (0 0 1)	~10^8 (XTEM)	Zhao et al. [6]
0.82	MOCVD	InP (0 0 1)	~10^{11}–10^{12} (XRD-FWHM)	Zhao et al. [7]
0.83	GSMBE	InP (0 0 1)	≤10^7 (XTEM)	Present work
0.53	SSMBE	GaAs (0 0 1)	~10^6 (XTEM)	Lubyshev et al. [8]
0.6	SSMBE	GaAs (0 0 1)	~10^8 (XTEM)	Valtuefia et al. [9]
0.75–1	SSMBE	GaAs (0 0 1)	~10^9–10^{10} (plan-view TEM)	Chang et al. [10]
1	SSMBE	GaAs (0 0 1)	~10^8–10^9 (XTEM)	Chang et al. [11]
0.8	Not mentioned	GaAs (0 0 1)	~10^5 (EPD)	Zimmermann et al. [12]
0.63	SSMBE	GaAs (0 0 1)	~10^8 (XTEM)	Song et al. [13]
0.85	SSMBE	GaAs (0 0 1)	Not mentioned	Jurczak et al. [14]
0.83	GSMBE	GaAs (0 0 1)	~10^9–10^{10} (XTEM)	Present work
0.53	SSMBE	Si (1 1 1)	Not mentioned	Gao et al. [15]

Table 1. Parameters of $In_xGa_{1-x}As$ ($0.53 \le x \le 1$) as reported by various researchers.

been optimized to acquire high-performance InGaAs PDs with relatively high-lattice mismatch. Additionally, some strategies were used to reduce the residual strain and decrease the TDD, such as composition overshoot and the insertion of digital alloy (DA) intermediate layer in the MBL.

All the material growth in this chapter was performed in a VG Semicon V80H gas source molecular beam epitaxy (GSMBE) system. The best background chamber pressure achieved in this system was about 3×10^{-11} Torr, and the pressure during growth is typically in the 10^{-5} Torr range. The elemental In, gallium (Ga), and aluminum (Al) sources were used as group III sources, and the elemental silicon (Si) and Be were used as n-type and p-type dopant sources, respectively. Their fluxes were controlled by changing the cell temperatures. Arsine and phosphine cracking cells were used as group V sources, and their fluxes were controlled by adjusting the pressure. The cracking temperature was approximately 1000°C measured by using a thermocouple. The growth rates were all adjusted to about 1 μm/h for InP, InAlAs, and InGaAs.

2. Optimize the growth of wavelength extended InGaAs PDs on InP

2.1. The effects of growth temperature on the characteristics of InP-based $In_{0.83}Ga_{0.17}As$ PDs

The growth temperature (T_g) is one of the most important parameters of MBE. To acquire high-device performance wavelength-extended InGaAs PD on InP substrate, the growth temperature should be optimized for each layer. Generally, a high T_g can reduce the incorporation of background impurities into the epilayers during the growth process, whereas the higher T_g will enhance the In segregation on the surface of InGaAs with a high In content [16], which will cause a poor surface and nonuniformity of compositions in the epilayer and then increase the dark current of PDs. In this work, T_g was optimized for the InAlAs buffer layer and InGaAs absorption layer, respectively [17].

Four $In_{0.83}Ga_{0.17}As$ PD structures with continuously graded InAlAs buffers were grown on (1 0 0)-oriented Fe-doped InP epi-ready substrate. Each structure consisted of a 1.95-μm N^+ (N = 3×10^{18} cm^{-3}) InAlAs buffer layer and a 2-μm-thick n^- (n = 3×10^{16} cm^{-3}) $In_{0.84}Ga_{0.16}As$ absorption layer followed by a P^+ (P = 7×10^{18} cm^{-3}) 600-nm-thick $In_{0.84}Al_{0.16}As$ cap. The In profiles and T_g at stages A, B, C, and D of the four PD structures were shown in **Figure 1**.

The results show that the growth temperature of the absorption layer should not be very low, whereas the InAlAs MBL with the T_g graded from 500 to 460°C exhibits better crystalline quality than that grown at either 460 or 500°C. The sample number 3 with InAlAs MBL grown with temperature graded from 500 to 460°C, and InGaAs absorption layer grown at 500°C exhibits better crystalline quality. It shows smoother surface with lower root-mean-square (RMS) of atomic force microscopy (AFM), narrower full width at half maximum (FWHM) of X-ray diffraction (XRD), and stronger photoluminescence (PL) intensity than others, as shown in **Figure 2**. Therefore, this temperature strategy for growth of wavelength-extended InGaAs PDs on InP substrate will be always used hereafter except for special reminding.

Sample no.	Growth temperature (°C)			
	A	B	C	D
1	500	500	500	500
2	460	460	500	500
3	500	460	500	500
4	500	460	460	460

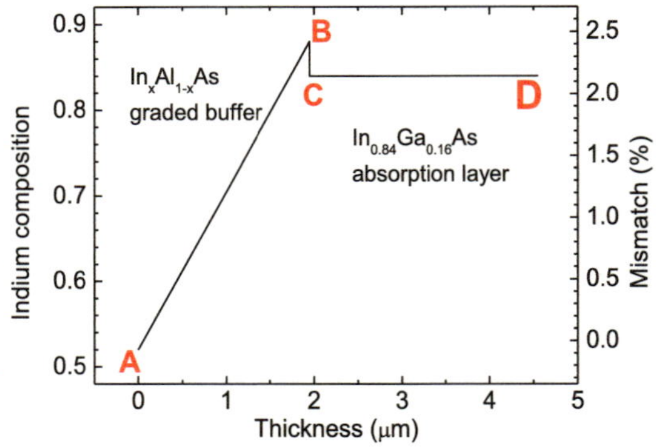

Figure 1. Growth temperatures at points A–D of the $In_{0.84}Ga_{0.16}As$ PD structures. The inset shows the indium composition vs. the grown thickness of samples 1–4. Reprinted with permission from Elsevier.

2.2. The optimization and comparison of PD structures on different metamorphic buffer layers

To acquire high-quality wavelength-extended InGaAs materials with relatively high-lattice mismatch, an appropriate InAlAs MBL is needed. Generally, for the growth of lattice-mismatched material system, the MBLs should be thick enough so that the lattice parameter can finally relax to be a freestanding status, and the misfit dislocation can annihilate ultimately during the growth process of the MBL. However, it is impractical and high cost to grow an excessively thick MBL for device structure by MBE technique due to a limited growth rate of around 1 μm h^{-1}. Therefore, proper schemes of thin MBL should be explored.

In our previous work, the effect of various kinds of MBL schemes such as a thick uniform buffer, continuously graded buffer, and step-graded buffer on the structural characteristics and device performances of InP-based InGaAs PDs has been compared [18–20].

Two $In_{0.83}Ga_{0.17}As$ PD wafers with continuously (sample S1) and step-graded (sample S2) InAlAs MBLs were grown on semi-insulated (S.I.) (1 0 0)-oriented InP epi-ready substrate. Each wafer consisted of a 1.9-μm N^+ ($N = 3 \times 10^{18}$ cm^{-3}) InAlAs MBL and a 1.5-μm-thick n^- ($n = 3 \times 10^{16}$ cm^{-3}) $In_{0.83}Ga_{0.17}As$ absorption layer followed by a P^+ ($P = 7 \times 10^{18}$ cm^{-3}) 530-nm-thick InAlAs cap layer. The typical schematic structures of the two PDs were shown in **Figure 3**.

The growth of sample S1 with a continuously graded $In_xAl_{1-x}As$ MBL started with a 200-nm InP buffer layer grown at 460°C, following a 0.1-μm-thick N^+ $In_{0.52}Al_{0.48}As$ layer grown at 500°C, the N^+ continuously graded $In_xAl_{1-x}As$ buffer layer was grown with growth temperature decreased linearly from 500 to 460°C, and the In composition x was graded from 0.52

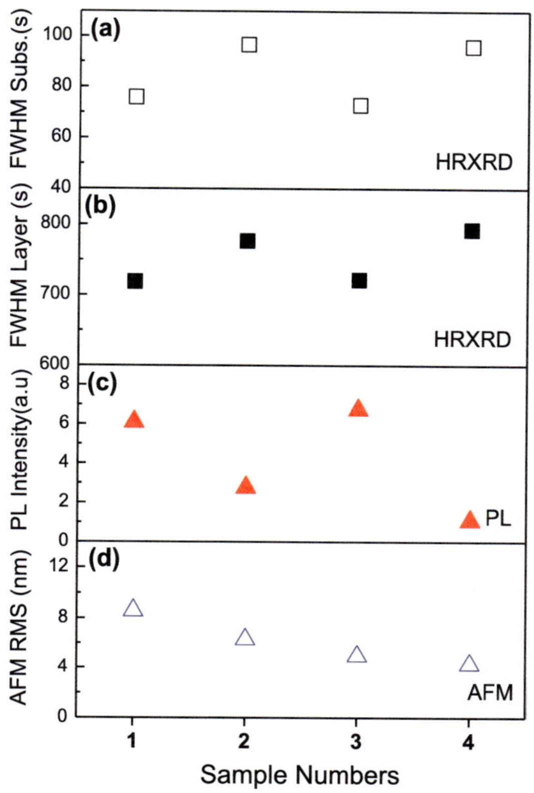

Figure 2. Summary of the measured data of samples with different temperature strategies. (a) HRXRD FWHM of substrate peak, (b) HRXRD FWHM of layer peak, (c) PL intensity of InGaAs absorption layer and (d) AFM RMS of samples 1–4. Reprinted with permission from Elsevier.

to 0.83 through the simultaneously linear increase of In source temperature and decrease of Al source temperature in a relatively small temperature region. This caused the In and Al beam fluxes to change with cell temperatures exponentially in opposite directions so that the composition grading profile in the buffer approximates a linear grade. The thickness of the InAlAs MBL was about 1.9 μm. After that, the growth temperature was increased to 490°C for the following growth.

The growth of sample S2 with a step-graded $In_xAl_{1-x}As$ MBL started with a 200-nm InP buffer layer grown at 460°C, following a 0.1-μm-thick N^+ $In_{0.52}Al_{0.48}As$ layer grown at 500°C, and then four N^+ $In_xAl_{1-x}As$ buffer layers with In composition of 0.60, 0.68, 0.76, and 0.83 were grown in sequence at 490, 480, 470, and 460°C, respectively. The In composition x was step graded from 0.52 to 0.83. Therefore, the thickness of the step-graded InAlAs MBL was also about 1.9 μm with each step layer of about 475 nm. After that, the growth temperature was kept at 490°C for the following growth.

To reduce the conduction band discontinuity and the dislocation density between the InGaAs absorption layer and InAlAs buffer and cap layers due to the small lattice mismatch between

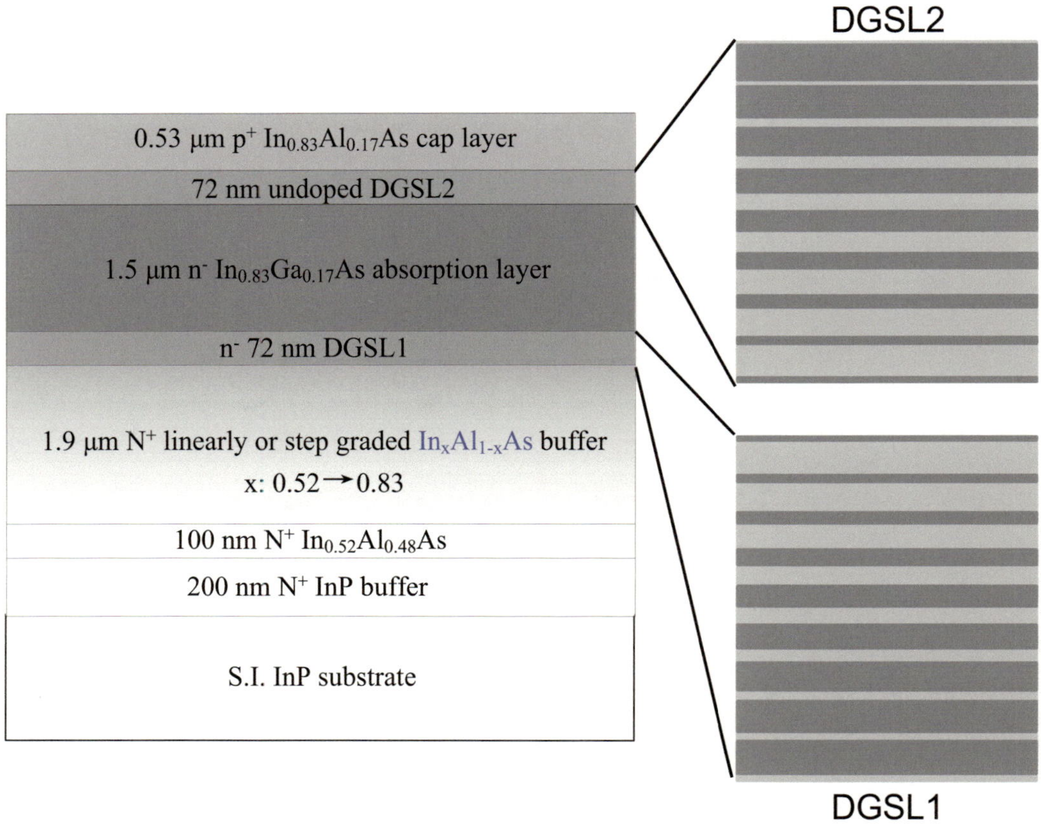

Figure 3. The schematic structure of $In_{0.83}Ga_{0.17}As$ photodetectors.

InGaAs and InAlAs at the interfaces, digital-graded superlattices (DGSLs) were grown at the interfaces, respectively. First, DGSL1 consisted of nine groups of $In_{0.83}Ga_{0.17}As/In_{0.83}Al_{0.17}As$ thin films which alternate with each other. The thickness of each group is about 8 nm. The thickness ratios of $In_{0.83}Ga_{0.17}As/In_{0.83}Al_{0.17}As$ in each group are 1:9, 2:8, …, finally coming to …, 8:2, 9:1, respectively. Similarly, DGSL2 was comprised of nine groups of $In_{0.83}Al_{0.17}As/In_{0.83}Ga_{0.17}As$ thin films which also alternate with each other but in a reverse order, and the thickness ratios of $In_{0.83}Al_{0.17}As/In_{0.83}Ga_{0.17}As$ in each group are 1:9, 2:8, …, finally coming to …, 8:2, 9:1, respectively. It is similar to the chirped $In_{0.53}Ga_{0.47}As/In_{0.52}Al_{0.48}As$ superlattice-graded bandgap layer used in some literatures [21–23].

Figure 4(a) shows the two different growth profiles of the In composition of $In_xAl_{1-x}As$ buffer for the InP-based $In_{0.83}Ga_{0.17}As$ PDs. Through a comprehensive comparison, sample S1 with continuously graded $In_xAl_{1-x}As$ MBL shows better properties on both material and devices than that of sample S2 with step-graded $In_xAl_{1-x}As$ MBL. As shown in the cross-sectional transmission electron micrographs (XTEMs) of **Figure 5**, almost none evident TD was found in the absorption layer of sample S1. By contrast, some obvious TDs exist in the absorption layer of sample S2. Unsurprisingly, sample S1 shows superior device performance compared

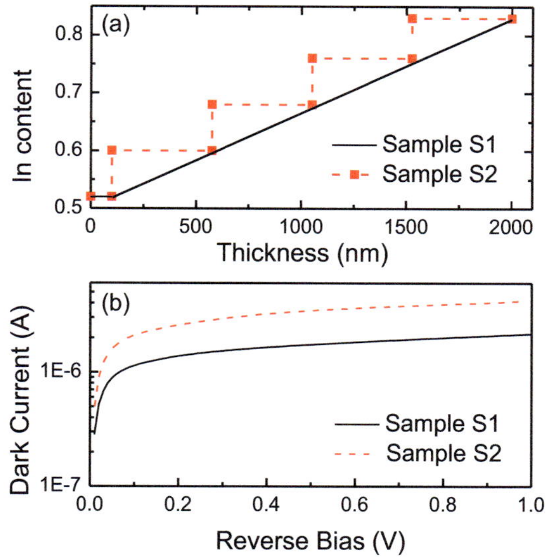

Figure 4. (a) The profiles of the indium composition vs. the grown thickness, and (b) I-V characteristics at 300 K of samples S1 and S2.

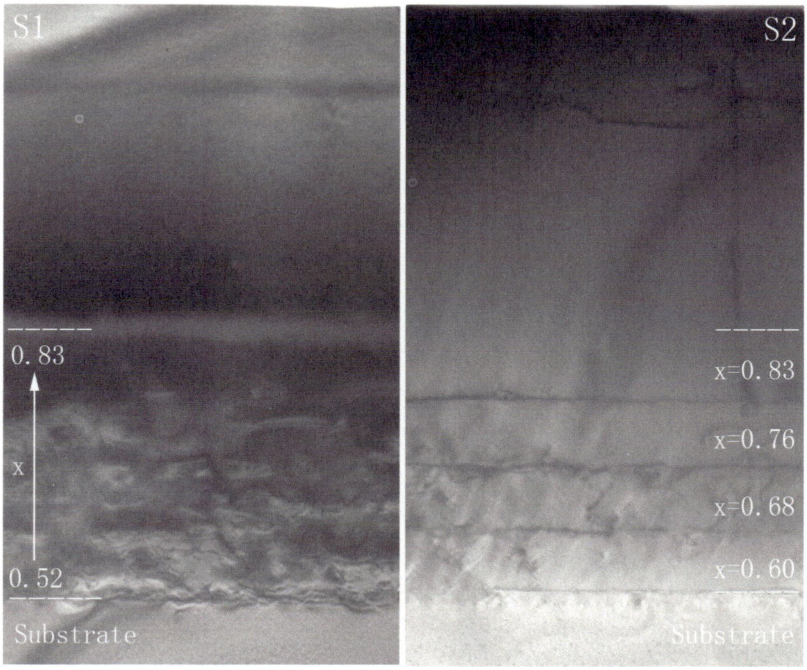

Figure 5. XTEM images of samples S1 and S2.

to sample S2. By measuring the mesa-type devices in diameter size of 200 μm, the room temperature (RT) dark current of the sample S1 was smaller than that of sample S2 at low reverse bias condition, as shown in **Figure 4(b)**. At reverse bias of −10 mV, the dark currents are 259 nA

(8.25×10^{-4} A/cm^2) and 473 nA (1.51×10^{-3} A/cm^2) for samples S1 and S2, respectively. Therefore, continuously graded $In_xAl_{1-x}As$ MBL scheme is more suitable for the growth of $In_{0.83}Ga_{0.17}As$ PDs on InP substrate.

In the previous continuously graded InAlAs MBLs, all the In contents were graded from 0.52. The initial lattice constant matches to the InP substrate. Strain energies in the continuously graded buffer increase gradually with increasing In content. Then, dispersed misfit dislocations will be generated to release the strain energy slowly in the buffer. Little overlap and interaction would occur among the TDs. Thus, an enough thick buffer is needed to slide and annihilate the TDs during the growth process. Here, we propose another method to promote the overlap of dispersed TDs by introducing an abrupt initial $In_yAl_{1-y}As$ ($y > 0.52$) layer of InAlAs MBL on InP substrate. This abrupt initial layer, mismatched with the InP substrate, is similar to a step buffer. But it was combined with a continuously graded $In_xAl_{1-x}As$ buffer with x graded from y to the composition needed. This buffer scheme, so-called one-step continuously graded buffer, is expected to possess the advantages of both step and continuously graded schemes. The strain is released rapidly by introducing the initial abrupt buffer, and TDs were restricted between the step buffer and the InP substrate. We show that this method results in lower TDD in the final-graded buffer and the InGaAs absorption layer in the wavelength-extended PD structure. This buffer strategy has been also adopted and demonstrated for the growth of $InAs_xP_{1-x}$ on InP by MOCVD method [23].

To investigate the effect of this buffer scheme on the properties of InP-based $In_{0.83}Ga_{0.17}As$ PDs with a lattice mismatch of 2.1%. Samples with three different initial In compositions were grown by GSMBE. As a comparison, the In composition grading profile and growth method of sample E were remained the same as before [19, 24]. For the deposition of wafers with one-step continuously graded $In_yAl_{1-y}As$ buffer, the In contents y of the one-step buffer were set to be 0.68 and 0.77 with the thicknesses of 0.5 and 0.7 μm, respectively. Then, the one-step buffer was followed by a continuously graded $In_xAl_{1-x}As$ buffer with In composition graded from 0.68 or 0.77 to 0.83, respectively. The growth temperatures were graded from 520 to 490°C and 505 to 490°C, respectively. They were defined as samples F and G, respectively. Since the total thickness of the InAlAs MBL was maintained to be 1.9 μm, the higher the initial indium content, the lower the mismatch grading rate of the continuously graded $In_yAl_{1-y}As$ buffer, as shown in **Figure 6(a)**. The mismatch grading rates for samples E, F, and G were 1.08, 0.702, and 0.312% μm^{-1}, respectively.

By using this buffer scheme, it was found that RT PL intensity of $In_{0.83}Ga_{0.17}As$ layer was enhanced in sequence from E to G as shown in **Figure 6(b)**, indicating a promoted optical quality of the absorption layer. For the mesa type devices in the diameter size of 200 μm, the dark current was deceased along with the increase of initial In content y from 0.52 to 0.68 and 0.77 under low reverse bias condition. At a reverse bias of −10 mV, the dark current halves for the sample G with respect to that of the sample E at 300 K, as shown in **Figure 6(c)**. This validated the suppression effects of one-step continuously graded $In_yAl_{1-y}As$ MBL on the TDs.

2.3. Strain gradient and composition overshoot in the metamorphic buffer layer

An ideal buffer cannot only restrict the misfit dislocations in the inactive region and reduce the TDD in the absorption layer but also immensely reduce the residual lattice strain in the

http://dx.doi.org/10.5772/intechopen.70259

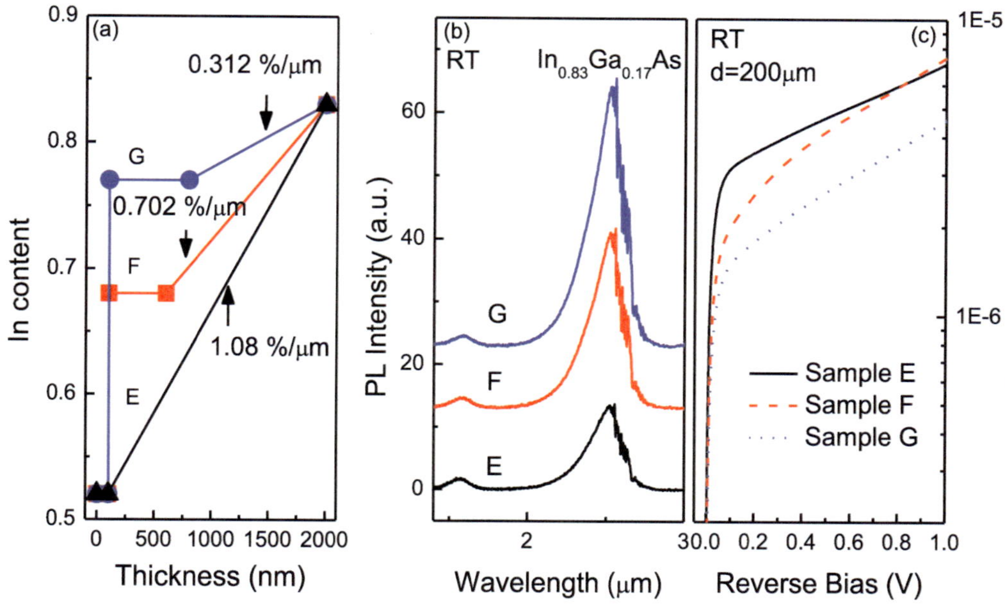

Figure 6. (a) In content vs. the grown thickness, (b) RT PL spectra, and (c) measured I-V characteristics of samples E, F, and G. Reprinted with permission from Elsevier.

epilayer. The strain gradient is proportional to the compositional grading rate in a linearly graded MBL. In this section, 100-nm thin $In_{0.8}Ga_{0.2}As$ layers were grown atop on the 1.6 μm and 0.6 μm linearly graded $In_xGa_{1-x}As$ MBLs on InP, which were defined as samples A and B, respectively. Since the lattice mismatch between the $In_{0.8}Ga_{0.2}As$ layer and the InP substrate is about +1.86%. The strain gradients correspond to be around +1.2% μm^{-1} and +3.1% μm^{-1}, for samples A and B, respectively. The In composition profiles versus growth thickness for the two samples were shown in **Figure 7**. The effects of strain gradient on the surface, structural, and optical properties of the $In_{0.80}Ga_{0.20}As$ metamorphic structures were investigated [25].

The results show that although lower strain gradient caused a slightly rougher surface of the structure shown in **Figure 8**, but it also brought about a larger degree of lattice relaxation and a less residual strain in the top $In_{0.8}Ga_{0.2}As$ layer. **Figure 9** shows the measurement of high-resolution XRD (HRXRD) reciprocal space mapping (RSM). Some results were extracted from the RSM and listed in **Table 2**.

The peaks of the top $In_{0.80}Ga_{0.20}As$ layer of sample A are stronger than that of sample B in both (0 0 4) and (2 2 4) reflections, indicating higher lattice quality of sample A. The lattice relaxation degree is 82.5% for sample A and 77.8% for sample B, respectively. Correspondingly, the residual strain of sample A is smaller than that of sample B. Therefore, it can be concluded that the use of lower strain gradient in the linearly graded MBL is beneficial to the lattice relaxation and the release of residual strain. In addition, a nearly fully strained thin top layer of the structure was observed, indicating a two-step relaxation procedure of the linearly graded MBL as predicted by the Tersoff's model [26]. However, it is still not sufficient to achieve a full relaxation even in the linearly graded $In_xGa_{1-x}As$ MBL with a mismatch

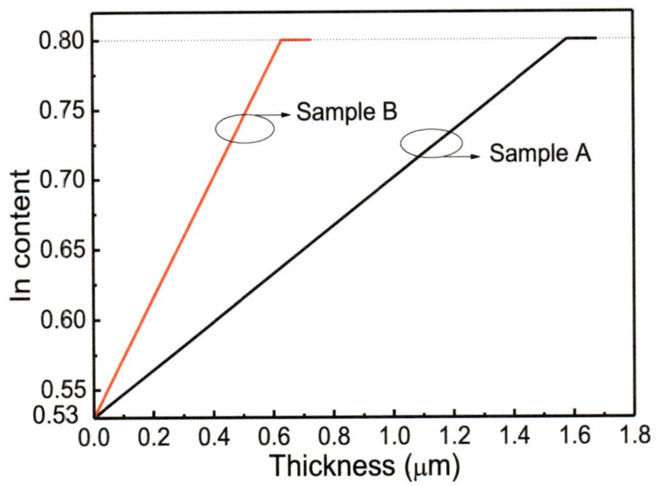

Figure 7. The indium composition versus growth thickness for the two samples. Reprinted with permission from IOP.

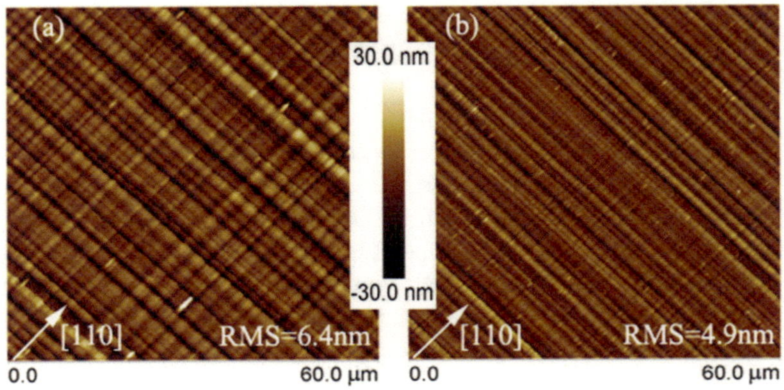

Figure 8. AFM images of (a) sample A and (b) sample B. Reprinted with permission from IOP.

grading rate of 1.2% μm^{-1} in our experiments. Therefore, to further increase the lattice relaxation degree, composition "overshoot" in the MBL should be introduced.

It was proposed that a dislocation-free portion will be formed when the thickness of the buffer is excess of a value of in the linearly graded buffer. The value of Z_c can be expressed as [27]:

$$Z_c = W - (2\lambda / bc\,\varepsilon')^{1/2} \tag{1}$$

where W is the total thickness of the buffer, λ is the energy per unit length of the dislocation, c is the appropriate elastic constant for biaxial strain, b is the misfit component of the Burgers vector of dislocation, and ε' is the strain gradient. The residual strain in this dislocation-free portion can be given by:

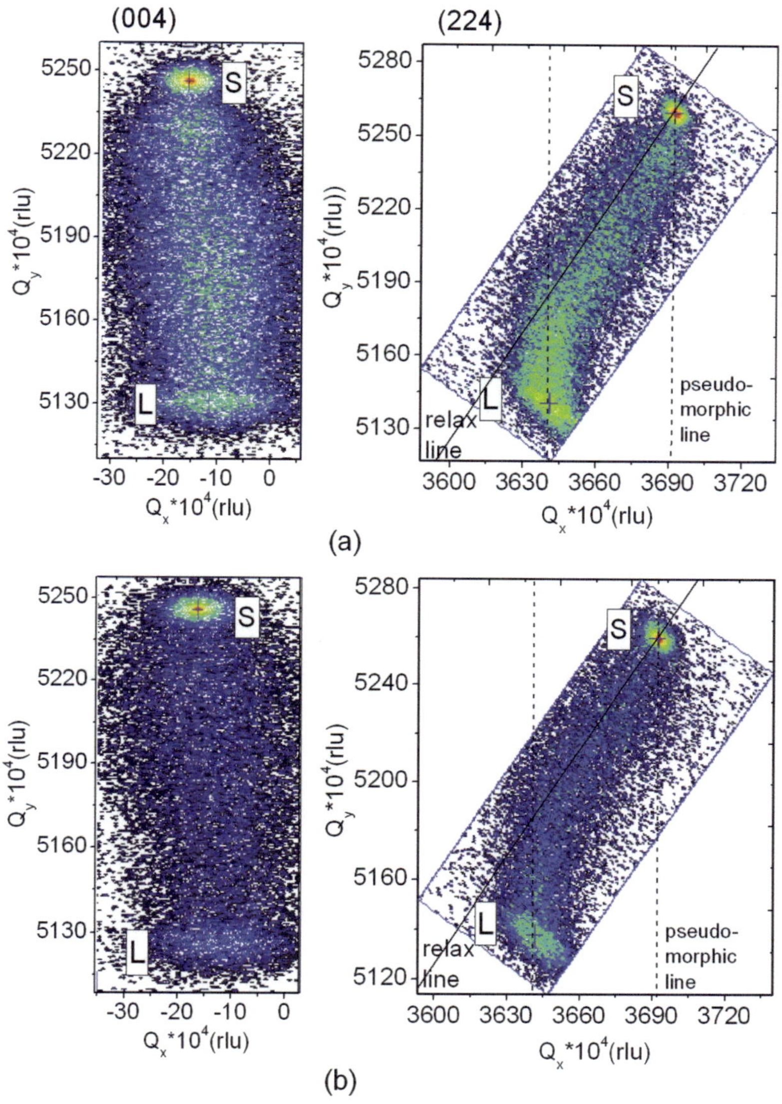

Figure 9. HRXRD RSMs of (a) sample A and (b) sample B. Reprinted with permission from IOP.

Samples	Bulk mismatch (%)	Indium content	Relaxation degree (%)	Parallel mismatch (%)	Perpendicular mismatch (%)	Residual strain (10^{-3})	
						XRD RSM	Tersoff's model
M	1.907	0.806	82.5	1.573	2.322	−3.27	−4.4
N	1.885	0.803	77.8	1.466	2.364	−4.11	−7.0

Table 2. Results extracted from the RSM measurements (reprinted with permission from IOP).

$$\bar{\varepsilon} = W\varepsilon' - Z_c\varepsilon' = (2\lambda\,\varepsilon'/bc)^{1/2} \qquad (2)$$

Therefore, in the linearly graded buffer, an "overshoot" of the mismatch by an amount is needed to make the graded buffer be lattice matched to the following layers. Therefore, to judge this hypothesis, two sets of $In_{0.78}Ga_{0.22}As/In_{0.78}Al_{0.22}As$ quantum wells (QWs) and $In_{0.84}Ga_{0.16}As$ PD structures with and without compositional overshoot in the linearly graded InAlAs MBLs were grown. The effects on the material properties and the lattice relaxation degree have been investigated in detail [28].

As shown in **Figure 10**, the QW structures consist of a 100-nm-thick InP buffer, a 1.7-μm-thick linearly graded $In_xAl_{1-x}As$ MBL, and a three periods of $In_{0.78}Ga_{0.22}As/In_{0.78}Al_{0.22}As$ QWs. The thicknesses of well and barrier layers are 10 and 12 nm respectively, as shown in **Figure 10(a)** and **(b)**. The In composition x in the $In_xAl_{1-x}As$ buffer of this structure was graded from 0.52 to 0.78 for sample 1 and from 0.52 to 0.82 for sample 2. For the $In_{0.84}Ga_{0.16}As$ PD structures, the In composition x in the $In_xAl_{1-x}As$ buffer was designed grading from 0.52 to 0.84 for sample 3 and from 0.52 to 0.88 for sample 4, as shown in **Figure 10(c)** and **(d)**, respectively. Therefore, compositional overshoot of about 0.04 was introduced in the end of the $In_xAl_{1-x}As$ MBLs of samples 2 and 4.

From the AFM, PL, and HRXRD RSM measurement results shown in **Figure 11**, **Figure 12**, and **Table 3**, respectively, we have concluded that the use of compositional overshoot in the InAlAs MBLs not only reduces the surface roughness but also enhances the optical quality and the lattice relaxation degree for both QW and PD structures.

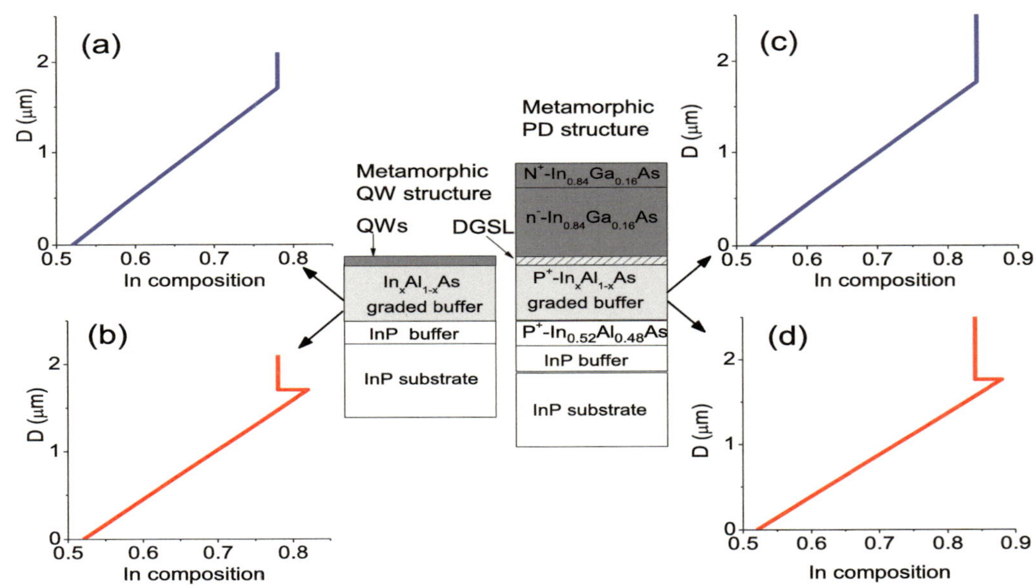

Figure 10. The indium composition profiles in the InAlAs buffer and schematic structures of samples (a) 1, (b) 2, (c) 3, and (d) 4, respectively.

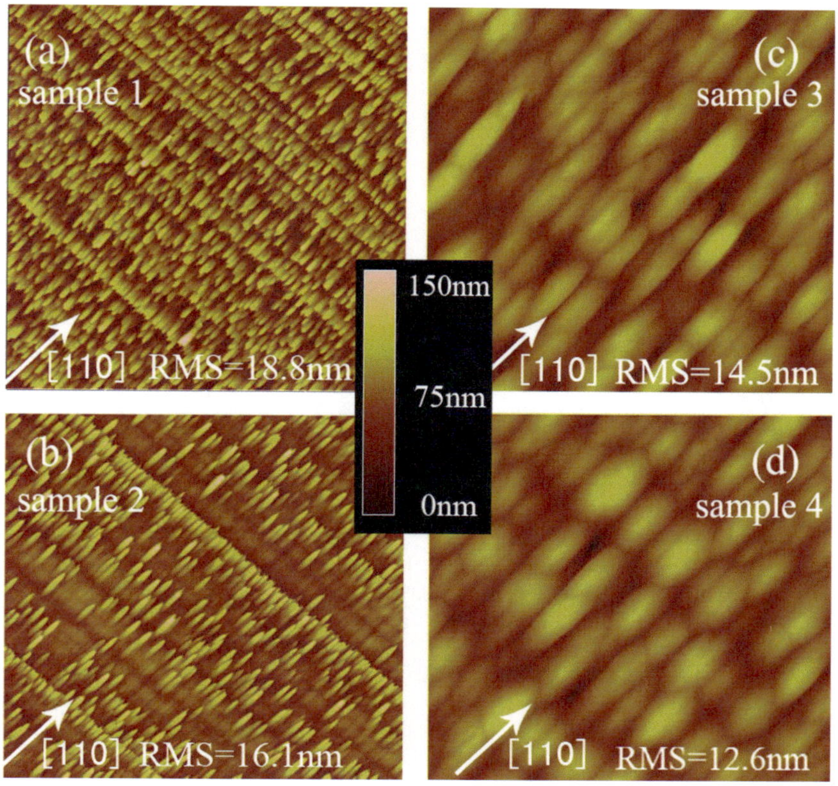

Figure 11. AFM images of (a) sample 1, (b) sample 2, (c) sample 3, and (d) sample 4.

2.4. Adoption of DA for dislocation restriction in the metamorphic buffer layer

Recently, it was confirmed that the insertion of $InAs/In_{0.52}Al_{0.48}As$ DA intermediate layer in the InAlAs MBL can restrain the TDs effectively and improve the structural and optical qualities of the lattice-mismatched system [29].

By alternating the epitaxy of two or more thin layers, DA intermediate layers can be formed. They are expected to reduce the strain energy, restrain the three dimensional (3D) growth, and increase the critical thickness. If the total thickness of one period is set to be d, the needed composition z of the DA layer can be achieved by adjusting the thickness of each thin layer in each period (d_i) through the equation:

$$\sum_{i=0}^{n} \frac{d_i \times m}{d} = z \tag{3}$$

n is the total number of thin layers and m is the composition in each thin layer. These layers are so thin that they would intermix with each other significantly.

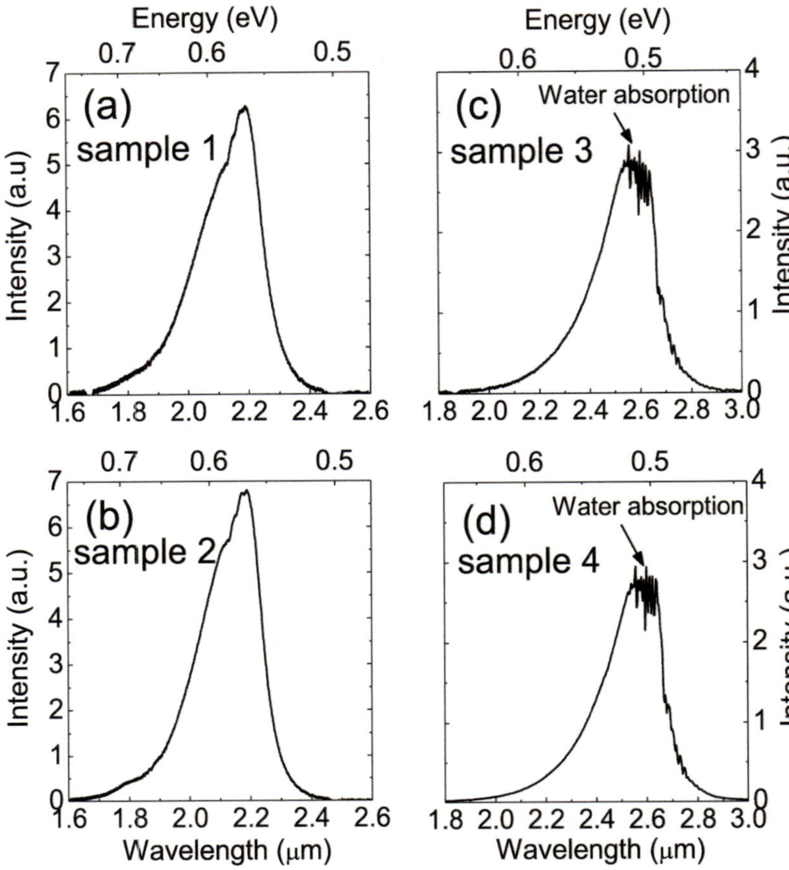

Figure 12. The RT PL spectra of (a) sample 1, (b) sample 2, (c) sample 3, and (d) sample 4.

Samples	Indium content	Cubic mismatch (%)	Parallel mismatch (%)	Perpendicular mismatch (%)	Relax. Degree (%)	Residual strain (10^{-3})
1	0.774	1.675	1.459	1.510	87.2	−2.13
2	0.781	1.730	1.716	2.006	96.5	−0.60
3	0.838	2.125	2.069	2.324	97.4	−0.54
4	0.840	2.136	2.095	2.288	99.4	−0.45

Table 3. Results extracted from RSM measurements of samples 1–4.

As shown schematically in **Figure 13(a)**, two $In_{0.8}Al_{0.2}As/In_{0.8}Ga_{0.2}As$ QW structures were grown on linearly graded InAlAs MBL on InP. As a reference, the In composition x in sample A was continuously graded from 0.52 to 0.8. For sample B, two $In_yAl_{1-y}As$ DA intermediate layers of about 100 nm were inserted into the graded InAlAs MBL, with a separation of every

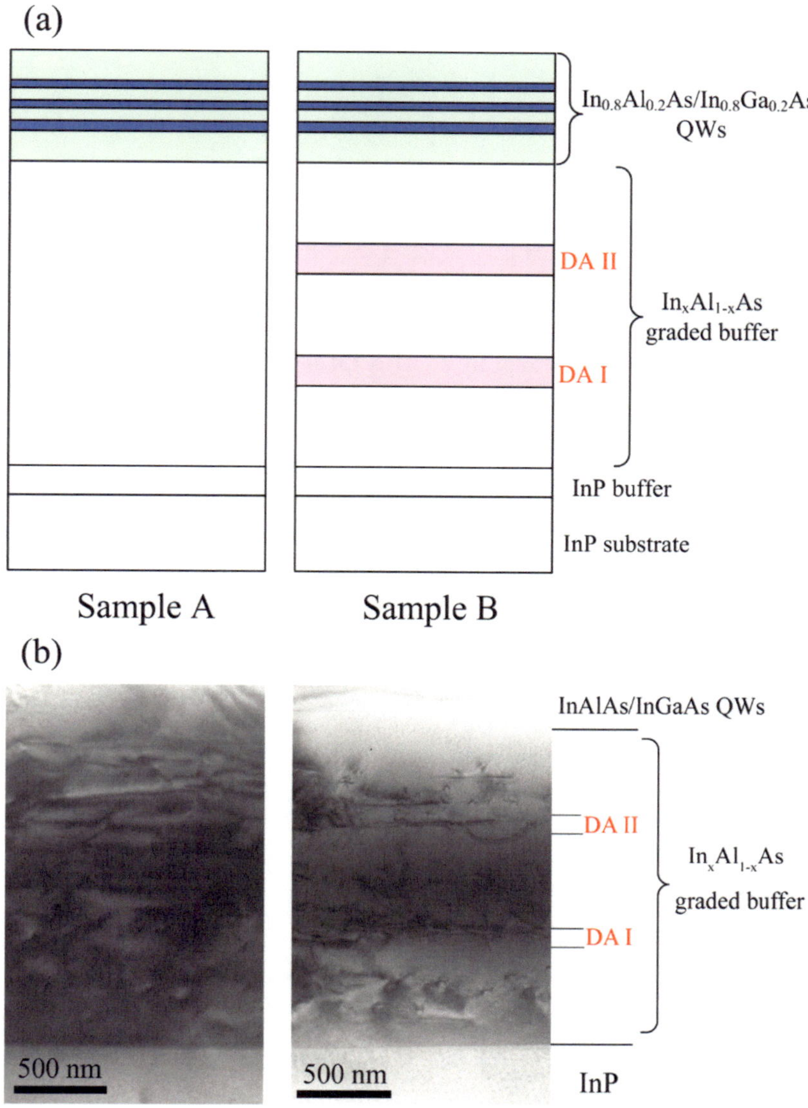

Figure 13. (a) Schematic design structure and (b) XTEM images of the samples. Reprinted with permission from IOP.

500 nm $In_xAl_{1-x}As$ buffer. The In composition of the intermediate layer, y, was designed to be the same as the composition in the inserted positions of the buffer, i.e., y = 0.62 and 0.71 for DA intermediate layers I and II, respectively. The DA layers were formed by the alternating epitaxy of very thin InAs and $In_{0.52}Al_{0.48}As$ layers. The thickness d of a period including one InAs layer and one $In_{0.52}Al_{0.48}As$ layer was 1 nm. The thicknesses of InAs (d_1) and $In_{0.52}Al_{0.48}As$ (d_2) in each period were adjusted to achieve the equivalent In composition y and obtained from the equation:

$$\begin{cases} \dfrac{d_1 + d_2 \times 0.52}{d} = y \\ d_1 + d_2 = d \end{cases}$$

d_1 and d_2 are so small that a considerable material intermixing is expected to occur between the two layers since that the thinnest layer is a submonolayer, and the thickest layer is only about two monolayers (MLs). After the growth of the MBL, three 10-nm $In_{0.8}Ga_{0.2}As$ QWs sandwiched by $In_{0.8}Al_{0.2}As$ barriers were grown. The thickness was 12 nm for the $In_{0.8}Al_{0.2}As$ barrier between the two wells and 100 nm for the first and last barriers.

The measurement results were shown in Figures **13(b)–16**. Although the insertion of InAs/$In_{0.52}Al_{0.48}As$ DA intermediate layers caused a slightly decrease of lattice relaxation degree listed in **Table 4**, the TDD has been decreased immensely in the following InAlAs MBLs as shown in the XTEM images of **Figure 13(b)**, and both the PL and AFM RMS have been improved markedly, indicating the positive effect of the DAs on the structural and optical qualities. However, to maximize these effects, the composition, number, and thickness of the DA should be further optimized.

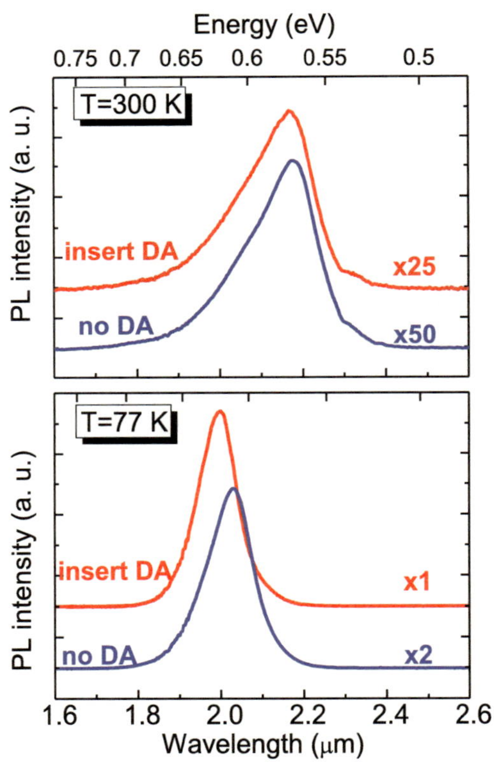

Figure 14. PL spectra of the samples at 300 and 77 K. Reprinted with permission from IOP.

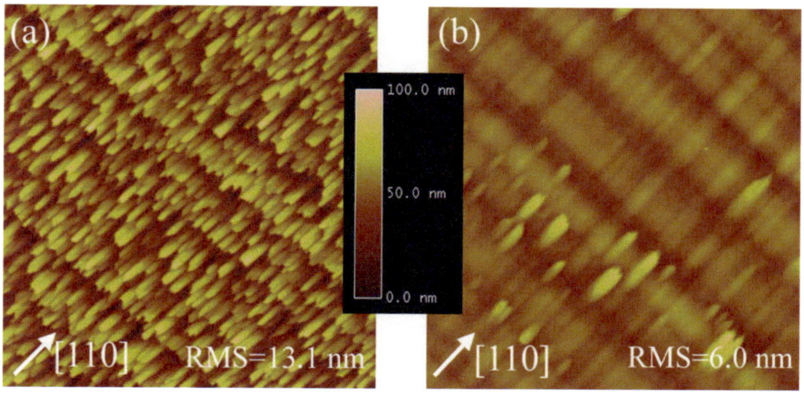

Figure 15. 20 × 20 μm² AFM surface images of (a) sample A and (b) sample B. Reprinted with permission from IOP.

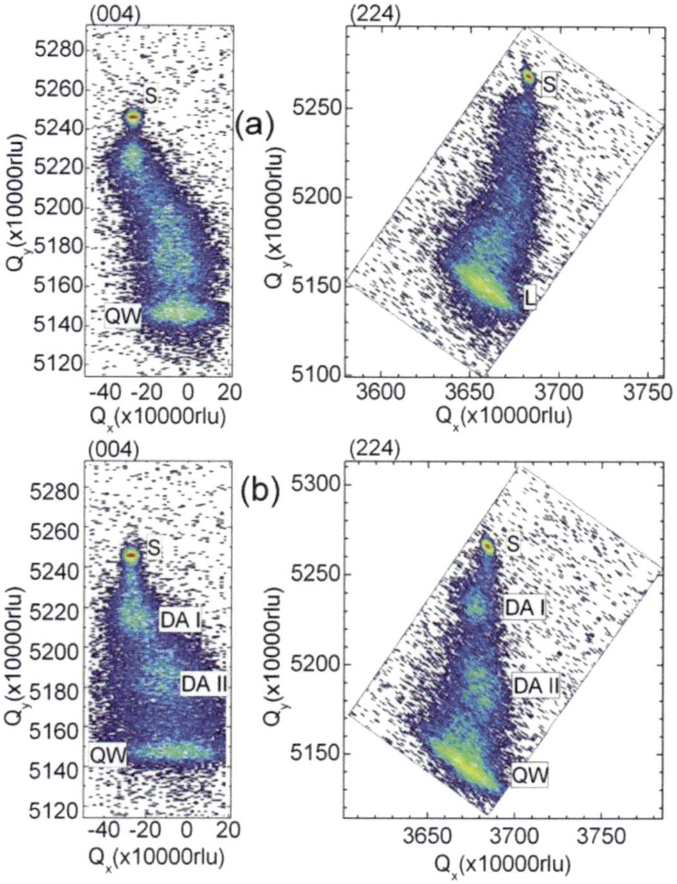

Figure 16. HRXRD RSMs of (a) sample A and (b) sample B. Reprinted with permission from IOP.

Sample	Epitaxy layers	Indium content (y)	Degree of relaxation, R (%)	Residual strain, ε (10^{-3})
A	QW structure	0.760	82.7	−3.5
	QW structure	0.736	70.2	−4.3
B	DA II	0.653	67.1	−2.9
	DA I	0.580	54.9	−1.8

Table 4. Results of XRD RSM measurements (reprinted with permission from IOP).

3. Trials to move $In_{0.83}Ga_{0.17}As$ PD from InP to GaAs substrate

3.1. Comparison of GaAs- and InP-based $In_{0.83}Ga_{0.17}As$ PDs with different lattice mismatches

Currently, most of the wavelength-extended InGaAs PD structures were grown on the InP substrate [2]. Even if a larger lattice mismatch will be introduced, GaAs may still be an attractive substrate for fabrication of InGaAs PDs with large size epitaxial wafers as well as FPAs with more pixels for the advantages of robustness, lower cost, and larger size. However, there are few attempts to transfer $In_xGa_{1-x}As$ ($x > 0.53$) PDs from InP to GaAs substrate.

In our recent researches, two $In_{0.83}Ga_{0.17}As$ PD structures with continuously graded InAlAs MBL were grown on (1 0 0)-oriented S-doped InP or GaAs epi-ready substrates by GSMBE [30]. In this way, the lattice mismatch will increase from +2.1 to +5.9% for $In_{0.83}Ga_{0.17}As$ on GaAs compared to that on InP. The growth condition of InP-based $In_{0.83}Ga_{0.17}As$ PD structure was the same as that in the previous study. For the deposition of the GaAs-based $In_{0.83}Ga_{0.17}As$ PD structure, the InAlAs MBL started with a 0.1-μm-thick highly Si-doped N^+ $In_{0.1}Al_{0.9}As$ layer grown at 530°C, followed by a 1.9-μm-thick compositionally graded $In_xAl_{1-x}As$ layer with In composition x graded from 0.1 to 0.87, and the substrate temperature graded from 530 to 460°C. The mismatch grading rate of 3.1% μm^{-1} for sample A was larger than that of 1.1% μm^{-1} for sample B. Then, a 0.65-μm $In_{0.83}Al_{0.17}As$ template was grown at 460°C in the end of the graded buffer. The growth temperatures of the InGaAs absorption layer and the InAlAs cap layer were 490°C, and the doping level in each layer of both structures was kept uniform. They were renamed as samples S and P for GaAs-based and InP-based $In_{0.83}Ga_{0.17}As$ PD structures, respectively. Their features have been evaluated on both material qualities and device performances.

As the AFM images shows in **Figure 17**, typical anisotropic features of the surface are detected along the [1 1 0] or [1 −1 0] direction in both samples. While the oval-like defects on the surface of sample S seem more distinct, and the undulations of the pattern are larger than those of sample P. The RMS roughness values are 11.3 nm and 4.8 nm for samples S and P, respectively, as shown in **Figure 17**. The relatively smaller RMS value of surface roughness indicates a better crystalline quality of InP-based $In_{0.83}Ga_{0.17}As$ PD structure. The results of measured symmetric (0 0 4) and asymmetric (2 2 4) reflection RSMs (not shown here) showed that

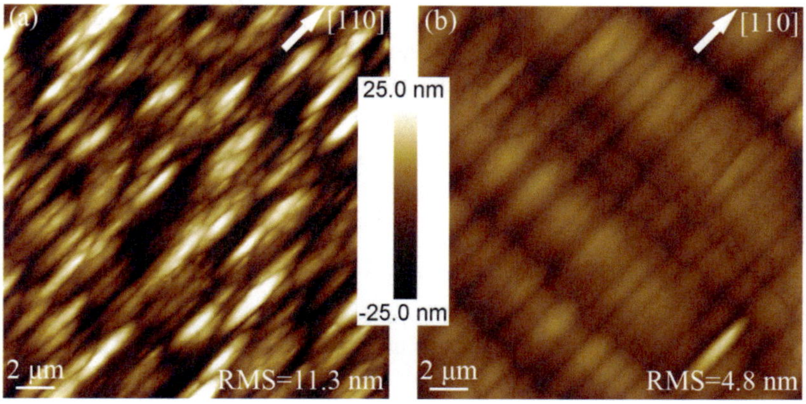

Figure 17. AFM images of (a) sample S, and (b) sample P. The scan area is 20 × 20 μm². Reprinted with permission from Elsevier.

though high lattice relaxation degrees were acquired in both structures, a relatively higher residual strain was observed in the GaAs-based $In_{0.83}Ga_{0.17}As$ PD structure than that in the InP-based structure. This implies that the continuously graded $In_xAl_{1-x}As$ MBL with relatively thin thickness could not well accommodate the relatively higher lattice mismatch between $In_{0.83}Ga_{0.17}As$ and GaAs.

From XRD measurement shown in **Figure 18(a)**, the intensity ratio of epilayer/substrate for sample S is much smaller than that of sample P. Moreover, the FWHM value of the $In_{0.83}Ga_{0.17}As$

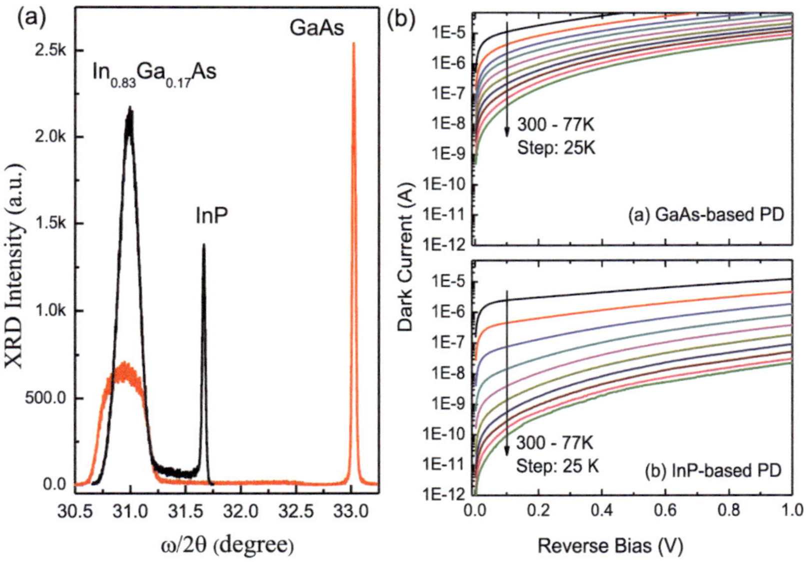

Figure 18. (a) High-resolution (0 0 4) $\omega - 2\theta$ XRD scan curves, and (b) temperature-dependent reverse I-V characteristics of the two samples. Reprinted with permission from Elsevier.

layer of sample S is much larger than that of sample P with values of 1638 and 644 arcsec, respectively. These suggests that the crystal quality of the $In_{0.83}Ga_{0.17}As$ grown on the GaAs substrate is not so good as that grown on the InP substrate with a higher lattice mismatch by using the same type of MBL. This verified one of the conclusions in our previous work [31] that full relaxation and favorable optical property of the InGaAs layer could not occur for the PD wafers with a mismatch grading rate of 2.4% μm^{-1} or above but occur for the wafers with a mismatch grading rate of about 1.2% μm^{-1} or lower.

Correspondingly, quite different temperature dependent I-V characteristics of the two PD chips with diameter of 300 μm were observed in the temperature range from 77 to 300 K, as shown in **Figure 18(b)**. At reverse bias of −10 mV, the dark currents of 2.28 μA at 300 K and 2.17 nA at 77 K for sample S are much larger than that of 674 nA at 300 K and 3.99 pA at 77 K for sample P, respectively, while the zero bias R_0A of the GaAs-based PD is comparable with that of the InP-based PD at 300 K [32], as shown in **Table 5**. This indicates that via appropriate structural design, GaAs may act as a feasible substrate for replacement of InP for $In_{0.83}Ga_{0.17}As$ PDs in some low-end application area around RT.

To analyze the cause of the different electrical performances of the two samples, XTEM measurement was performed firstly, as shown in **Figure 19**. From the XTEM bright-field images, it is obvious that the majority of misfit dislocations of sample P are mainly localized at the early stage of the continuously graded $In_xAl_{1-x}As$ MBL. Threading dislocations have been prevented from propagating into the $In_{0.83}Ga_{0.17}As$ absorption layer by the continuously graded InAlAs MBL on InP under this strain gradient condition. However, it seems that the effect of the continuously graded InAlAs MBL on GaAs is not so remarkable as that on InP. It can be easily seen that many TDs such as 60° dislocations and 60° dislocation pairs generated in the interface of InAlAs MBL and GaAs substrate of sample S have penetrated through the InAlAs buffer and came into the active region. Roughly calculated from **Figure 19(a)**, the TDD in the absorption layer is more than 10^9 cm^{-2} in sample S, much higher than that of sample P with the TDD estimated to be less than 10^8 cm^{-2}.

Second, deep level transient spectroscopy (DLTS) was measured in the temperature range of 77–300 K using DLS-83D DLTS test system and analyzed using standard techniques for the two samples. **Figure 20** shows the DLTS temperature scan signals using reverse bias voltage $V_0 = -2.0$ V, filling pulse height $V_P = 0.5$ V, and filling pulse duration $t_p = 20$ μs. It is evident that no clear peak can be found in the scan signal of InP-based PD. This indicates a low enough density of electrically active defect in the active region of sample P. By contrast, a large electron trap peak around 275 K is clearly observed in that of sample S. Temperature scans with other three lock-in frequencies have also been made and put together to acquire emission

Sample	Relax degree (%)	Residual strain ε (10^{-3})	R@1310 nm (A/W)	η_e@1310 nm (%)	R_0A (Ω cm²)	$D^*_{\lambda p}$ (cm Hz$^{1/2}$/W)
S	97.5	1.5	0.48	45	3.02	1.30×10^{10}
P	97.7	0.5	0.56	53	9.07	2.25×10^{10}

Table 5. The measured material and device results of samples A and B at 300 K (reprinted with permission from Elsevier).

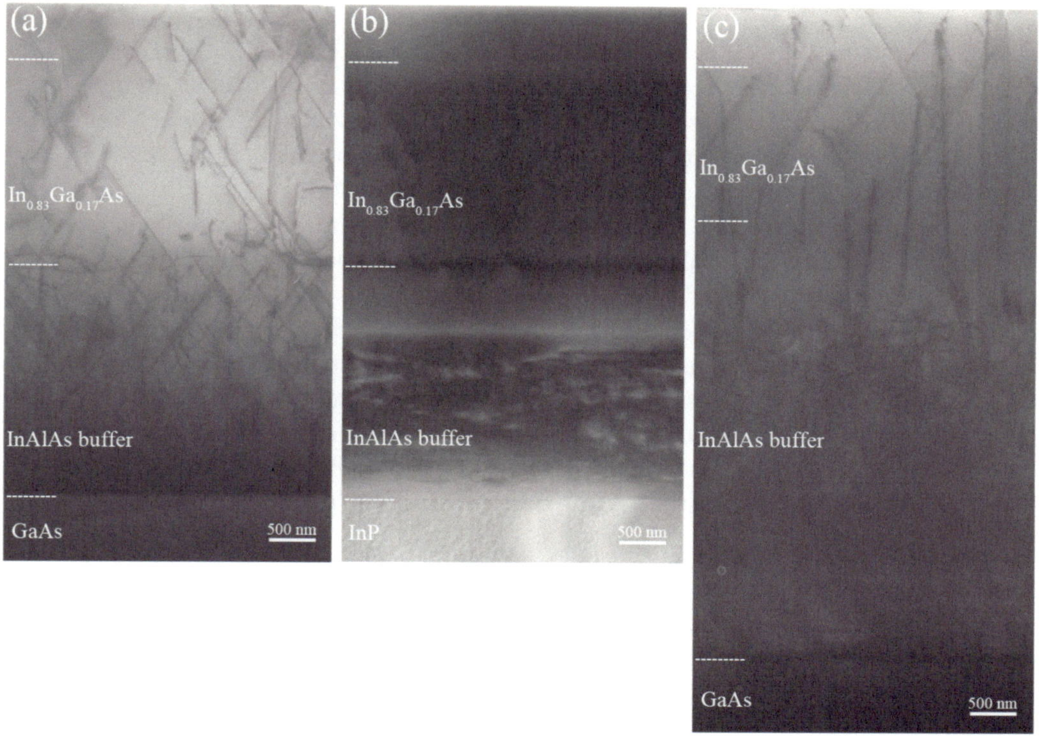

Figure 19. XTEM images of (a) sample S, (b) sample P, and (c) sample S0. Reprinted with permission from Elsevier.

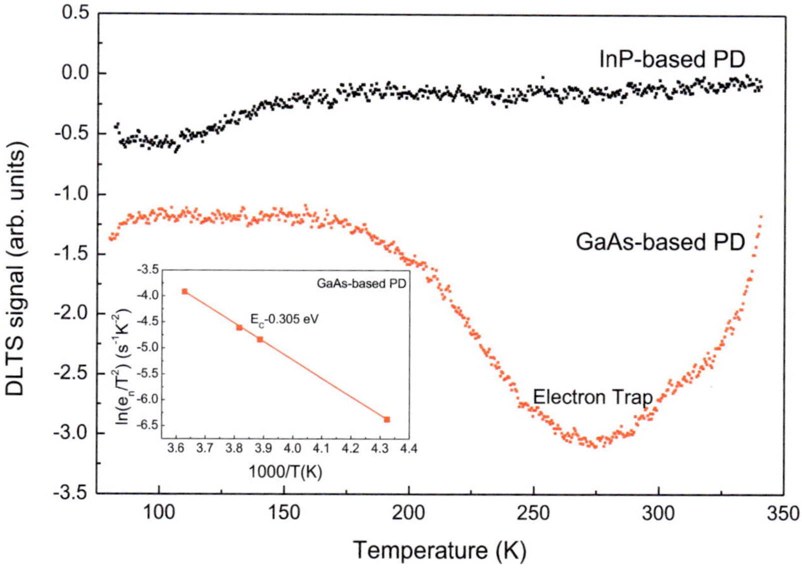

Figure 20. DLTS spectra for the two samples measured at a rate window of 680 Hz. Inset: Arrhenius plot of the deep state observed in GaAs-based PD sample using different lock-in frequencies of 41 Hz, 240 Hz, 320 Hz and 680 Hz. Reprinted with permission from Elsevier.

rate-temperature (e_n-T) data pairs. Then, the standard least-squares fitting for the plot (Arrhenius plot) was applied to extract the trap activation energy and the capture cross-section from the signal peaks. The results are shown in the inset of **Figure 20**. The trap activation energy $E_c - E_T$ = 0.305 eV and the capture cross-section $\sigma_n = 2.25 \times 10^{-18}$ cm^{-2}. This trap will definitely have a deleterious effect on the electrical performance of the PD. Previous literature has reported some deep centers located above the maximum of valence band in lattice-matched $In_{0.53}Ga_{0.47}As$/InP PDs, which were believed to be related with Fe impurities diffused from the Fe-doped substrates to the InGaAs layers [33]. Obviously, the trap occurred here cannot be resulted from Fe impurities because no Fe-doped substrate has been used in this experiment. However, similar traps have been observed in lattice-mismatched $In_{1-x}Ga_xAs$/GaAs hetero-structures [34] and $In_{0.78}Ga_{0.22}As$/InP PDs [35]. Therefore, this 275 K electron trap signal could be associated with the dislocations and point defects in the $In_{0.83}Ga_{0.17}As$ layer because of the even higher lattice mismatch on GaAs compared to that on InP. Since the trap level observed here locates near the center of the bandgap, it is considered that the trap-assisted tunneling current could be the main source of the large dark current in the GaAs-based PD at low temperature range.

On the other hand, the dark current could be expressed by using the thermal activation energy E_a and temperature T as $I_d \sim \exp(-E_a/kT)$ at a fixed reverse bias voltage, where $E_a = E_g/n$, E_g, and n are the band gap energy and the ideal factor, respectively. Therefore, to see the temperature-dependent characteristics of the dark current, Arrhenius plots of the dark currents at −10 mV are made for both samples, as shown in **Figure 21**. For the InP-based PD, near room temperature E_a is almost equivalent to the band gap energy of $In_{0.83}Ga_{0.17}As$ (E_g = 0.48 eV).

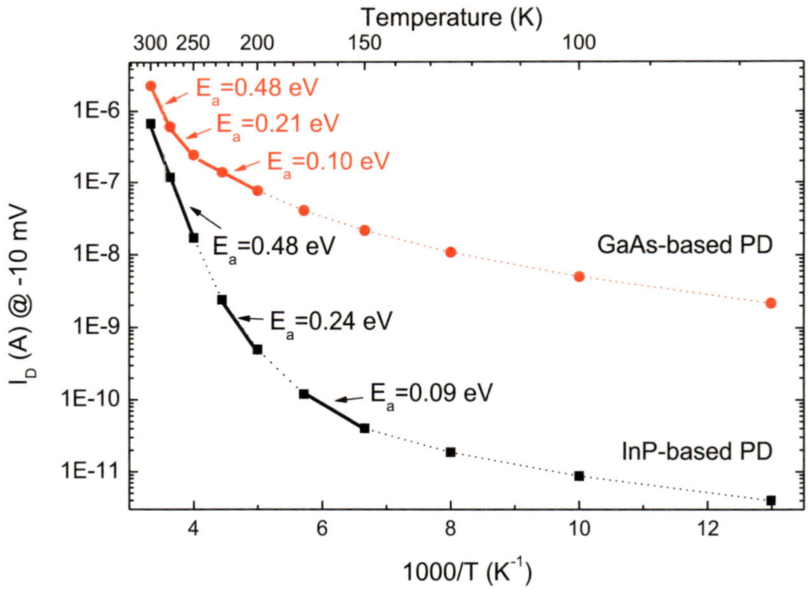

Figure 21. Arrhenius plots of the dark current at −10 mV versus reciprocal temperatures of GaAs- and InP-based PDs. The activation energies in specific temperature regions are also shown in the figure. Reprinted with permission from Elsevier.

http://dx.doi.org/10.5772/intechopen.70259

It indicates that the dark current is dominated by the diffusion current (I_{diff}) with temperature ranging from about 225 to 300 K. As the temperature drops, E_a is about 0.24 eV in the temperature range of 175–225 K, that is, the thermal activation energy is about half of the band gap energy of $In_{0.83}Ga_{0.17}As$. It means that the dark current is dominated by the generation-recombination current (I_{g-r}) in this range. As the temperature reduces lower than 175 K, E_a becomes smaller, and thus, the components of trap-assisted tunneling current (I_{tat}) and/or band-to-band tunneling (I_{btb}) start to play an important role. However, it was found that the temperature dependence of E_a in the GaAs-based PD is much stronger than that of InP-based PD. That is, I_{diff} and I_{g-r} dominate the dark current of the GaAs-based PD in a much smaller temperature region than that of InP-based PD. Alternatively, the majority component is tunneling current at temperatures lower than 250 K, when E_a is much less than the band gap of $In_{0.83}Ga_{0.17}As$. This explained the higher dark current of the GaAs-based PD at the low-temperature region shown in **Figure 18(b)**. The large component proportion of tunneling current could be due to a high TDD in the active region caused by the relatively high lattice mismatch between $In_{0.83}Ga_{0.17}As$ and GaAs [30]. This agrees well with the DLTS results.

To further optimize the GaAs-based PD structure and understand the strain relaxation mechanism with such a relatively high lattice mismatch, it is necessary to increase the buffer thickness and thus decrease the strain gradient of the continuously graded $In_xAl_{1-x}As$ MBL. Therefore, as a comparison, an $In_{0.83}Ga_{0.17}As$ PD structure (renamed as sample S0) with a 5-μm-thick continuously graded $In_xAl_{1-x}As$ MBL and 1.5-μm-thick $In_{0.83}Ga_{0.17}As$ absorption layer has been also grown on a GaAs substrate. So that the mismatch grading rate in the $In_xAl_{1-x}As$ MBL was lowered to about 1.2% μm^{-1}, close to that used for the $In_xAl_{1-x}As$ MBL on InP substrate. By comparison with samples S and P, the XRD and PL results of sample S0 are really better than those of sample S but still worse than that of sample P. Roughly estimated from the XTEM image of **Figure 19(c)**, the average TDD in the absorption layer of sample S0 is only slightly less than that of sample S but still much larger than that of sample P. In addition, some differences were observed between the two XTEM images of samples S and S0. The 60° dislocations and 60° dislocation pairs, which are dominant in sample S, have been mostly replaced by Lomer (90°) dislocations in sample S0. It would be resulted from reactions of 60° dislocations from different glide systems to form Lomer dislocations on the top of the InAlAs MBL due to a two-dimensional growth mode in the thicker buffer of sample S0 [36, 37]. This means that the lower mismatch grading rate of the MBL has improved the characteristics of the PD structure very limited except for the evolution of the TDs.

Therefore, it is supposed that the composition continuously graded buffer may not be a good strategy for growth of material systems with relatively high lattice mismatch due to a quite low growth rate of the MBE technique. A more suitable thin buffer layer should be exploited, and the metamorphic strategy will be updated accordingly.

3.2. Optimization of InAlAs metamorphic buffer on GaAs with relatively high lattice mismatch

From the discussions above, GaAs-based high In content InGaAs PDs may be not a good choice for practical special application of remote sensing at the low temperature range.

However, it is still valuable to design and develop an appropriate buffer scheme for the device development from material system with relatively high-lattice mismatch.

Since that the way strain is introduced at the initial stage of the MBL has been proved to play a critical role in the final TDD [10]. The strain energy should be released as quickly as possible, and the multiplication of TD must be avoided occurring at the final stage of the buffer layer. Therefore, if we take the accessory advantage of this relatively high lattice mismatch between high In content InGaAs and GaAs, we can promote the nucleation of dispersed TDs at the initial stage of the MBL. Considering that the high In content InAlAs and/or InGaAs will quickly achieve a high relaxation on GaAs due to the small critical thickness. Specifically, for the growth of $In_{0.83}Ga_{0.17}As$ PD on GaAs, we can use the fixed-composition $In_{0.83}Al_{0.17}As$ as MBL to accelerate the release of strain energy and the nucleation of dispersed misfit dislocations at the initial stage of the buffer and thus restrain the misfit dislocation at the interface between $In_{0.83}Al_{0.17}As$ and GaAs. The uniform composition buffer will also reduce the strain gradient at the later stage of the MBL with lattice nearly fully relaxed. We show that this method results in a lower TDD and a smoother surface of the final absorption and cap layer.

In this work, four $In_{0.83}Ga_{0.17}As$ PD structures with different buffer schemes were grown by GSMBE on S.I. (1 0 0)-oriented GaAs epi-ready substrates [38]. Each structure consisted of a 2.5-μm N^+ InAlAs MBL and a 1.5-μm n^- $In_{0.83}Ga_{0.17}As$ absorption layer followed by a 530 nm P^+ InAlAs cap. The detailed buffer schemes for samples GS1, GS2, GS3, and GS4 were listed in **Table 6**. Growth condition was exactly the same as that in our previous study [30]. The strategies of substrate temperature graded from 530 to 460°C, and the uniform temperature of 490 °C were adopted for the deposition of continuously graded $In_xAl_{1-x}As$ and fixed-composition $In_{0.83}Al_{0.17}As$ buffer, respectively. In addition, InAs wetting layer was inserted between the fixed-composition $In_{0.83}Al_{0.17}As$ layer and the GaAs substrate for samples GS3 and GS4 to investigate the effect of interfacial layer on the TD behaviors at the interface.

As shown in **Figure 22**, compared to sample GS1, the cross-hatch pattern in the 2D AFM image of sample GS2 is less pronounced, and smaller diameters of 3D mounds align along the [1 1 0] direction in the 3D AFM image, indicating a smaller residual strain fields on the surface [39]. Thus, the RMS roughnesses of samples GS2–GS4 are a little smaller than that of sample GS1, as summarized in **Table 6**. As well known, when the thickness of the epilayer is beyond the critical value, the misfit strain could relax by introducing misfit dislocation arrays as well as surface undulation [40, 41]. The larger RMS value and larger size of 3D mounds on the structural surface of sample GS1 reflect that the strain-reliving surface roughening is more serious in sample GS1 than samples GS2–GS4.

Figure 23(a) shows the normalized HRXRD rocking curves for (0 0 4) reflections of all samples, which is also related to the TDD and other crystal imperfections [42]. A broader peak located at about 31° of sample GS1 reflects the continuously grading profile of $In_xAl_{1-x}As$ ($x = 0.1 \rightarrow 0.86$) MBL which ultimately combined with the $In_{0.83}Ga_{0.17}As$ peak. While much lower FWHM of $In_{0.83}Ga_{0.17}As$ layers for samples GS2–GS4 mainly due to the thicker fixed-composition $In_{0.83}Al_{0.17}As$ MBL. Though all PD structures exhibited high degree of lattice relaxation, shown in **Table 6**, the improvement of structural surface and crystal quality was believed to be associated with the amelioration of misfit dislocation formation at the

http://dx.doi.org/10.5772/intechopen.70259

Sample	Buffer scheme	XRD FWHM (arcsec)	Relaxation degree (%)	In composition	AFM RMS (nm)
GS1	2.5 μm $In_xAl_{1-x}As$ buffer, x grades from 0.1 to 0.86	1782	94.8	0.84	11.6
GS2	2.5 μm $In_{0.83}Al_{0.17}As$	544	97.5	0.82	8.1
GS3	5 MLs InAs QDs + 2.5 μm $In_{0.83}Al_{0.17}As$	410	98.3	0.83	8.8
GS4	50 nm InAs + 2.5 μm $In_{0.83}Al_{0.17}As$	511	98.8	0.82	8.6

Table 6. Buffer schemes and measured results for four samples (reprinted with permission from Elsevier).

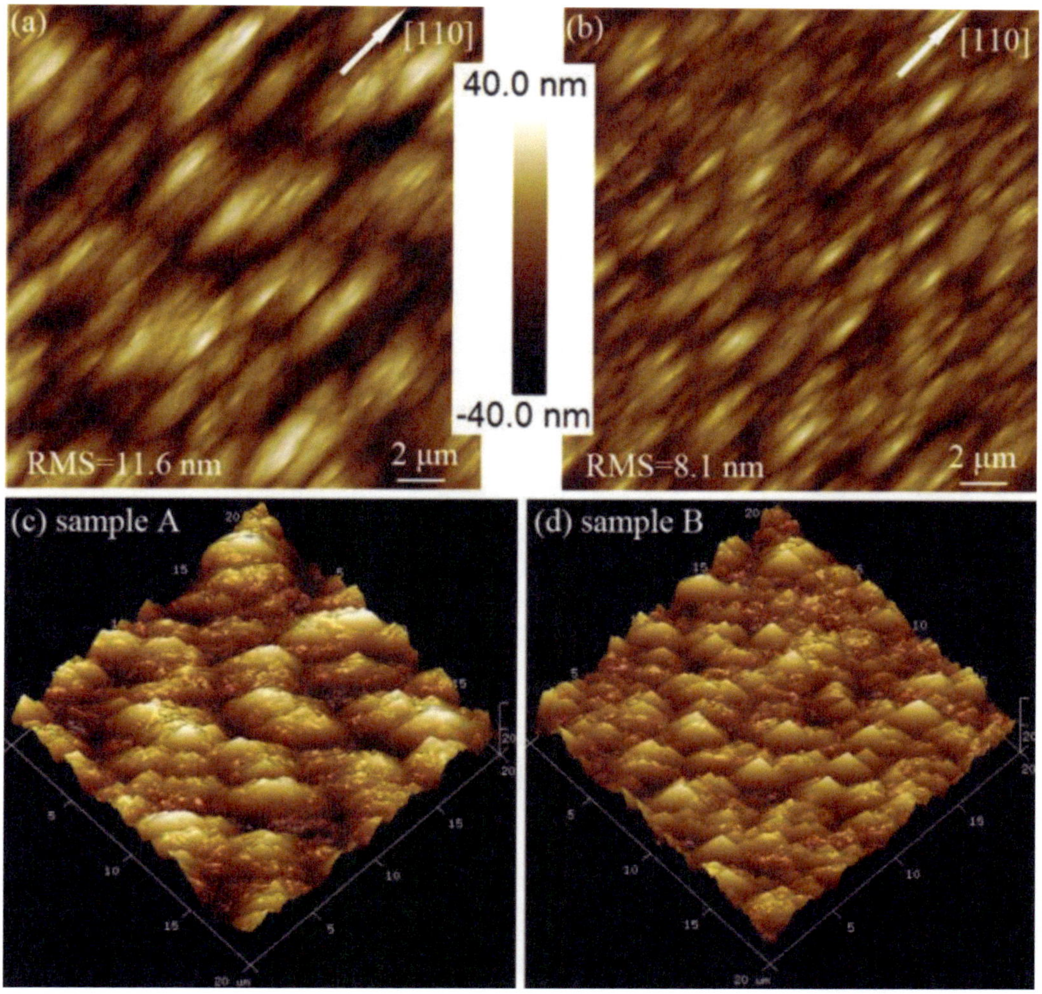

Figure 22. AFM images of samples A and B. The scan area is 20 × 20 μm^2. Reprinted with permission from Elsevier.

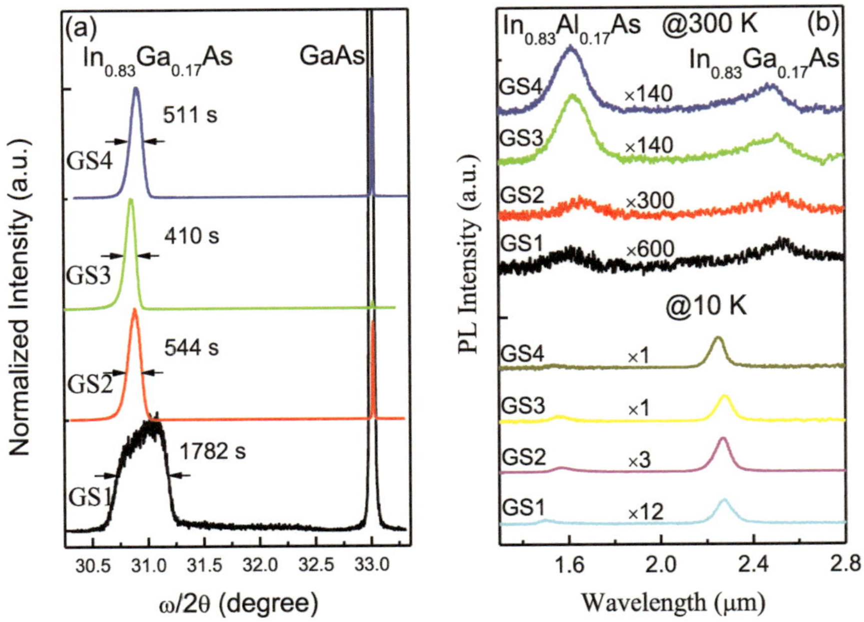

Figure 23. (a) (0 0 4) $\omega - 2\theta$ high resolution XRD patterns. (b) PL spectra at 300 K and 10 K of samples GS1–GS4. Reprinted with permission from Elsevier.

film/substrate interface [43]. This was further demonstrated by sample GS3 with a narrowest FWHM value of 410 arcs by inserting 5 MLs InAs QDs at the $In_{0.83}Al_{0.17}As$/GaAs interface. This indicates that the growth of the QD is beneficial to the dislocation nucleation at the interface and the confinement of misfit dislocations in the inactive region, while the effect is slightly weakened when InAs was thickened and degenerated, resulting in a degraded interfacial quality.

From the PL spectra measured at 300 and 10 K shown in **Figure 23(b)**, we can see that the PL intensity of the $In_{0.83}Ga_{0.17}As$ absorption layers increased in sequence for samples GS1–GS3, meaning that the crystal defect has been suppressed through the substitution of continuously graded $In_xAl_{1-x}As$ MBL by fixed-composition $In_{0.83}Al_{0.17}As$ MBL. The XTEM images of the two typical PD structures were shown in **Figure 24**. For sample GS1, it is notable that the misfit dislocation networks were separated by 0.1 μm from the interface between $In_{0.1}Al_{0.9}As$ and GaAs because of the $In_{0.1}Al_{0.9}As$ layer, and highly intensive dislocation arrays accumulated in the lower area of the $In_xAl_{1-x}As$ grading layer, as shown in **Figure 24(a)**. Many vertical dislocations with long dislocation length, thread from the tangling area through the buffer along the [1 0 0] direction and propagate to the top active region of the PD structure. Strain in this structure released dominantly by surface undulation, which deteriorates the surface morphology of the PD structure and leads to a large RMS roughness. The surface ripple troughs–induced misfit dislocations were generated by gliding of dislocation half loops from the surface through the epilayer [44].

However, the dislocations in the structure of sample GS2 are short, and most of them are not perpendicular to the sample surface, as shown in **Figure 24(b)**. We supposed that $In_{0.83}Al_{0.17}As$ QDs would form at the beginning of growth process of the fixed-composition $In_{0.83}Al_{0.17}As$

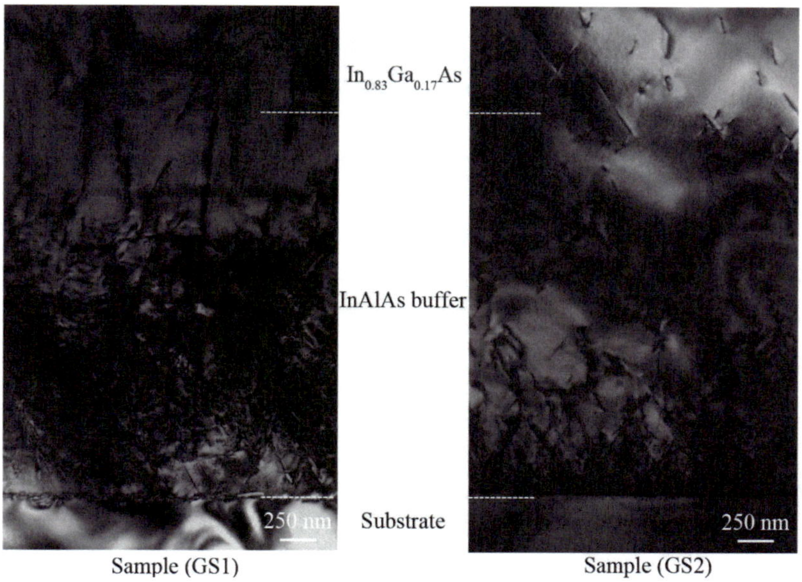

Figure 24. XTEM images of (a) sample GS1 and (b) sample GS2. Reprinted with permission from Elsevier.

layer because of the relatively high-lattice mismatch of 5.9% with respect to GaAs. Therefore, the relaxation process in the fixed-composition buffer of sample GS2 may start with a layer-to-island transition (Stranski-Kranstanov growth mode) after the deposition of a couple MLs of $In_{0.83}Al_{0.17}As$ on GaAs. Then, the large compressive strain can be released by generation and reaction of misfit dislocation networks. Since the initial growth mechanism will also affect the dislocation generation mechanism in the hetero-epitaxy of $In_xGa_{1-x}As$ on GaAs except for the lattice mismatch [10]. The formation of $In_{0.83}Al_{0.17}As$ QDs at the initial stage of the $In_{0.83}Al_{0.17}As$ MBL has indeed played an important role by the formation of high density of nucleation sites [13], which can act as core of misfit dislocations. In this way, the relaxation process is promptly due to the substantial lattice mismatch and most of the misfit dislocations are inhibited to the core areas, which locate close to the $In_{0.83}Al_{0.17}As$/GaAs interface. Therefore, TDs in the buffer of sample GS2 cannot propagate a long distance due to the island growth near the interface. They annihilated much easier by interaction with the point defects existing in the high In content $In_{0.83}Al_{0.17}As$ layer with respect to the $In_xAl_{1-x}As$ MBL of sample GS1. It is expected that the intermediate layer of 5 MLs InAs QD will enhance the dislocation nucleation process at the $In_{0.83}Al_{0.17}As$/GaAs interface because of a higher lattice mismatch between InAs and GaAs. Through appropriate buffer design, fixed-composition $In_{0.83}Al_{0.17}As$ buffer may be a better choice for the growth of metamorphic $In_{0.83}Ga_{0.17}As$ PDs on GaAs substrate.

4. Conclusion

In conclusion, due to the inventions of metamorphic techniques such as, linearly, step, and one-step continuously graded buffer, compositional overshoot, and DA in the growth of lattice-mismatched material system, the development of InP- and GaAs-based wavelength-extended $In_xGa_{1-x}As$ (x > 0.53) photodetectors with relatively high lattice mismatch has achieved remarkable success in SWIR band especially for the applications at higher operation temperatures and robust circumstances. Though GaAs-based $In_{0.83}Ga_{0.17}As$ PD shows large potential in some low-end application area around RT, the high densities of TDs and electron traps in the active region due to the relatively high-lattice mismatch have hindered its development. By designing of abrupt interface and appropriate buffer, the initial dislocation nucleation process may pave a way of development of lattice-mismatched device structures.

Acknowledgements

The authors wish to acknowledge the support of the National Key Research and Development Program of China under Grant No. 2016YFB0402400, the National Natural Science Foundation of China under grant Nos. 61405232, 61675225, and 61605232, and the Youth Innovation Promotion Association CAS under Grant No. 2013155.

Author details

Xing-you Chen*, Yi Gu and Yong-gang Zhang

*Address all correspondence to: xychen@mail.sim.ac.cn

State Key Laboratory of Functional Materials for Informatics, Shanghai Institute of Microsystem and Information Technology, Chinese Academy of Sciences, China

References

[1] Hoogeveen RWM, JvdA R, Goede APH. Mint: Extended wavelength InGaAs infrared (1.0-2.4 μm) detector arrays on SCIAMACHY for space-based spectrometry of the Earth atmosphere. Infrared Physics Technology. 2001;**42**:1-16. DOI: 10.1016/S1350-4495(00)00061-X

[2] Zhang YG, Gu Y. Gas source MBE grown wavelength extending InGaAs photodetectors. In: Betta GFD, editor. Advances in Photodiodes. Croatia: InTech; 2011. pp. 349-376. DOI: 10.5772/13910

[3] Tangring I, Wang SM, Zhu XR, Larsson A, Lai ZH, Sadeghi M. Mint: Manipulation of strain relaxation in metamorphic heterostructures. Applied Physics Letters. 2007;**90**: 071904. DOI: 10.1063/1.2435609

[4] Song YX, Wang SM, Lai ZH, Sadeghi M. Mint: Enhancement of optical quality in metamorphic quantum wells using dilute nitride buffers. Applied Physics Letters. 2010;**97**:091903. DOI: 10.1063/1.3483839

[5] Ji L, Lu SL, Zhao YM, Tan M, Dong JR, Yang H. Mint: Compositionally undulating step-graded $InAs_yP_{1-y}$ buffer layer growth by metal-organic chemical vapor deposition. Journal of Crystal Growth. 2013;**363**:44-48. DOI: 10.1016/j.jcrysgro.2012.09.035

[6] Zhao L, Sun JG, Guo ZX, Miao GQ. Mint: TEM dislocations characterization of $In_xGa_{1-x}As$/InP (100) (x = 0.82) on mismatched InP substrate. Materials Letters. 2013;**106**:222-224. DOI: 10.1016/j.matlet.2013.04.116

[7] Zhao L, Guo ZX, Wei QL, Miao GQ, Zhao L. Mint: The relationship between the dislocations and microstructure in $In_{0.82}Ga_{0.18}As$/InP heterostructures. Scientific Reports. 2016;**6**:35139. DOI: 10.1038/srep35139

[8] Lubyshev D, Liu WK, Stewart TR, Cornfeld AB, Fang XM, Xu X, Specht P, Kisielowski C, Naidenkova M, Goorsky MS. Mint: Strain relaxation and dislocation filtering in metamorphic high electron mobility transistor structures grown on GaAs substrates. Journal of Vacuum Science and Technology B. 2001;**19**:1510-1514. DOI: 10.1116/1.1376384

[9] Valtuena JF, Sacedon A, Alvarez AL, Izputa I, Calle F, Calleja E, MacPherson G, Goodhew PJ, Pacheco FJ, Garcia R and Molina SI. Mint: Influence of the surface morphology on the relaxation of low-strained $In_xGa_{1-x}As$ linear buffer structures. Journal of Crystal Growth. 1997;**182**:281-291. DOI: 10.1016/S0022-0248(97)00377-1

[10] Chang SZ, Chang TC, Lee SC. Mint: The growth of highly mismatched $In_xGa_{1-x}As$ ($0.28 \le x \le 1$) on GaAs by molecular-beam epitaxy. Journal of Applied Physics. 1993;**73**:4916. DOI: 10.1063/1.353809

[11] Chang CA, Serrano CM, Chang LL, Esaki L. Mint: Studies by cross-sectional transmission electron microscope of InAs grown by molecular beam epitaxy on GaAs substrates. Applied Physics Letters. 1980;**37**:538. DOI: 10.1063/1.91977

[12] Zimmermann L, John J, de Weerd M, Slaman M, Nemeth S, Merken P, Borghs S, Van Hoof C. Mint: InGaAs on GaAs extended wavelength linear detector arrays. Proceedings of SPIE. 2001;**4288**:77-84. DOI: 10.1117/12.429396

[13] Song YX, Wang SM, Tangring I, Lai ZH, Sadeghi M. Mint: Effects of doping and grading slope on surface and structure of metamorphic InGaAs buffers on GaAs substrates. Journal of Applied Physics. 2009;**106**:123531. DOI: 10.1063/1.3273492

[14] Jurczak P, Sablon KA, Gutierrez M, Liu HY, Wu J. Mint: 2.5-μm InGaAs photodiodes grown on GaAs substrates by interfacial misfit array technique. Infrared Physics & Technology. 2017;**81**:320-324. DOI: 10.1016/j.infrared.2017.02.001

[15] Gao FL, Wen L, Zhang XN, Guan YF, Li JL, Zhang, SG, Li GQ. Mint: Structural properties of $In_{0.53}Ga_{0.47}As$ epitaxial films grown on Si (111) substrates by molecular beam epitaxy. Thin Solid Films. 2015;**589**:32-37. DOI: 10.1016/j.tsf.2015.04.085

[16] Toikkanen L, Hakkarainen T, Schramm A, Tukiainen A, Laukkanen P, Guina M. Mint: Metamorphic growth of tensile strained GaInP on GaAs substrate. Journal of Crystal Growth. 2010;**312**:3105-3110. DOI: 10.1016/j.jcrysgro.2010.07.059

[17] Gu Y, Zhang YG, Wang K, Fang X, Li C, Zhou L, Li AZ, Li Hsby. Mint: Effects of growth temperature and buffer scheme on characteristics of InP-based metamorphic InGaAs photodetectors. Journal of Grystal Growth. 2013;**378**:65-68. DOI: 10.1016/j.jcrysgro.2012.12.049

[18] Gu Y, Zhang YG, Wang K, Fang X, Li C, Cao YY, Li AZ, Li YY. Mint: InP-based InAs/InGaAs quantum wells with type-I emission beyond 3 μm. Applied Physics Letters. 2011;**99**:081914. DOI: 10.1063/1.3629999

[19] Xi SP, Gu Y, Zhang YG, Chen XY, Zhou L, Li AZ, Li Hsby. Mint: Effects of continuously graded or step-graded $In_xAl_{1-x}As$ buffer on the performance of InP-based $In_{0.83}Ga_{0.17}As$ photodetectors. Journal of Grystal Growth. 2015;**425**:337-340. DOI: 10.1016/j.jcrysgro.2015.03.040

[20] Du B, Gu Y, Zhang YG, Chen XY, Xi SP, Ma YJ, Ji WY, Shi YH. Mint: Effects of continuously or step-continuously graded buffer on the performance of wavelength extended InGaAs photodetectors. Journal of Grystal Growth. 2016;**440**:1-5. DOI: 10.1016/j.jcrysgro.2016.01.016

[21] Jang JH, Cueva G, Hoke WE, Lemonias PJ, Fay P, Adesida I. Mint: Metamorphic graded bandgap InGaAs-InGaAlAs-InAlAs double heterojunction P-i-I-N photodiodes. Journal of Lightwave Technology. 2002;**20**:507-514. DOI: 10.1109/50.989001

[22] Lenox C, Nie H, Kinsey G, Yuan P, Holmes AL, Streetman BG, Campbell JC. Mint: Improved optical response of superlattice graded InAlAs/InGaAs p-i-n photodetectors. Applied Physics Letters. 1998;**73**:3405. DOI: 10.1063/1.122757

[23] Jandl A, Bulsara MT, Fitzgerald EA. Mint: Materials properties and dislocation dynamics in InAsP compositionally graded buffers on InP substrates. Journal of Applied Physics. 2014;**115**:153503. DOI: 10.1063/1.4871289

[24] Zhang YG, Gu Y, Tian ZB, Li AZ, Zhu XR, Wang K. Mint: Wavelength extended InGaAs/InAlAs/InP photodetectors using n-on-p configuration optimized for back illumination. Infrared Physics & Technology. 2009;**52**:52-56. DOI: 10.1016/j.infrared.2008.12.001

[25] Fang X, Gu Y, Chen XY, Zhou L, Cao YY, Li Hsby, Zhang YG. Mint: InP-based $In_xGa_{1-x}As$ metamorphic buffers with different mismatch grading rates. Journal of Semiconductors. 2013;**34**:073005. DOI: 10.1088/1674-4926/34/7/073005

[26] Tersoff J. Mint: Dislocations and strain relief in compositionally graded layers. Applied Physics Letters. 1993;**62**:693. DOI: 10.1063/1.108842

[27] Boschetti A, Bassi D, Iacob E, Iannotta S, Ricci L, Scotoni M. Mint: Resonant photoacoustic simultaneous detection of methane and ethylene by means of a 1.63 μm diode laser. Applied Physics B-Lasers and Optics. 2002;**74**:273-278. DOI: 10.1007/s003400200790

[28] Fang X, Gu Y, Zhang YG, Zhou L, Zhou L, Wang K, Li Hsby, Liu KH, Cao YY. Mint: Effects of compositional overshoot on InP-based InAlAs metamorphic graded buffer. Journal of Infrared Millimeter Waves. 2013;**32**:481-490. DOI: 10.3724/SP.J.1010.2013.00481

[29] Gu Y, Zhang YG, Wang K, Fang X, Liu KH. Mint: InAlAs graded metamorphic buffer with digital alloy intermediate layers. Japanese Journal of Applied Physics. 2012;**51**:080205. DOI: 10.1143/JJAP.51.080205

[30] Chen XY, Zhang YG, Gu Y, Zhou L, Cao YY, Fang X, Li Hsby. Mint: GaAs-based $In_{0.83}Ga_{0.17}As$ photodetector structure grown by gas source molecular beam epitaxy. Journal of Grystal Growth. 2014;**39**:75-80. DOI: 10.1016/j.jcrysgro.2013.11.083

[31] Zhang YG, Gu Y, Wang K, Li AZ, Li C. Mint: Properties of gas source molecular beam epitaxy grown wavelength extended InGaAs photodetector structures on a linear graded InAlAs buffer. Semiconductor Science and Technology. 2008;**23**:125029. DOI: 10.1088/0268-1242/23/12/125029

[32] Zhou L, Zhang YG, Chen XY, Gu Y, Li Hsby, Cao YY, Xi SP. Mint: Dark current characteristics of GaAs-based 2.6 μm InGaAs photodetectors on different types of InAlAs buffer layers. Journal of Physics D: Applied Physics. 2014;**47**:085107. DOI: 10.1088/0022-3727/47/8/085107

[33] Guillot G, Bremond G, Benyattou T, Ducroquet F, Wirth B, Colombet M, Louati A, Bencherifa A. Mint: Identification of the Fe acceptor level in $Ga_{0.47}In_{0.53}As$. Semiconductor Science Technology. 1990;**5**:391-394. DOI: 10.1088/0268-1242/5/5/003

[34] Pal C, Gombia E, Mosca R, Bosacchi A, Franchi S. Mint: Deep levels in virtually unstrained InGaAs layers deposited on GaAs. Journal of Applied Physics. 1998;**84**:2965-2967. DOI: 10.1063/1.368404

[35] Ji XL, Liu BQ, Tang HJ, Yang XL, Li X, Gong HM, Shen B, Han P, Yan F. Mint: 2.6 μm MBE grown InGaAs detectors with dark current of SRH and TAT. AIP Advances. 2014;**4**:087135. DOI: 10.1063/1.4894142

[36] Henager CH, Hoagland RG. Mint: A rebound mechanism for misfit dislocation creation in metallic nanolayers. Scripta Materialia. 2004;**50**:701-705. DOI: 10.1016/j.scriptamat.2003.09.002

[37] Wang Y, Ruterana P, Desplanque L, ElKazzi S, Wallart X. Mint: Growth mode dependence of misfit dislocation configuration at lattice mismatched III-V semiconductor interfaces. Europhysics Letters. 2012;**97**:68011. DOI: 10.1209/0295-5075/97/68011

[38] Chen XY, Gu Y, Zhang YG, Xi SP, Guo ZX, Zhou L, Li AZ, Li Hsby. Mint: Optimization of InAlAs buffers for growth of GaAs-based high indium content InGaAs photodetectors. Journal of Grystal Growth. 2015;**425**:346-350. DOI: 10.1016/j.jcrysgro.2015.02.102

[39] Samavedan SB, Fitzgerald EA. Mint: Novel dislocation structure and surface morphology effects in relaxed Ge/Si-Ge(graded)/Si structures. Journal of Applied Physics. 1997;**81**:3108. DOI: 10.1063/1.364345

[40] Tersoff J. Mint: Stress-induced roughening in epitaxial growth. Applied Surface Science. 1996;**102**:1-2. DOI: 10.1016/0169-4332(96)00002-5

[41] France RM, Geisz JF, Steiner MA, To B, Romero MJ, Olavarria WJ, King RR. Mint: Reduction of crosshatch roughness and threading dislocation density in metamorphic GaInP buffers and GaInAs solar cells. Journal of Applied Physics. 2012;**111**:103528. DOI: 10.1063/1.4721367

[42] Ayers JE. Mint: The measurement of threading dislocation densities in semiconductor crystals by X-ray diffraction. Journal Crystal Growth. 1994;**135**:71-77. DOI: 10.1016/0022-0248(94)90727-7

[43] Andrews AM, Speck JS, Romanov AE, Bobeth M, Pompe W. Mint: Modeling cross-hatch surface morphology in growing mismatched layers. Journal of Applied Physics. 2002; **91**:1933. DOI: 10.1063/1.1428091

[44] Cullis AG, Pidduck AJ, Emeny MT. Mint: Misfit dislocation sources at surface ripple troughs in continuous heteroepitaxial layers. Physical Review Letters. 1995;**75**:2368-2371. DOI: 10.1103/PhysRevLett.75.2368